MATHEMATICAL FOUNDATION OF COMPUTER SCIENCE

Dr. Bhavanari Satyanarayana

Dr. Tumurukota Venkata Pradeep Kumar

Dr. Shaik Mohiddin Shaw

BSP **BS Publications**

CRC Press
Taylor & Francis Group
Boca Raton London New York

CRC Press is an imprint of the
Taylor & Francis Group, an **informa** business

CRC Press
Taylor & Francis Group
6000 Broken Sound Parkway NW, Suite 300
Boca Raton, FL 33487-2742

First issued in paperback 2023

ISBN 13: 978-1-03-265419-5 (pbk)
ISBN 13: 978-0-367-36681-0 (hbk)
ISBN 13: 978-0-367-36723-7 (ebk)

DOI: 10.1201/9780367367237

Publisher's Note
The publisher has gone to great lengths to ensure the quality of this reprint but points out that some imperfections in the original copies may be apparent.

Print edition not for sale in South Asia (India, Sri Lanka, Nepal, Bangladesh, Pakistan or Bhutan)

Library of Congress Cataloging-in-Publication Data
A catalog record has been requested

Visit the Taylor & Francis Web site at
http://www.taylorandfrancis.com

and the CRC Press Web site at
http://www.crcpress.com

 **BS Publications**

Mathematical Foundation of Computer Science

Preface

When something can be read without effort, great effort has gone into its writing.

By- Enrique Jardiel Poncela

We have great pleasure in presenting this book "Mathematical Foundation of Computer Science" for B.Tech students.

Book has been written in a simple, lucid and easy to understandable style. Number of problems have been worked out as examples. Hints and answers are given for the problems in the exercise to help the students for self-learning. Each chapter has been planned as independent unit so that various topics connected which can also be read separately.

Suggestions for further improvement of the book are most welcome and will be gratefully acknowledged.

No originality of the subject was claimed by the authors. The authors presented the subject in their own way for the better understanding of the target group of the readers. The authors referred several existing books. Some of them were listed in the book.

-Authors

Acknowledgement

The Authors are pleased to express their thanks to the following for their constant encouragement in each of their endeavors:

*Dr. P. V. Arunachalam, Former Vice-Chancellor of Dravidian University, Kuppam, A.P., India.

*Dr. D. Ramakotaiah, Former Vice – Chancellor of Acharya Nagarjuna University, A.P.

*Dr. A. Radhakrishna former Professor of Mathematics, Kakatiya University, Warangal,

Telangana, India.

*Dr. L. Radha Krishna, Visiting Professor, Mathematics, Bangalore University, Bangalore, Karnataka, India.

*Dr. Kuncham Syamprasad, Manipal University, Manipal.

*Dr. Kedukodi Babushri Srinivas, Manipal University, Manipal.

*Dr. P. Trimurthy and Dr. Y. Venkateswara Reddy of Acharya Nagarjuna University.

Dr. Bhavanari Satyanarayana places on record his deep sense of gratitude to his parents: Bhavanari Ramakotaiah (a teacher in an elementary school in the village Madugula, Guntur(dt), A.P., India) and Bhavanari Anasuyamma, without whose constant encouragement and help it would not have been possible for him to pursue higher studies in Mathematics. Also, he thanks his wife Bhavanari Jayalakshmi, and his children, Mallikharjun (M.Tech), Satyasri, MBBS (China), and Satya Gnyana Sri (B.Tech) for their constant patience with him and help in creating better output.

-Authors

Contents

Chapter 3: Number Theory

Chapter 4: Mathematical Induction

Chapter 5: Set Theory

Chapter 6: Functions

Chapter 10: Graph Theory - IV

Chapter 11: Algebraic Structures

Chapter 14: Binomial Theorem

Chapter 15: Recurrence Relations

Chapter 16: Some Methods of Solving Recurrence Relations

CHAPTER - 1

Propositional Calculus

Introduction

Logic means reasoning. One of the important aims of logic is to provide rules through which one can determine the validity of any particular argument. Logic concern with all types of reasonings such as legal arguments, mathematical proofs, conclusions in a scientific theory based upon a set of given hypothesis. The rules are called as rules of inference. The rules should be independent of any particular argument or discipline or language used in the argument.

We shall mean, by formal logic, a system of rules and procedures used to decide whether or not a statement follows from some given set of statements.

In order to avoid ambiguity, we use symbols. The symbols are easy to write and easy to manipulate. Hence, the logic that we study is named as "Symbolic logic". For example, observe the following statements.

   (i)   All men are mortal

   (ii) Socrates is a man

Therefore (iii) Socrates is mortal

According to the logic, if any three statements have the following form

   (i)   All M are P   (ii)   S is M

Therefore (iii) S is P

then (iii) follows from (i) and (ii). The argument is correct, no matter whether the meanings of statements (i), (ii), and (iii) are correct. All that required is that they have the forms (i), (ii), and (iii). In Aristotelian logic, an argument of this type is called **syllogism**.

The formulation of the syllogism is contained in Aristotle's organon. It had a great fascination for medieval logicians, for almost all their work centered about ascertaining its valid moods. The three characteristic properties of a syllogism are as follows:

(i) It consists of three statements. The first two statements are called as **premises**, and the third statement is called as **conclusion**. The third one (**conclusion**) being a logical consequence of the first two (the **premises**).

(ii) Each of the three sentences has one of the four forms given in the Table.

Classification	Examples
Universal and affirmative judgment	All X is Y All monkeys are tree climbers All integers are real numbers All men are mortal
Universal and negative judgment	No X is Y No man is mortal No monkey is a tree climber No negative number is a positive number
Particular and affirmative judgment	Some X is Y Some men are mortal Some monkeys are tree climbers Some real numbers are integers
Particular and negative judgment	Some X is not Y Some men are not mortal Some monkeys are not tree climbers Some real numbers are not integers

So a **syllogism** is an argument consisting of two propositions called **premises** and a third proposition called the **conclusion**.

1.1 Statements and Notations

In any language, a sentence, in practice, is constructed by means of words. So a meaningful sequence of words is a sentence.

A statement is a sentence for which we can say whether it is true or false.

Consider the two sentences

(i) "Rama is a man"

(ii) "open the door"

We can say that "Rama is a man" is true. Hence it is a sentence.

It is clear that "open the door" is not a statement. Thus "open the door" is a sentence which is not a statement.

Example 1.1

Let us consider the following sentences.

(i)   Socrates is a man

(ii)   Rama killed Ravana

(iii) Open the door

(iv) $\sqrt{3}$ is an irrational number

(v)   How are you?

(vi)   Delhi is the Capital of India

(vii) Guntur is the Capital of Andhra Pradesh

The sentences (i), (ii), (iv), (vi), (vii) are statements. The statements (i), (ii), (iv), (vi) are true statements. The statement (vii) is false. The sentences (iii) and (v) are not statements.

Note:

"True" or "False" are called the truth value of the statement considered. "True" is denoted by "T" or "I". "False" is denoted by "F" or "O". For any given statement "P" we assign the truth value "T" or "I" if P is true. We assign the truth value "F" or "O" if P is false.

1.1.1   Subject and Predicate

Consider the statement "Rama is a man". In this statement "Rama" is the subject of the statement. The other part "is a man" is called predicate.

Note:

A simple statement is a statement that contains only one subject and one predicate. Simple statements are the basic units of our language to frame the rules of inference. These statements are also called as primary statements. These are the statements that cannot be further broken down into smaller statements. Such statements are also called as atomic statements. "$3 + 2 = 5$" and "$7 + 5 = 12$" are two primary (or simple or atomic) statements.

1.1.2   Notation

Observe the following:

(i)   p: Socrates is a man

(ii)   q: Guntur is the Capital City of Andhra Pradesh

In statements (i), (ii), p and q are the symbols used. Here "p" is a statement in symbolic logic that corresponds to the English statement "Socrates is a man".

Here "Socrates" is the subject and "is a man" is the predicate. The statement P (that is, "Socrates is a man") contains only one subject and only one predicate. Hence it is a primary statement.

Similarly the statement "q" that represents the English statement "Guntur is the Capital City of Andhra Pradesh" is a primary statement. Note that the symbols "p" and "q" were used as the names of the statements. We use this symbolic notation for statements throughout this unit.

From the above discussion, we understand that the basic unit of our objective language is called as a primary statement (or variable). We assume that these primary statements cannot be further broken down or analyzed into simple statements.

Example 1.2

Determine the truth value of each of the following statements:

  (i) $6 + 2 = 7$ and $4 + 4 = 8$

  (ii) Four is even

  (iii) $4 + 3 = 7$ and $6 + 2 = 8$

[JNTU–H, Nov 2010 (Set–1)]

Solution:

  (i)   As $6 + 2 = 8$, the statement $6 + 2 = 7$ has truth value "False".
        $4 + 4 = 8$ is true.

  (ii) The truth value of "Four is even" is true.

  (iii) The truth value of the statement "$4 + 3 = 7$" is true. The truth value of the statement "$6 + 2 = 8$" is true.

1.2   Connectives and Truth Tables

By using connectives like "not', "or", "and", we may combine two or more primary statements. The words "or", "and" are called as connectives.

Example 1.3

Consider two statements

  (i)   Rama is a boy.

  (ii) Sita is a girl.

These two statements (i) and (ii) are primary statements. By using connective "and" we can combine these two statements. (iii) Rama is a boy and Sita is a girl.

The statement (iii) is called as compound statement.

The sentences constructed by using two or more primary (or simple) statements and certain sentential connectives are called as compound statements. The simple statements used to form compound statements are called as the components of the compound statement.

To form compound statement we use simple sentences and the connectives "and", "or", "if....then....", "if and only if", etc.

Example 1.4

Suppose that p and q are two statements given by

p: Rama is a boy

q: Sita is a girl

Here p and q are two simple statements. We may combine p and q to get different compound statements:

(i)   p and q: Rama is a boy and Sita is a girl.

(ii) p or q: Rama is a boy or Sita is a girl.

(iii) If p then q: If Rama is a boy then Sita is a girl.

(iv) p if and only if q: "Rama is a boy" if and only if "Sita is a girl".

1.2.1   Negation of a Statement

Associated with every given statement 'p' there is another statement called its negation. The negation of 'p' is nothing but "not p". It is denoted by "~p" or "$\tilde{p}$" or $\bar{p}$ or "⌐p"

It is important to understand that

If p is true then "~p" is false

If p is false then "~p" is true

Example 1.5

(i)   The negation of

      p: Socrates is a man is

      ~p: Socrates is not a man.

(ii) q: Delhi is the Capital of India

      ⌐q (or ~q): Delhi is not the Capital of India

(iii) r: London is a City

      ⌐r (or ~r): London is not a City

Note:

The truth value of a compound statement or negation of a statement depends on the truth values of the statements used to form a compound statement or to form a negation of a statement.

The depending of the truth values of the statements involved will be shown in the form of a table called "Truth Table".

Truth Table for negation is given below

p	~p
T	F
F	T

1.2.2   Conjunction (or meet)

Let p and q be two statements. The conjunction of the statements p, q is the statement "p and q" (denoted by $p \wedge q$). The truth value of $p \wedge q$ is true only if "p is true" and "q is true". In all other cases $p \wedge q$ has the truth value "False" (or F).

The truth table for "Conjunction" is given below

p	q	$p \wedge q$
T	T	T
T	F	F
F	T	F
F	F	F

Truth table for $p \wedge q$

The symbol " $\wedge$ "may be read as "meet". The meaning of the Truth table for $p \wedge q$ is as follows:

(i)   If p is true and q is true then $p \wedge q$ is true;

(ii)   If p is true and q is false then $p \wedge q$ is false;

(iii)   If p is false and q is true then $p \wedge q$ is false;

(iv)   If p is false and q is false then $p \wedge q$ is false.

Example 1.6

Translate the statement

"Rama and Mallikarjun went up the hill"into symbolic form.

Solution: suppose that

p: Rama went up to the hill

q: Mallikarjun went upto the hill

Then $p \wedge q$: Rama went upto the hill and Mallikarjun went upto hill.

Therefore the statement "Rama and Mallikarjun went up the hill" is symbolised as "$p \wedge q$".

1.2.3 Disjunction (or join)

Let p and q be two statements. The disjunction of the statements p, q is the statement "p or q" (denoted by $p \vee q$). The truth value of $p \vee q$ is false only if "p is false" and "q is false". In all other cases $p \vee q$ is true. The truth table for "Disjunction" is given below:

p	q	$p \vee q$
T	T	T
T	F	T
F	T	T
F	F	F

Truth table for p ∨ q

The symbol "$\vee$" may be read as "join".

If p stands for 'I shall purchase a book' and q stands for 'I shall purchase a pencil, then $p \vee q$ stands for the compound statement "I shall purchase a book and a pencil".

The meaning of the truth table for $p \vee q$ is as follows:

(i)   If p is true and q is true then $p \vee q$ is true

(ii) If p is true and q is false then $p \vee q$ is true

(iii) If p is false and q is true then $p \vee q$ is true

(iv) If p is false and q is false then $p \vee q$ is false

Example 1.7

Translate the statement

"Rama or Mallikarjun went up the hill" into symbolic form.

Solution: Suppose that

p: Rama went up to the hill

q: Mallikarjun went upto hill

Then $p \vee q$: Rama went upto the hill or Mallikarjun went upto the hill.

Therefore the statement "Rama or Mallikarjun went upto hill" is symbolised as " $p \vee q$ ".

Example 1.8

Write the following statements in symbolic form
  (i)   Anil and Sunil are rich
  (ii)   Neither Ramu nor Raju is poor
  (iii) It is not true that Ravi and Raju are both rich

[JNTU–H, Nov 2010, Set–1]

Solution:
  (i)   Suppose that
        p: Anil is rich
        q: Sunil is Rich
        Then the given statement can be written in the symbolic form as $p \wedge q$.

  (ii)   write
        p: Ramu is poor
        q: Raju is poor
        ~p: Ramu is not poor
        ~q: Raju is not poor
        The other form of the given statement is "Ramu is not poor" and "Raju is not poor", so the answer is $(\sim p) \wedge (\sim q)$.

  (iii) write
        p: Ravi is rich
        q: Raju is rich
        $p \wedge q$: Ravi is rich and Raju is rich

        $p \wedge q$: Both Ravi and Raju are rich

        $\sim(p \wedge q)$: It is not true that both Ravi and Raju are rich.

Example 1.9

What is the negation of the statement:

  "2 is even and –3 is negative"

[JNTU–H, Nov 2008 (Set–4)]
[JNTU–H, June 2010 (Set–2)]

Solution: Write

     p: 2 is even

     q: −3 is negative

Then $p \wedge q$: 2 is even and −3 is negative.

The negation of $p \wedge q$ is $\neg(p \wedge q)$

     $\neg(p \wedge q) = (\neg p) \vee (\neg q)$ (by demorgan laws)

     $\neg p$: 2 is not even

     $\neg q$: −3 is not negative

Hence the negation of the given statement is $(\neg p) \vee (\neg q)$.

That is

2 is not even (or) −3 is not negative

Example 1.10

What is the compound statement that is true when exactly two of the three statements p, q and R are true?

[JNTU–H, June 2010(Set–2)]

[JNTU–H, Nov 2008 (Set–4)]

Solution: We note that

     $(p \wedge q) \wedge (\neg r)$ is true if "p and q are true" and "r is false".

     $(p \wedge r) \wedge (\neg q)$ is true if "p and r are true" and "q is false".

     $(q \wedge r) \wedge (\neg p)$ is true if "q and r are true" and "p is false".

Hence $\left[(p \wedge q) \wedge (\neg r)\right] \vee \left[(p \wedge r) \wedge (\neg q)\right] \vee \left[(q \wedge r) \wedge (\neg p)\right]$ is the compound statement

that is true when exactly two of three statements p, q and r are true.

1.2.4   Notation

For any two statements p and q

  (i)   The compound statement $\sim (p \wedge q)$ is denoted by "$p \uparrow q$"

  (ii)   The compound statement $\sim (p \vee q)$ is denoted by "$p \downarrow q$"

  (iii)   Note that the symbols "$\uparrow$" and "$\downarrow$" are also connectives.

1.2.5   Draw Truth Table for "p ↑ q"

Solution:

p	q	p ∧ q	p ↑ q
T	T	T	F
T	F	F	T
F	T	F	T
F	F	F	T

Truth table for "p ↑ q"

1.2.6   Draw the Truth Table for "p ↓ q"

Solution:

p	q	p ∨ q	p ↓ q
T	T	T	F
T	F	T	F
F	T	T	F
F	F	F	T

Truth table for "p ↓ q"

1.2.7   Statement Formulas and Truth Tables

We know that simple (or primary or atomic) statements contains no connectives. Those statements which contains one or more simple statements and some connectives are called as compound (or composite or molecular) statements.

For example, let p and q be two simple statements. Then ⅂p, p ∧ q, p ∨ q, p ∧ (⅂q), (⅂p) ∨ (⅂q) are some compound statements.

These compound statements are also called as statement formulas derived from the simple statements p and q. In this case, p and q are called as the components of the statement formulas. The truth value of a statement formula depend on the truth value of the primary statements involved in it.

Here ⅂p means negation of p.

⅂(p ∧ q) means negation of (p ∧ q)

Example 1.11

Construct the truth tables for the following statement formulas

(i)   ~(~p) [that is ⅂(⅂p)]
(ii)   q ∧ (⅂q)
(iii)   p ∨ (⅂q)
(iv)   (p ∨ q) ∨ ⅂p

Solution:

(i)

p	~p	~(~p)
T	F	T
F	T	F

Truth table for ~(~p)

Note that the truth value of both p and ~(~p) are same in all cases

(ii)

q	⌐q	q∧(⌐q)
T	F	F
F	T	F

Truth table for q∧(⌐q)

(iii)

p	q	⌐q	p∨(⌐q)
T	T	F	T
T	F	T	T
F	T	F	F
F	F	T	T

Truth table for p∨ (⌐q)

(iv)

p	q	p ∨ q	⌐p	(p ∨ q) ∨ ⌐p
T	T	T	F	T
T	F	T	F	T
F	T	T	T	T
F	F	F	T	T

Truth table for (p ∨ q) ∨ ⌐p

1.2.8   Implication (or Conditional Statement)

Let p and q be two statements. The implication of the statements p, q is a statement "if p, then q" (denoted by p → q). The truth value of "p → q" is false only if "p is true" and "q is false". In all other cases "p → q" has truth value "true" (or T).

The truth table for "implication" is given below.

p	q	p → q
T	T	T
F	T	T
T	F	F
F	F	T

Truth table for p → q

(The symbol "→"may be read as "conditional".) The meaning of the Truth table for
p → q is as follows:

(i)   If p is true and q is true then p → q is true

(ii) If p is false and q is true then p → q is true

(iii) If p is true and q is false then p → q is false

(iv) If p is false and q is false then p → q is true

Example 1.12

Construct the truth table for $(p \land \neg q) \to r$

Solution:

p	q	r	$p \land \neg q$	$(p \land \neg q) \to r$
T	T	T	F	T
T	T	F	F	T
T	F	T	T	T
T	F	F	T	F
F	T	T	F	T
F	T	F	F	T
F	F	T	F	T
F	F	F	F	T

Truth table for $(p \land \neg q) \to r$

Example 1.13

Construct the truth table for $p \to p \lor q$

Solution:

p	q	$p \lor q$	$p \to p \lor q$
T	T	T	T
T	F	T	T
F	T	T	T
F	F	F	T

Truth table for $p \to p \lor q$

1.2.9   Biconditional (or Double Implication)

Let p and q be two statements. The double implication of the statements p, q is a statement "p if and only if q" (denoted by p $\rightleftarrows$ q). The truth value of p $\rightleftarrows$ q is true if "p" and "q" have the same truth values and if false if they have opposite truth values.

The truth table for "Double implication" is given below

p	q	p $\rightleftarrows$ q
T	T	T
F	T	F
T	F	F
F	F	T

Truth table for p $\rightleftarrows$ q

The symbol "$\rightleftarrows$" may be read as "if and only if". The meaning of the truth table for p $\rightleftarrows$ q is as follows:

(i)   If "p" is true and "q" is true then p $\rightleftarrows$ q is true

(ii)   If "p" is false and "q" is true then p $\rightleftarrows$ q is false

(iii)   If "p" is true and "q" is false then p $\rightleftarrows$ q is false

(iv)   If "p" is false and "q" is false then p $\rightleftarrows$ q is true

1.2.10   Construction of the Truth Table for p $\rightleftarrows$ q

Solution: We know that p $\rightleftarrows$ q is nothing but $(p \rightarrow q) \wedge (q \rightarrow p)$

p	q	p $\rightarrow$ q	q $\rightarrow$ p	p $\rightleftarrows$ q
T	T	T	T	T
T	F	F	T	F
F	T	T	F	F
F	F	T	T	T

Truth table for p $\rightleftarrows$ q

1.2.11   Well Formed Formulas

We know that a statement formula is an expression which is a string consisting of variables, parentheses and connectives.

Now we provide a recursive definition for "statement formula". It is often called as a well-formed formula.

A well-formed formula can be generated by the following formula.

Rule-1: A statement symbol (or variable) is a well-formed formula

Rule-2: if x is a well-formed formula then ~x is a well formed formula

Rule-3: if x and y are well formed formulas then $(x \lor y)$, $(x \land y)$, $(x \rightarrow y)$, $(x \rightleftarrows y)$ are also well formed formulas.

A string consisting of statement symbols, parenthesis, connectives is a well formed formula if it can be obtained by finitely many applications of the rules 1, 2 and 3 mentioned above.

For example, p, $\rceil$p, $p \lor q$, $p \land q$, $(\rceil p) \lor q$, $p \rightarrow q$, $(p \rightarrow q) \rightarrow r$, $(p \rightarrow q) \lor (r \rightarrow q)$, $p \rightleftarrows (q \land r)$ are well formed formulas.

Example 1.14

Find the truth table for the propositional (statement) formula

$$(p \rightleftarrows \rceil q) \rightleftarrows (q \rightarrow p)$$

[JNTU–H, Nov 2008 (set–4)]

[JNTU–H, June 2010 (set–2)]

Solution: The truth table for the given statement formula is given below. (as there are only two variables, the truth table consists exactly four rows).

p	q	~q	$p \rightleftarrows \sim q$	$q \rightarrow p$	$(p \rightleftarrows \sim q) \rightleftarrows (q \rightarrow p)$
T	T	F	F	T	F
T	F	T	T	T	T
F	T	F	T	F	F
F	F	T	F	T	F

Example 1.15

Construct the truth table for the following statement $(\sim p \rightleftarrows q) \rightleftarrows (q \rightleftarrows r)$

[JNTU Nov 2010, Set–4]

Solution: The truth table for the given statement is as follows:

p	q	r	~p	~q	$\sim p \rightleftarrows \sim q$	$q \rightleftarrows r$	$(\sim p \rightleftarrows \sim q) \rightleftarrows (q \rightleftarrows r)$
T	T	T	F	F	T	T	T
T	T	F	F	F	T	F	F
T	F	T	F	T	F	F	T
T	F	F	F	T	F	T	F
F	T	T	T	F	F	T	F

Contd...

p	q	r	~p	~q	~ p $\rightleftarrows$ ~ q	q $\rightleftarrows$ r	(~ p $\rightleftarrows$ ~ q) $\rightleftarrows$ (q $\rightleftarrows$ r)
F	T	F	T	F	F	F	T
F	F	T	T	T	T	F	F
F	F	F	T	T	T	T	T

Truth table for the expression $(\sim p \rightleftarrows \sim q) \rightleftarrows (q \rightleftarrows r)$

1.2.12 The Operation ⊕ or Δ

There is one more operation on statements frequently used denoted by ⊕ or Δ. It may be called as the operation "ring sum". The truth table for this operation ⊕ is given below

p	q	p ⊕ q(or p Δ q)
T	T	F
T	F	T
F	T	T
F	F	F

Example 1.16

If p and q are two statements, then show that the statement $(p \oplus q) \vee (p \downarrow q)$ is equivalent to $p \uparrow q$.

[JNTUH, June 2010, Set–4]

Solution: The equivalence of two compound statements is shown in the following truth table:

Truth Table

p (1)	q (2)	p ⊕ q (3)	p ↓ q (4)	(p ⊕ q) ∨ (p ↓ q) (5)	p ↑ q (6)
T	T	F	F	F	F
T	F	T	F	T	T
F	T	T	F	T	T
F	F	F	T	T	T

Since values in columns (5) and (6) are same, therefore the two given compound statements are equivalent.

Example 1.17

If p and q are two statements, then show that the statement $(p \uparrow q) \oplus (p \wedge q)$ is equivalent to $(p \vee q) \wedge (p \downarrow q)$.

Solution: The truth table of two compound statements is as follows:

Truth table

p (1)	q (2)	p↑q (3)	p∧q	(p↑q)⊕(p∧q) (4)	p∨q (5)	p↓q (6)	(p∨q)∧(p↓q) (7)
T	T	F		F	T	F	F
T	F	T		F	T	F	F
F	T	T		F	T	F	F
F	F	T		F	F	T	F

Since values in columns (4) and (7) are same, therefore the two given are equivalent.

1.3   Tautology and Contradiction

Tautology and contradiction are two important concepts in the study of logic.

1.3.1   Tautology

Tautology is an expression which has truth value 'T' for all possible values of the statement variables involved in the expression.

Example 1.18

Show that $p \vee (\sim p)$ is a tautology.

Solution:

p	~p	p∨(~p)
T	F	T
F	T	T

Table for $p \vee (\neg p)$

Observe the Table for $p \vee (\neg p)$. It is clear that in all cases, the truth value of $p \vee (\neg p)$ is true. Hence $p \vee (\neg p)$ is a tautology.

Example 1.19

Prove that $(p \vee q) \vee (\neg p)$ is a tautology.

Solution:

p	q	p∨q	~p	(p∨q)∨(~p)
T	T	T	F	T
T	F	T	F	T
F	T	T	T	T
F	F	F	T	T

Truth table for $(p \vee q) \vee (\sim p)$

Observe the table for $(p \vee q) \vee (\sim p)$. It is clear that in all cases, the truth value of $(p \vee q) \vee (\sim p)$ is true. Hence $(p \vee q) \vee (\sim p)$ is a tautology.

Example 1.20

Show the implication: $[(p \rightarrow q) \rightarrow q] \rightarrow p \vee q$

[JNTU Nov 2008, Set no.3]

Solution: We have to show that "$[(p \rightarrow q) \rightarrow q] \rightarrow p \vee q$" is a tautology

p	q	p → q	(p → q) → q	p ∨ q	[(p → q) → q] → p ∨ q
T	T	T	T	T	T
T	F	F	T	T	T
F	T	T	T	T	T
F	F	T	F	F	T

Truth table for $[(p \rightarrow q) \rightarrow q] \rightarrow p \vee q$

Observe the truth table for "$[(p \rightarrow q) \rightarrow q] \rightarrow p \vee q$". In all cases the truth value of the statement is T. Therefore given statement "$[(p \rightarrow q) \rightarrow q] \rightarrow p \vee q$" is a tautology.

Example 1.21

Show that the proposition is a tautology. $[(p \vee \sim q) \wedge (\sim p \vee \sim q] \vee q$

[JNTU Nov 2008, Set no.3]

Solution: The truth table for given statement is as follows:

Let $E = [(p \vee \sim q) \wedge (\sim p \vee \sim q] \vee q$

p	q	~p	~q	(p ∨ ~q)	(~p ∨ ~q)	(p ∨ ~q) ∧ (~p ∨ ~q)	E
T	T	F	F	T	F	F	T
T	F	F	T	T	T	T	T
F	T	T	F	F	T	F	T
F	F	T	T	T	T	T	T

Truth table for $[(p \vee \sim q) \wedge (\sim p \vee \sim q] \vee q$

Observe the truth table for the given statement $[(p \vee \sim q) \wedge (\sim p \vee \sim q)] \vee q$. In all the cases, the truth values of the statement are T. Therefore the given statement is a tautology.

Example 1.22

Determine which of the following is not a Tautology.

[JNTU June 2010 Set no.1]

(i) $\sim (p \to q) \to p$

(ii) $\sim (p \to q) \to \sim q$

(ii) $\sim p \wedge (p \vee q) \to q$

(iii) $(p \to q) \wedge (q \to p)$

Solution:

(i)   The truth tables for the given statement "$\sim (p \to q) \to p$" is as follows:

p	q	p → q	~(p → q)	(~(p → q) → p)
T	T	T	F	T
T	F	F	T	T
F	T	T	F	T
F	F	T	F	T

Truth table for $\sim (p \to q) \to p$

Observe the truth table for "$\sim (p \to q) \to p$". In all the cases, the truth value of the statement "$\sim (p \to q) \to p$"is true. Therefore "$\sim (p \to q) \to p$"is a tautology.

(ii)   The truth table for the given statement "$\sim (p \to q) \to \sim q$" is as follows:

p	q	~q	p → q	~(p → q)	~(p → q) →~ q
T	T	F	T	F	T
T	F	T	F	T	T
F	T	F	T	F	T
F	F	T	T	F	T

Truth table for $\sim (p \to q) \to \sim q$

Observe the truth table for "$\sim (p \to q) \to \sim q$". In all cases, the truth value of the statement "$\sim (p \to q) \to \sim q$"are T. Hence "$\sim (p \to q) \to \sim q$" is a tautology.

(iii)   The truth table for the given statement "$\sim p \wedge (p \vee q) \to q$" is as follows

p	q	~p	(p ∨ q)	~p ∧ (p ∨ q)	~p ∧ (p ∨ q) → q
T	T	F	T	F	T
T	F	F	T	F	T
F	T	T	T	T	T
F	F	T	F	F	T

Truth table for $\sim p \wedge (p \vee q) \to q$

Observe the truth table for "$\sim p \wedge (p \vee q) \to q$". In all the cases, the truth values of the statement "$\sim p \wedge (p \vee q) \to q$" are T. Hence the statement "$\sim p \wedge (p \vee q) \to q$" is a tautology.

(iv) The truth table for the given statement "$(p \to q) \wedge (q \to p)$" is as follows:

p	q	p → q	q → p	(p → q)∧(q → p)
T	T	T	T	T
T	F	F	T	F
F	T	T	F	F
F	F	T	T	T

Truth table for (p → q) ∧ (q → p)

Observe the truth table for "$(p \to q) \wedge (q \to p)$". In all the cases, the truth values of the statement "$(p \to q) \wedge (q \to p)$" are not true. Hence the given statement "$(p \to q) \wedge (q \to p)$" is not a tautology.

1.3.2   Contradiction (or Fallacy)

A contradiction (or Fallacy) is an expression which has truth value 'F' for all possible values of the statement variables involved in that expression.

Example 1.23

Show that $p \wedge (\sim p)$ is a contradiction.

Solution:

p	~p	p∧ ~ p
T	F	F
F	T	F

Truth table for p∧ ~ p

Observe the truth table for "$p \wedge (\sim p)$". It is clear that in all cases, the truth value of "$p \wedge (\sim p)$" is false. Hence $p \wedge (\sim p)$ is a contradiction.

Example 1.24

Prove that $(p \wedge q) \rightleftarrows (\sim p \vee \sim q)$ is a contradiction

Solution:

p	q	p∧q	~p	~q	(~ p∨ ~ q)	p∧q ⇄ (~ p∨ ~ q)
T	T	T	F	F	F	F
T	F	F	F	T	T	F
F	F	F	T	T	T	F
F	F	F	T	T	T	F

Truth table for $(p \wedge q) \rightleftarrows (\sim p \vee \sim q)$

Observe the truth table for "$(p \wedge q) \rightleftarrows (\sim p \vee \sim q)$". It is clear that in all cases, the truth value of "$(p \wedge q) \rightleftarrows (\sim p \vee \sim q)$" is false. Hence $(p \wedge q) \rightleftarrows (\sim p \vee \sim q)$ is a contradiction.

1.3.3   Contingency

A proposition (statement) that is neither a tautology nor a contradiction is called a contingency.

Example 1.25

Prove that the proposition (statement) "$(p \rightarrow q) \rightarrow (p \wedge q)$" is a contingency.

[JNTU Nov 2008 (Set–1)]

Solution: Given statement is $(p \rightarrow q) \rightarrow (p \wedge q)$.

The truth table of given statement "$(p \rightarrow q) \rightarrow (p \wedge q)$" is given below.

p	q	p → q	(p ∧ q)	(p → q) → (p ∧ q)
T	T	T	T	T
T	F	F	F	T
F	T	T	F	F
F	F	T	F	F

Truth table for $(p \rightarrow q) \rightarrow (p \wedge q)$

Observe the truth table for the statement "$(p \rightarrow q) \rightarrow (p \wedge q)$". Clearly all the truth values of given statement is neither "T" nor "F". Therefore it is neither a tautology nor a contradiction. Hence the given statement is a contingency.

Example 1.26

Construct the truth table and show that $((p \wedge \sim q) \rightarrow r) \rightarrow (p \rightarrow (q \vee r))$ is a tautology.

Solution: Let E denote the given expression. First we form the truth table.

p	q	r	~q	p∧ ~ q	(p∧ ~ q) → r	p → (q ∨ r)	E
T	T	T	F	F	T	T	T
T	T	F	F	F	T	T	T
T	F	T	T	T	T	T	T
T	F	F	T	T	F	F	T
F	T	T	F	F	T	T	T
F	T	F	F	F	T	T	T
F	F	T	T	F	T	T	T
F	F	F	T	F	T	T	T

From the truth table we conclude that E (the given statement formula) is a tautology

Example 1.27

(In this example, we denote the truth value T by '1' and the truth value F by '0'). Consider the statement q where p, q and r are three propositions.

Show that the given statement formula is a contingency.

Truth table for $(p \vee q) \wedge \bar{r}$

p	q	r	p∨q	$\bar{r}$	$(p \vee q) \wedge \bar{r}$
0	0	0	0	1	0
0	0	1	0	0	0
0	1	0	1	1	1
0	1	1	1	0	0
1	0	0	1	1	1
1	0	1	1	0	0
1	1	0	1	1	1
1	1	1	1	0	0

From the truth table, we understand that $(p \vee q) \wedge \bar{r}$ is a contingency.

Example 1.28

Show that $[p \wedge (p \vee q)] \wedge \bar{p}$ is a contradiction.

Solution: Now we write down the truth table

p	q	p∨q	p∧(p∨q)	$\bar{p}$	$[p \wedge (p \vee q)] \wedge \bar{p}$
0	0	0	0	1	0
0	1	1	0	1	0
1	0	1	1	0	0
1	1	1	1	0	0

Observing the table, we can conclude that $[p \wedge (p \vee q)] \wedge \bar{p}$ is always false. Hence $[p \wedge (p \vee q)] \wedge \bar{p}$ is a contradiction.

Example 1.29

Show that $[p \wedge (p \vee q)] \vee \bar{p}$

Solution: Now we write down the truth table.

p	q	p ∨ q	p ∧ (p ∨ q)	$\bar{p}$	[p ∧ (p ∨ q)] ∨ $\bar{p}$
0	0	0	0	1	1
0	1	1	0	1	1
1	0	1	1	0	1
1	1	1	1	0	1

Observing the table we can conclude that $[p \wedge (p \vee q)] \vee \bar{p}$ is always true. Hence $[p \wedge (p \vee q)] \vee \bar{p}$ is a tautology.

Example 1.30

Show that the following statement is a tautology $\bar{p} \wedge (p \to q) \to \bar{q}$.

[JNTUH Nov 2010, Set No.2]

Solution: Given statement is $\bar{p} \wedge (p \to q) \to \bar{q}$

Now we construct the truth table for given statement

p	q	$\bar{p}$	$\bar{q}$	p → q	$\bar{p} \wedge (p \to q)$	$\bar{p} \wedge (p \to q) \to \bar{q}$
T	T	F	F	T	F	T
T	F	F	T	F	F	T
F	T	T	F	T	T	F
F	F	T	T	T	T	T

Given statement is not a tautology.

1.4   Equivalence of Statements /Formulas

1.4.1   Statements/Formulas

Let A and B be two statements involving the variables $p_1, p_2, ...p_n$ then we say A and B are equivalent if the truth value of A is equal to the truth value of B for every 2^n possible sets of truth values assigned to $p_1, p_2, ...p_n$ and is denoted by $A \Leftrightarrow B$. In other words $A \Leftrightarrow B$ is a tautology.

Example 1.31

Verify that $(p \to q) \Leftrightarrow (\sim p \vee q)$

Solution: Given that $A : (p \to q)$, $B : (\sim p \vee q)$ are two statements. We have to verify $A \Leftrightarrow B$.

The truth table for given statements is given below.

p	q	~p	p→q	~p∨q
T	T	F	T	T
T	F	F	F	F
F	T	T	T	T
F	F	T	T	T

Truth table for A & B

Observe the truth table for statements A & B. The truth values of A and B are equal in all cases. Therefore, the statement A is equivalent to statement B.

That is $A \Leftrightarrow B$ is a tautology.

Hence $(p \to q) \Leftrightarrow (\sim p \vee q)$

Example 1.32

Show that $\sim(\sim p)$ is equivalent to p.

Solution: Let A: p, B: $\sim(\sim p)$ be two statements.

The truth table for the statements A & B is given below

p	~p	~(~p)
T	F	T
T	F	T
F	T	F
F	T	F

Truth table for statements p and $\sim(\sim p)$

Observe the truth table for statements p and $\sim(\sim p)$ truth values of statement p and statement $\sim(\sim p)$ are identical (or equal for all cases). Therefore statement p is equivalent to statement $\sim(\sim p)$.

1.4.2 Equivalent Formulas

1. $\left. \begin{array}{l} p \vee p \Leftrightarrow p \\ p \wedge p \Leftrightarrow p \end{array} \right\}$ Idempotent laws

2. $\left. \begin{array}{l} p \vee (q \vee r) \Leftrightarrow (p \vee q) \vee r \\ p \wedge (q \wedge r) \Leftrightarrow (p \wedge q) \wedge r \end{array} \right\}$ Associative laws

3. $\left.\begin{array}{l} p \vee q \Leftrightarrow q \vee p \\ p \wedge q \Leftrightarrow q \wedge p \end{array}\right\}$ Commutative laws

4. $\left.\begin{array}{l} p \vee (q \wedge r) \Leftrightarrow (p \vee q) \wedge (p \vee r) \\ p \wedge (q \vee r) \Leftrightarrow (p \wedge q) \vee (p \wedge r) \end{array}\right\}$ Distributive laws

5. $P \vee F \Leftrightarrow P$
   $P \wedge F \Leftrightarrow F$

6. $P \vee T \Leftrightarrow T$
   $P \wedge T \Leftrightarrow P$

7. $\left.\begin{array}{l} P \vee \sim P \Leftrightarrow T \\ P \wedge \sim P \Leftrightarrow F \end{array}\right\}$ Complement laws

8. $\left.\begin{array}{l} p \vee (p \wedge q) \Leftrightarrow p \\ p \wedge (p \vee q) \Leftrightarrow p \end{array}\right\}$ Absorption laws

9. $\left.\begin{array}{l} \sim(p \vee q) \Leftrightarrow \sim p \wedge \sim q \\ \sim(p \wedge q) \Leftrightarrow \sim p \vee \sim q \end{array}\right\}$ De Morgan's law

Example 1.33

Show that the following statements are logically equivalent (without using truth table).

$$\sim(p \vee (\sim p \wedge q)) \Leftrightarrow (\sim p \wedge \sim q)$$

[JNTU Nov 2010 Set No.2]

Solution: Given statements are A : $\sim(p \vee (\sim p \wedge q))$, B : $(\sim p \wedge \sim q)$ we have to show the statements A and B are equivalent. Consider statement A : $\sim[p \vee (\sim p \wedge q)]$

$\Leftrightarrow \sim[(p \vee \sim p) \wedge (p \vee q)]$ [by distributive law]

$\Leftrightarrow \sim[T \wedge (p \vee q)]$ [since $p \vee \sim p$ is a tautology]

$\Leftrightarrow \sim(p \vee q)$ (law 5)

$\Leftrightarrow \sim p \wedge \sim q$ [by Demorgan's law]

Example 1.34

Show that the following statements are logically equivalent (without using truth table)

$$(p \rightarrow q) \wedge (p \rightarrow r) \Leftrightarrow p \rightarrow (q \wedge r)$$

[JNTU Nov. 2010 Set no.4]

Solution: Consider the statement

$(p \to q) \wedge (p \to r)$

$\Leftrightarrow (\sim p \vee q) \wedge (p \to r)$ [See example 1.4.2 since $(p \to q) \Leftrightarrow (\sim p \vee q)$]

$\Leftrightarrow (\sim p \vee q) \wedge (\sim p \vee r)$ [since $(p \Rightarrow r) \Leftrightarrow (\sim p \vee r)$]

$\Leftrightarrow [\sim p \vee (q \wedge r)]$ [by Distributive law]

$\Leftrightarrow [p \to (q \wedge r)]$ [since $(p \to q) \Leftrightarrow (\sim p \vee q)$]

Therefore, the statements " $(p \to q) \wedge (p \to r)$ " and " $p \to (q \wedge r)$ " are equivalent.

Example 1.35

Show that the value of $\left[(p \downarrow p) \downarrow (q \downarrow q) \right]$ is logically equivalent to $p \wedge q$.

[JNTU June 2010 Set No.3]

Solution: We know that " $p \downarrow q$ " is defined as " $\sim (p \vee q)$ "

That is $(p \downarrow q) \Leftrightarrow \sim (p \vee q)$ .....(1.1)

Consider the statement

$\left[(p \downarrow p) \downarrow (q \downarrow q) \right] \Leftrightarrow \ \sim \left[(p \downarrow p) \vee (q \downarrow q) \right]$

$\Leftrightarrow \ \sim \left[\sim (p \vee p) \vee \sim (q \vee q) \right]$

(by the definition of " $\downarrow$ ")

$\Leftrightarrow \ \sim \left[\sim p \vee \sim q \right]$

$(p \vee p \Leftrightarrow p$ and $q \vee q \Leftrightarrow q$)

$\Leftrightarrow \sim \left[\sim (p \wedge q) \right]$

[Demorgan law: $\sim (p \wedge q) = (\sim p \vee \sim q)$]

$\Leftrightarrow p \wedge q$

Therefore, the given statements $\left[(p \downarrow p) \downarrow (q \downarrow q) \right]$ and $p \wedge q$ are equivalent.

Example 1.36

Show that " $p \to q$ " and " $\sim q \to \sim p$ " are equivalent (use truth table).

Solution:

p	q	p → q	~q	~p	~q → ~p
T	T	T	F	F	T
T	F	F	T	F	F
F	T	T	F	T	T
F	F	T	T	T	T

Truth table for both " p → q " and " ~ q → ~ p "

Observe the truth table for "p → q" and "~ q → ~ p". The truth values for both the statements "p → q" and "~ q → ~ p" are identical in all cases. Hence "p → q" and "~ q → ~ p" are equivalent statements.

Example 1.37

Let n be a find positive integer

Consider the statements

p: n is an even number

q: n + 1 is an odd number

prove that p and q are equivalent

Solution: Suppose that p is true. Then n is an even number. Since n is even, we know that (n + 1) is an odd number. Hence q is true.

Similarly if (n + 1) is an odd number then n = (n + 1) −1 is even. Hence $q \to p$ is true. Therefore $p \Leftrightarrow q$. That is p and q are two different equivalent statements.

Example 1.38

Prove that $\sim (p \vee q) \Leftrightarrow \sim p \wedge \sim q$
Solution:

p	q	~ (p ∨ q)	~p ∧ ~q
T	T	F	F
T	F	F	F
F	T	F	F
F	F	T	T

Hence $\sim (p \vee q) \Leftrightarrow \sim p \wedge \sim q$

1.5   Duality Law and Tautological Implication

1.5.1   Duality Law

Let A and B be any two formulas. Then A and B are said to be dual each other, if either one can be obtained from the other by replacing " $\wedge$ " by " $\vee$ " and " $\vee$ " by " $\wedge$ ".

Note:

   (i)   The connectives " $\wedge$ " and " $\vee$ " are dual each other.

  (ii)   The dual of the variable "T" is "F" and the dual of F is T.

Example 1.39

Write the dual of the following.

   (i) $(p \vee q) \wedge r$

  (ii) $(p \wedge q) \vee T$

 (iii) $\sim (p \vee q) \wedge (p \vee \sim (q \wedge \sim s))$

Solution: The duals of the given formulas (statements) as follows:

   (i)   " $(p \wedge q) \vee r$ " is the dual of " $(p \vee q) \wedge r$ "

  (ii) $(p \vee q) \wedge T$ is the dual of $(p \wedge q) \vee T$

 (iii) $\sim (p \wedge q) \vee (p \wedge \sim (q \vee \sim s))$ is the dual of $\sim (p \vee q) \wedge (p \vee \sim (q \wedge \sim s))$

1.5.2   Tautological Implications

A statement A is said to tautologically imply a statement B if and only if $A \rightarrow B$ is a tautology.

We denote this fact by $A \Rightarrow B$ which is read as "A implies B".

Note:

   (i)   " $\Rightarrow$ " is not a connective, " $A \Rightarrow B$ " is not a statement formula.

  (ii) $A \Rightarrow B$ states that " $A \rightarrow B$ is a tautology" or "A tautologically implies B".

The connectivities $\wedge$, $\vee$ and $\rightleftarrows$ are symmetric in the sense that

$$p \wedge q \Leftrightarrow q \wedge p$$

$$p \vee q \Leftrightarrow q \vee p \quad \text{and}$$

$$p \rightleftarrows q \Leftrightarrow q \rightleftarrows p \text{ but } p \rightarrow q \text{ is not equivalent to } q \rightarrow p.$$

1.5.3   Converse

For any statement formula $p \rightarrow q$, the statement formula $q \rightarrow p$ is called its converse.

1.5.4   Inverse

For any statement formula $p \rightarrow q$, the statement formula $\sim p \rightarrow \sim q$ is called its inverse.

1.5.5   Contrapositive

For any statement formula $p \rightarrow q$, the statement formula "$\sim q \rightarrow \sim p$" is called its Contrapositive.

Write down Contrapositive of the following statement.

If Rama have Rs.100/- he will spend Rs. 50/- for his friend Krishna.

Solution: write p: "Rama have Rs. 100/-"

q: "Rama spends Rs. 50/- for krishna"

Given statement is "$p \rightarrow q$".

The Contrapositive of "$p \rightarrow q$" is "$\sim q \rightarrow \sim p$".

Now $\sim q$: "Rama does not spend Rs.50/- for Krishna"

$\sim p$: Rama does not have Rs. 100/-

The required statement is as follows

If "Rama does not spend Rs. 50/- for Krishna" then "Rama does not have Rs. 100/-".

1.5.6   Some Implications

The following implications have important applications. All of them can be proved by truth table or by any other methods.

$$p \wedge q \Rightarrow p \qquad\qquad .....(1.2)$$

$$p \vee q \Rightarrow q \qquad\qquad .....(1.3)$$

$$p \Rightarrow (p \vee q) \qquad\qquad .....(1.4)$$

$$\sim p \Rightarrow (p \rightarrow q) \qquad\qquad .....(1.5)$$

$$Q \Rightarrow (p \rightarrow q) \qquad\qquad .....(1.6)$$

$$\sim (p \rightarrow q) \Rightarrow p \qquad\qquad .....(1.7)$$

$$\sim (p \rightarrow q) \Rightarrow \sim q \qquad\qquad .....(1.8)$$

$$p \wedge (p \to q) \Rightarrow \sim p \qquad \qquad \qquad .....(1.9)$$

$$\sim p \wedge (p \vee q) \Rightarrow q \qquad \qquad \qquad .....(1.10)$$

$$(p \to q) \wedge (q \to r) \Rightarrow (p \to r) \qquad \qquad .....(1.11)$$

$$(p \vee q) \wedge (p \to r) \wedge (q \to r) \Rightarrow r \qquad .....(1.12)$$

$$(\sim q) \wedge (p \to q) \Rightarrow (\sim p) \qquad \qquad .....(1.13)$$

Example 1.40

Show that $(\sim q) \wedge (p \to q) \Rightarrow (\sim p)$

Solution: we have to show that the statement "$\sim q \wedge (p \to q)$" is tautological implication to statement "$\sim p$"

Suppose that $\sim q \wedge (p \to q)$ has truth value "T". This means, both "$\sim q$" and "$p \to q$"has the truth value T.

This shows that "q" has truth value "F" and $p \to q$ has truth value T. Since the truth value of q is F, by the implication (or conditional) table. We conclude that the truth value of p is F.

Therefore the truth value of "$\sim p$" is T. Hence "$\sim q \wedge (p \to q) \Rightarrow \sim p$"

Example 1.41

Construct the truth tables of converse, inverse and Contrapositive of the proposition (p→q).

[JNTU, June 2010, Set No.3]

Solution: Given proposition is "$p \to q$"

   (i)   The converse of the given proposition is "$q \to p$".
   (ii)   The inverse of the given proposition is "$\sim p \to \sim q$".
   (iii) The Contrapositive of the given proposition is "$\sim q \to \sim p$".

   (i)   The truth table of the converse of the proposition "$p \to q$" is as follows:

p	q	(p → q)	q → p
T	T	T	T
T	F	F	T
F	T	T	F
F	F	T	T

Truth table for proposition "$q \to p$"

(Converse of proposition "$p \to q$")

(ii) The truth table for the inverse of the proposition "$p \rightarrow q$" is as follows:

p	q	~p	~q	p → q	~p →~ q
T	T	F	F	T	T
T	F	F	T	F	T
F	T	T	F	T	F
F	F	T	T	T	T

Truth table for the proposition "$\sim p \rightarrow \sim q$"

(Inverse proposition of proposition "$p \rightarrow q$")

(iii) The truth table for the Contrapositive of the proposition "$p \rightarrow q$" is as follows:

p	q	~p	~q	p → q	~q →~ p
T	T	F	F	T	T
T	F	F	T	F	F
F	T	T	F	T	T
F	F	T	T	T	T

Truth table for the proposition "$\sim q \rightarrow \sim p$"

(Contrapositive of the proposition "$p \rightarrow q$")

1.6   Normal Forms

Suppose that A $(P_1, P_2,...P_n)$ be a statement formula where $P_1, P_2,...P_n$ are the atomic variables. Each P_i have truth value T (or 1) or F (or 0). Hence the n-table $(P_1, P_2,...P_n)$ have 2^n values. We can form the truth table for $A(P_1, P_2,...P_n)$ with 2^n rows.

If for all 2^n values of $(P_1, P_2,...P_n)$ the value of A $(P_1, P_2,...P_n)$ is T (or 1) then the statement formula is said to be identically true. In this case we also say that A $(P_1, P_2,...P_n)$ is a tautology.

If for all 2^n values of $(P_1, P_2,...P_n)$ the value of $A(P_1, P_2,...P_n)$ is F (or 0) then the statement formula $A(P_1, P_2,...P_n)$ is said to be identically false. In this case, we also say that $A(P_1, P_2,...P_n)$ is a contradiction.

If the truth value of $A(P_1, P_2,...P_n)$ is True (T or 1) for atleast one value of $(P_1, P_2,...P_n)$ then $A(P_1, P_2,...P_n)$ is said to be satisfiable.

1.6.1 Decision Problem

The problem of determining (in a finite number of steps) whether the given statement formula is a tautology (or) a contradiction (or) satisfiable is called as a decision problem.

So every problem in the statement calculus has a solution (we can decide this by forming a truth table for the given statement formula.

Now we study different forms called as normal forms

(i)   Disjunctive Normal Form (DNF)

(ii) Conjunctive Normal Form (CNF)

(iii) Principal Disjunctive Normal Form (PDNF)

(iv) Principal Conjunctive Normal Form (PCNF)

1.6.2 DNF

Let $X_1, X_2,...X_n$ be n given atomic variables and $\overline{X}_1, \overline{X}_2,...\overline{X}_n$ (or $(\sim X_1, \sim X_2... \sim X_n)$ are the negations of $X_1, X_2,...X_n$ respectively.

Product of same elements from $\left\{X_1, X_2...X_n, \overline{X}_1, \overline{X}_2...\overline{X}_n\right\}$ is called as an elementary product.

Sum of some elements from $\left\{X_1, X_2...X_n, \overline{X}_1, \overline{X}_2...\overline{X}_n\right\}$ is called as elementary sum.

For example, $X_i, \overline{X}_i \wedge X_j, \overline{X}_1 \wedge X_3 X_4, X_1 \wedge X_2 \wedge...\wedge X_n, X_1 \wedge X_2 \wedge \overline{X}_3 \wedge X_4 \wedge \overline{X}_5$, are elementary products.

$\overline{X}_1, X_1 \vee X_2, \overline{X}_3 \vee X_4 \vee X_5, X_1 \vee X_2 \vee...\vee X_n, \overline{X}_1 \vee \overline{X}_2 \vee X_3 \vee \overline{X}_4 \vee X_5$ are elementary sums.

A formula which is equivalent to a given formula and which consists of a sum of elementary products is called as Disjunctive Normal Form (DNF) of the given statement formula.

1.6.3 How to Find DNF

Suppose a statement formula $A(P_1, P_2,...P_n)$ is given. If it is in the form: sum of elementary products then it is already in DNF. If it is not in the form of DNF then we use some known results or formulas or axioms. Most of the cases when '$\rightarrow$' presents, then we use the known result.

$$P \rightarrow Q \Leftrightarrow \rceil P \vee Q. \text{ [that is } \overline{P} \vee Q\text{].}$$

We use distributive laws, demorgan laws, Commutative and associative laws, etc.

Example 1.42

Find DNF of $X \wedge (X \rightarrow Y)$

Solution: $X \wedge (X \to Y)$

$\Leftrightarrow X \wedge (\daleth X \vee Y)$ [by a known result $X \to Y \Leftrightarrow \overline{X} \vee Y$]

$\Leftrightarrow [X \wedge (\daleth X)] \vee [X \wedge Y]$ [by distributive law]

Now $[X \wedge \daleth X] \vee [X \wedge Y]$ is the sum of two elementary product terms $X \wedge \daleth X$ and $X \wedge Y$. Hence $[X \wedge \daleth X] \vee [X \wedge Y]$ is the DNF of the given statement formula $X \wedge (X \to Y)$.

1.6.4   CNF

A formula which is equivalent to a given formula and which consists of a product of elementary sums is called as Conjunctive Normal Form (CNF) of the given statement formula.

Example 1.43

Find CNF for the statement formula $X \wedge (X \to Y)$

Solution:

$X \wedge (X \to Y)$

$\Leftrightarrow X \wedge (\overline{X} \vee Y)$

The formula $X \wedge (\overline{X} \vee Y)$ is a product of sums: X and $\overline{X} \vee Y$. So it is in CNF. Hence $X \wedge (\overline{X} \vee Y)$ is a CNF for $X \wedge (X \to Y)$.

1.6.5   Principal Disjunctive Normal Form (PDNF)

Let P_1, P_2, ..., P_n be n statement variables. The expression $P_1^* \wedge P_2^* \wedge ... \wedge P_n^*$ where P_i^* is either P_i or $\sim P_i$ is called a minterm. There are 2^n such minterms.

The express $P_1^* \vee P_2^* \vee ... \vee P_n^*$, P_i^* is either P_i or $\sim P_i$ is called a maxterm. There are 2^n such maxterms.

Let P, Q, R be the three variables.

Then the minterms are:

$$P \wedge Q \wedge R, P \wedge Q \wedge \sim R, P \wedge \sim Q \wedge R, P \wedge \sim \wedge \sim R, \sim P \wedge Q \wedge R, \sim P \wedge Q \wedge \sim R,$$

$$\sim P \wedge \sim Q \wedge R, \sim P \wedge \sim Q \wedge \sim R.$$

For a given formula, an equivalent formula consisting of disjunction's of minterms only is known as its Principal Disjunctive Normal Form (PDNF) or sum of products canonical form.

Example 1.44

Find PDNF for $(\overline{X} \vee Y)$.

Solution: $\overline{X} \vee Y \Leftrightarrow (\overline{X} \wedge 1) \vee (Y \wedge 1)$ (since A $\wedge$ 1 = A for all A)

$$\Leftrightarrow \left[\overline{X} \wedge (Y \vee \overline{Y})\right] \vee \left[Y \wedge (X \vee \overline{X})\right] \text{ (since } A \vee \overline{A} = 1)$$

$$\Leftrightarrow \left[(\overline{X} \wedge Y) \vee (\overline{X} \wedge \overline{Y})\right] \vee \left[(Y \wedge X) \vee (Y \wedge \overline{X})\right] \text{ (by distributive law)}$$

$$\Leftrightarrow (\overline{X} \wedge Y) \vee (\overline{X} \wedge \overline{Y}) \vee (X \wedge Y) \vee (\overline{X} \wedge Y)$$

$$\Leftrightarrow (\overline{X} \wedge Y) \vee (\overline{X} \wedge \overline{Y}) \vee (X \wedge Y) \text{ (since } A \vee A = A)$$

Hence $(\overline{X} \wedge Y) \vee (\overline{X} \wedge \overline{Y}) \vee (X \wedge Y)$ is the PDNF for $(\overline{X} \wedge Y)$

The following notation will be used in the next coming Black box method of finding PDNF. For convenience often we are denoting ~p by $\overline{p}$.

We also use the following notation.

Suppose there are three atomic variables p, q, r. Then observe the following:

Binary Notation			Expression		
0	0	0	$\overline{p}$	$\overline{q}$	$\overline{r}$
0	0	1	$\overline{p}$	$\overline{q}$	r
0	1	0	$\overline{p}$	q	$\overline{r}$
0	1	1	$\overline{p}$	q	r
1	0	0	p	$\overline{q}$	$\overline{r}$
1	0	1	p	$\overline{q}$	r
1	1	0	p	q	$\overline{r}$
1	1	1	p	q	r

It is clear that $\overline{p}\,\overline{q}\,\overline{r}$ is the product term (or related expression) for 000; $pq\,\overline{r}$ is the related expression for 110.

1.6.6   Black Box Method (to find PDNF)

Suppose that $A(X_1, X_2, ...X_n)$ be the given statement formula where $X_1, X_2, ...X_n$ are atomic variables and each atomic variable may attain its value either 0 or 1 (that is, False or True).

Form the truth table for $A(X_1, X_2, ...X_n)$ that contains 2^n rows.

This truth table determines the PDNF simply by taking each product term that occurs when $A(X_1, X_2, ...X_n)$ takes Value 1.

Example 1.45

Find PDNF for the statement P→Q by Black Box Method (or by using truth tables)

Solution: First we form the truth table for P→Q.

Or

P	Q	P → Q
T	T	T
T	F	F
F	T	T
F	F	T

P	Q	P → Q
1	1	1
1	0	0
0	1	1
0	0	1

Consider the column under P→Q.

These are three 1's in this column.

The 1's are in 1st row, 3rd row and forth row. Consider 1st row in this row P & Q have truth values 1 and 1 respectively. So the related product term is PQ.

Consider 3rd row the truth values of P and Q are 0 and 1 respectively.

So the related product term is $\bar{P}.Q$. Consider 4th row the truth values of P and Q are 0 and 0 respectively. So the related product term is $\bar{P}\,\bar{Q}$.

The PDNF is $PQ \vee \bar{P}Q \vee \bar{P}\bar{Q}$

In other words $(P \wedge Q) \vee (\bar{P} \wedge Q) \vee (\bar{P} \wedge \bar{Q})$.

Example 1.46

Find PDNF for $(\bar{X} \vee Y)$ (by Black Box Method)

(Compare this problem with example 1.6.8)

Solution: First we form truth table for $(\bar{X} \vee Y)$.

X	Y	$\bar{X}$	$\bar{X} \vee Y$
0	0	1	1
0	1	1	1
1	0	0	0
1	1	0	1

Truth table for $\bar{X} \vee Y$

Observe the column of $\left(\overline{X} \vee Y\right)$. These are three's the one's are in first row, second row and fourth row.

Consider row–1

The product term related to row–1 is $\overline{X}\,\overline{Y}$ (because the truth values of X and Y are 0 and 0 respectively)

Consider row–2

The product term related to row–2 is $\overline{X}\,Y$ (because the truth values of X and Y are 0 and 1 respectively)

Consider row–4

The product term related to row–4 is $X\,Y$ (because the truth values of X and Y are 1 and 1 respectively)

Therefore the PDNF is

$$\overline{X}\,\overline{Y} \vee \overline{X}\,Y \vee X\,Y \ \text{ or } \ \left(\overline{X}\wedge\overline{Y}\right)\vee\left(\overline{X}\wedge Y\right)\vee\left(X\wedge Y\right)$$

1.6.7 Principal Conjunctive Normal Form (PCNF)

For a given statement formula, an equivalent formula consisting of the conjunction of maxterms (product of sums) is known as Principal Conjunctive Normal Form (PCNF) (or product of sums canonical form).

1.6.8 How to Find PCNF (by using truth table or through PDNF)

Step 1: Suppose the given statement formula is F.

Step 2: Find $\overline{F}$ (the complement of F)

Step 3: Find the PDNF & $\overline{F}$ (we may use black box method or direct method)

Step 4: PCNF = $\overline{\left(\text{PDNF of }\left(\overline{F}\right)\right)}$

Example 1.47

Find the PCNF of $X \wedge \overline{Y}$.

Sol: We follow the method given in 1.6.13.

Step 1: Consider the given statement formula

$$F = \left(X \wedge \overline{Y}\right)$$

Step 2: Now we find $\overline{F}$.

$$\overline{F} = \overline{\left(X \wedge \overline{Y}\right)} = \overline{X} \vee \left(\overline{\overline{Y}}\right) = \overline{X} \vee Y$$

Step 3: In this step we find the PDNF for $\overline{F} = \overline{X} \vee Y$

X	Y	$\overline{X}$	$\overline{X} \vee \overline{Y}$
0	0	1	1
0	1	1	1
1	0	0	0
1	1	0	1

Truth table for $\left(\overline{X} \vee \overline{Y}\right)$

As in example 1.6.11, we get that $PDNF\left(\overline{F}\right) = \left(\overline{X} \wedge \overline{Y}\right) \vee \left(\overline{X} \wedge Y\right) \vee \left(X \wedge Y\right)$

Step 4: $PCNF(F) = \overline{PDNF\left(\overline{F}\right)}$

$$= \overline{\left(\overline{X} \wedge \overline{Y}\right) \vee \left(\overline{X} \wedge Y\right) \vee \left(X \wedge Y\right)}$$

$$= \overline{\left(\overline{X} \wedge Y\right)} \wedge \overline{\left(\overline{X} \wedge Y\right)} \wedge \overline{\left(X \wedge Y\right)}$$

(by demorgan laws)

$$= \left(\overline{\overline{X}} \vee \overline{\overline{Y}}\right) \wedge \left(\overline{\overline{X}} \vee \overline{Y}\right) \wedge \left(\overline{X} \vee \overline{Y}\right)$$

(by demorgan laws)

$$= \left(X \vee \overline{Y}\right) \wedge \left(X \vee \overline{Y}\right) \wedge \left(\overline{X} \vee \overline{Y}\right)$$

$$= \left(X \vee \overline{Y}\right) \wedge \left(\overline{X} \vee \overline{Y}\right)$$

(since $A \wedge A = A$)

Example 1.48

Construct the principal conjunctive normal form of the propositional formula.

$(\sim p \rightarrow r) \wedge (p \Leftrightarrow q)$.

[JNTUH–June 2010 Set No.1]

[JNTUH–Nov 2008 Set No.1]

Solution: Given propositional formula is $(\sim p \rightarrow r) \wedge (p \Leftrightarrow q)$ we have to find the PCNF (Principal Conjunctive Normal Formal) of $(\sim p \rightarrow r) \wedge (p \Leftrightarrow q)$.

Consider $(\sim p \to r) \wedge (p \Leftrightarrow q)$

$\Leftrightarrow (p \vee r) \wedge \left[(p \to q) \wedge (q \to p) \right]$ [Since $\sim p \to r \Leftrightarrow p \vee r$]

$\Leftrightarrow (p \vee r) \wedge \left[(\sim p \vee q) \wedge (\sim q \vee p) \right]$ [Since $p \to q \Leftrightarrow \sim p \vee q$ & $q \to p \Leftrightarrow \sim q \vee p$]

$\Leftrightarrow (p \vee r) \wedge (\sim p \vee q) \wedge (\sim q \vee p)$

$\Leftrightarrow \left[p \vee r \vee (q \wedge \sim q) \right] \wedge \left[(\sim p \vee q) \vee (r \wedge \sim r) \right] \wedge \left[\sim q \vee p \vee (r \wedge \sim r) \right]$

$\Leftrightarrow \left[(p \vee r \vee q) \wedge (p \vee r \vee \sim q) \right] \wedge \left[(\sim p \vee q \vee r) \wedge (\sim p \vee q \wedge \sim r) \right] \wedge$

$\left[(\sim q \vee p \vee r) \wedge (\sim q \vee p \vee \sim r) \right]$

$\Leftrightarrow (p \vee q \vee r) \wedge (p \vee \sim q \vee r) \wedge (p \vee \sim q \vee \sim r) \wedge (\sim p \vee q \vee r) \wedge (\sim p \vee q \vee \sim r)$

Therefore the principal conjunctive normal formal of the given propositional formula "$(\sim p \to r) \wedge (p \Leftrightarrow q)$" is $(p \vee q \vee r) \wedge (p \vee \sim q \vee r) \wedge (p \vee \sim q \vee \sim r) \wedge (\sim p \vee q \vee r) \wedge (\sim p \vee q \vee \sim r)$.

Example 1.49

Show that the principal conjunctive normal of the formula

$\left[p \to (q \wedge r) \right] \wedge \left[\bar{p} \to (\bar{q} \wedge \bar{r}) \right]$ is $\pi (1, 2, 3, 4, 5, 6)$

[JNTU–Nov 2008 Set No.2]

Solution: Given formula is

$\left[p \to (q \wedge r) \right] \wedge \left[\bar{p} \to (\bar{q} \wedge \bar{r}) \right]$

$\Leftrightarrow \left[\bar{p} \vee (q \wedge r) \right] \wedge \left[\bar{p} \to (\bar{q} \wedge \bar{r}) \right]$ [Since $p \to q \Leftrightarrow \sim p \vee q$]

$\Leftrightarrow \left[\bar{p} \vee (q \wedge r) \right] \wedge \left[p \vee (\bar{q} \wedge \bar{r}) \right]$ [Since $p \to q \Leftrightarrow \sim p \vee q$]

$\Leftrightarrow \left[(\bar{p} \vee q) \wedge (\bar{p} \vee r) \right] \wedge \left[(p \vee \bar{q}) \wedge (p \vee \bar{r}) \right]$ [By Demorgan laws]

$\Leftrightarrow \left[\bar{p} \vee q \vee (r \wedge \bar{r}) \right] \wedge \left[\bar{p} \vee r \vee (q \wedge \bar{q}) \right] \wedge \left[p \vee \bar{q} \vee (r \wedge \bar{r}) \right] \wedge \left[p \vee \bar{r} \vee (q \wedge \bar{q}) \right]$

$\Leftrightarrow \left[\bar{p} \vee q \vee r \right] \wedge \left[\bar{p} \vee q \vee \bar{r} \right] \wedge \left[\bar{p} \vee r \vee q \right] \wedge \left[\bar{p} \vee r \vee \bar{q} \right] \wedge \left[p \vee \bar{q} \vee r \right] \wedge \left[p \vee \bar{q} \vee \bar{r} \right]$

$\wedge \left[p \vee \bar{r} \vee q \right] \wedge \left[p \vee \bar{r} \vee \bar{q} \right]$

$\Leftrightarrow (\bar{p} \vee q \vee r) \wedge (\bar{p} \vee q \vee \bar{r}) \wedge (\bar{p} \vee \bar{q} \vee r) \wedge (p \vee \bar{q} \vee r) \wedge (p \vee q \vee \bar{r}) \wedge (p \vee \bar{q} \vee \bar{r})$ is

the principal conjunctive normal form of given statement formula.

Now by the known representation, we have

1. Can be represented as $\bar{p} \, \bar{q} \, r$ (001)

2.   Can be represented as $\bar{p}\,q\,\bar{r}$ (010)

3.   Can be represented as $\bar{p}\,q\,r$ (011)

4.   Can be represented as $p\,\bar{q}\,\bar{r}$ (100)

5.   Can be represented as $p\,\bar{q}\,r$ (101)

6.   Can be represented as $p\,q\,\bar{r}$ (110)

Therefore, the principal conjunctive normal form of the given formula can be represented as π (1, 2, 3, 4, 5, 6).

It completes the solution.

Example 1.50

Obtain the canonical product of sums of the propositional formula

$$\bar{x}\wedge(\bar{y}\vee z)$$

[JNTUH, June 2010, Set-4]

Solution:

First part:

$$\bar{x}\Leftrightarrow\bar{x}\vee 0$$

$$\Leftrightarrow\bar{x}\vee(y\wedge\bar{y})=(\bar{x}\vee y)\wedge(\bar{x}\vee\bar{y})$$

$$\Leftrightarrow\big[(\bar{x}\vee y)\vee 0\big]\wedge\big[(\bar{x}\vee\bar{y})\vee 0\big]$$

$$\Leftrightarrow\big[(\bar{x}\vee y)\vee(z\wedge\bar{z})\big]\wedge\big[(\bar{x}\vee\bar{y})\vee(z\wedge\bar{z})\big]$$

$$\Leftrightarrow\big[(\bar{x}\vee y\vee z)\wedge(\bar{x}\vee y\vee\bar{z})\big]\wedge\big[(\bar{x}\vee\bar{y}\vee z)\wedge(\bar{x}\vee\bar{y}\vee\bar{z})\big]$$

Second part:

$$(\bar{y}\vee z)\Leftrightarrow(\bar{y}\vee z)\vee 0$$

$$\Leftrightarrow(\bar{y}\vee z)\vee(x\wedge\bar{x})$$

$$\Leftrightarrow(\bar{y}\vee z\vee x)\wedge(\bar{y}\vee z\vee\bar{x})$$

Combining First part and Second part we get that

$$\bar{x}\wedge(\bar{y}\vee z)\Leftrightarrow\big[(\bar{x}\vee y\vee z)\wedge(\bar{x}\vee y\vee\bar{z})\big]\wedge\big[(\bar{x}\vee\bar{y}\vee z)\wedge(\bar{x}\vee\bar{y}\vee\bar{z})\big]$$

$$\wedge\big[(\bar{y}\vee z\vee x)\wedge(\bar{y}\vee z\vee\bar{x})\big]$$

$$\Leftrightarrow(\bar{x}\vee y\vee z)\wedge(\bar{x}\vee y\vee\bar{z})\wedge(\bar{x}\vee\bar{y}\vee z)\wedge(\bar{x}\vee\bar{y}\vee\bar{z})\wedge(\bar{y}\vee z\vee x)$$

The last expression is the required Canonical form.

Example 1.51

Obtain the canonical sum of product form for the propositional formulas:

$$\overline{x} \wedge (\overline{y} \vee z)$$

Solution: Given propositional formula is

$$\overline{x} \wedge (\overline{y} \vee z)$$

We have to obtain product of sums of given statement $\overline{x} \wedge (\overline{y} \vee z)$.

Consider $\quad \overline{x} \wedge (\overline{y} \vee z)$

$$\Leftrightarrow (\overline{x} \wedge \overline{y}) \vee (\overline{x} \wedge z)$$

$$\Leftrightarrow \left[\overline{x} \wedge \overline{y} \wedge (z \vee \overline{z})\right] \vee \left[\overline{x} \wedge \overline{z} \wedge (y \vee \overline{y})\right]$$

$$\Leftrightarrow \left[(\overline{x} \wedge \overline{y} \wedge z) \vee (\overline{x} \wedge \overline{y} \wedge \overline{z})\right] \vee \left[\overline{x} \wedge z \wedge y \ \vee \ \overline{x} \wedge z \wedge \overline{y}\right]$$

$$\Leftrightarrow (\overline{x} \wedge \overline{y} \wedge z) \vee (\overline{x} \wedge \overline{y} \wedge \overline{z}) \vee (\overline{x} \wedge y \wedge z)$$

is the required canonical sum of products form of the given propositional formula

1.7   The Theory of Inference for Statement Calculus

Logic provides the rules of inference, or principles of reasoning. The theory deal with these rules is known as inference theory. This theory is concerned with the inferring of a conclusion from the given hypothesis (or certain premises).

When we derived a conclusion from the set of given statements (or premises) by using the accepted rules, then such process of derivation is known as deduction or a formal proof. In the formal proof, at any stage, the rule of inference used in the derivation is acknowledged.

The conclusion which is arrived following the rules of inference is called as Valid Conclusion; and the argument used is called a Valid Argument.

The actual truth values of the given premises do not play any role in determining the validity of the argument.

Logic was discussed by its ancient founder Aristotle (384 BC – 322 BC) from two quite different points of view. On one hand he regarded logic as an instrument or organ for appraising the correctness or strength of the reasoning. On the other hand, he treated the principles and methods of logic as interesting and important topics of the study. The study of logic will provide the reader certain techniques for testing the validity of a given arguments. Logic provides the theoretical basis for many areas of computer science such as digital logic design, automata theory and computability, and artificial intelligence. In this chapter, we discuss the truth tables, validity of arguments using the rules of inference. Further, we study the various normal forms and logical equivalences using the rules.

1.7.1   Tautology

Suppose that A and B are two statement formulas. We say that "B logically follows from A" or "B is a valid conclusion (or consequence) of the premise A" if and only if A→B is a tautology (that is, A⇒B).

1.7.2   Validity using Truth Table

Let P_1, P_2, ..., P_n be the variables appearing in the premises H_1, H_2, ..., H_m and the conclusion C. Let all possible combinations of truth values are assigned to P_1, P_2, ..., P_n and let the truth values of H_1, H_2, ..., H_m and C are entered in the table. We say that C follows logically from premises H_1, H_2, ..., H_m if and only if $H_1 \wedge H_2 \wedge ... H_m \Rightarrow C$. This can be checked from the truth table using the following procedure:

1.   Look at the rows in which C has the value F.
2.   In every such row if at least one of the values of H_1, H_2, ..., H_m is F then the conclusion is valid.

Example 1.52

Show that the conclusion C: ~P follows from the premises

H_1: ~ P ∨ Q, H_2: ~ (Q ∧ ~ R) and H_3: ~ R.

Solution: Given that C: ~ P, H_1: ~ P ∨ Q, H_2: ~ (Q ∧ ~ R) and H_3: ~ R.

P	Q	R	H_1	H_2	H_3	C
T	T	T	T	T	F	F
T	T	F	T	F	T	F
T	F	T	F	T	F	F
T	F	F	F	T	T	F
F	T	T	T	T	F	T
F	T	F	T	F	T	T
F	F	T	T	T	F	T
F	F	F	T	T	T	T

The row in which C has the truth values F has the situation that at least one of H_1, H_2, H_3 has truth value F. Thus C logically follows form H_1, H_2, and H_3.

1.7.3   Rules of Inference

We now describe the process of derivation by which one demonstrates particular formula is a valid consequence of the given set of premises. The following are the three rules of inference.

Rule P: A premise may be introduced at any point in the derivation.

Rule T: A formula S may be introduced in a derivation if S is tautologically implied by any one or more of the preceding formulas in the derivation.

Rule CP: If we can derive S and R and a set of premises then we can derive R $\rightarrow$ S from the set of premises alone.

1.7.4 Some Implications

$$I_1 \quad : \quad p \wedge q \Rightarrow p$$
$$I_2 \quad : \quad p \wedge q \Rightarrow q$$
(Simplification)

$$I_3 \quad : \quad p \Rightarrow p \vee q$$
$$I_4 \quad : \quad q \Rightarrow p \vee q$$
(addition)

$$I_5 \quad : \quad \bar{p} \Rightarrow p \rightarrow q$$

$$I_6 \quad : \quad q \Rightarrow p \rightarrow q$$

$$I_7 \quad : \quad \overline{p \rightarrow q} \Rightarrow p$$

$$I_8 \quad : \quad \overline{p \rightarrow q} \Rightarrow \bar{q}$$

$$I_9 \quad : \quad p, q \Rightarrow p \wedge q$$

$$I_{10} \quad : \quad \bar{p}, p \vee q \Rightarrow q \qquad \text{[disjunctive syllogism]}$$

$$I_{11} \quad : \quad p, p \rightarrow q \Rightarrow q \qquad \text{[modus ponens]}$$

$$I_{12} \quad : \quad \bar{q}, p \rightarrow q \Rightarrow \bar{p} \qquad \text{[modus tollens]}$$

$$I_{13} \quad : \quad p \rightarrow q, q \rightarrow r \Rightarrow p \rightarrow q \qquad \text{[hypothetical syllogism]}$$

$$I_{14} \quad : \quad p \vee q, p \rightarrow r, q \rightarrow r \Rightarrow r \text{ [dilemma]}$$

1.7.5 Some Equivalences

$$E_1 \quad : \quad \bar{\bar{p}} \Leftrightarrow p \text{ [double negation]}$$

$$E_2 \quad : \quad p \wedge q \Leftrightarrow q \wedge p$$

$$E_3 \quad : \quad p \vee q \Leftrightarrow q \vee p$$

$$E_4 \quad : \quad (p \wedge q) \wedge r \Leftrightarrow p \wedge (q \wedge r)$$

$$E_5 \quad : \quad (p \vee q) \vee r \Leftrightarrow p \vee (q \vee r)$$

$$E_6 \quad : \quad p \wedge (q \vee r) \Leftrightarrow (p \wedge q) \vee (p \wedge r)$$

$$E_7 \quad : \quad p \vee (q \wedge r) \Leftrightarrow (p \vee q) \wedge (p \vee r)$$

$$E_8 \quad : \quad \overline{p \wedge q} \Leftrightarrow \bar{p} \vee \bar{q}$$

E_9	:	$\overline{p \vee q} \Leftrightarrow \overline{p} \wedge \overline{q}$
E_{10}	:	$p \vee p \Leftrightarrow p$
E_{11}	:	$p \wedge p \Leftrightarrow p$
E_{12}	:	$r \vee (p \wedge \overline{p}) \Leftrightarrow r$
E_{13}	:	$r \wedge (p \vee \overline{p}) \Leftrightarrow r$
E_{14}	:	$r \vee (p \vee \overline{p}) \Leftrightarrow T$
E_{15}	:	$r \wedge (p \wedge \overline{p}) \Leftrightarrow F$
E_{16}	:	$p \rightarrow q \Leftrightarrow \overline{p} \vee q$
E_{17}	:	$\overline{p \rightarrow q} \Leftrightarrow p \wedge \overline{q}$
E_{18}	:	$p \rightarrow q \Leftrightarrow \overline{q} \rightarrow \overline{p}$
E_{19}	:	$p \rightarrow (q \rightarrow r) \Leftrightarrow (p \wedge q) \rightarrow r$
E_{20}	:	$\overline{p \rightleftarrows q} \Leftrightarrow (p \rightleftarrows \overline{q})$
E_{21}	:	$(p \rightleftarrows q) \Leftrightarrow (p \rightarrow q) \wedge (q \rightarrow p)$
E_{22}	:	$p \rightleftarrows q \Leftrightarrow (p \wedge q) \vee (\overline{p} \wedge \overline{q})$

Example 1.53

Show that $r \wedge (p \vee q)$ is a valid conclusion from the premises $p \vee q$, $q \rightarrow r$, $p \rightarrow m$ and $\overline{m}$.

Solution:

$\{1\}$	(1) $p \rightarrow m$	Rule P
$\{2\}$	(2) $\overline{m}$	Rule P
$\{1,2\}$	(3) $\overline{p}$	Rule T, (1), (2) and I_{12}.
$\{4\}$	(4) $p \vee q$	Rule P
$\{1,2,4\}$	(5) q	Rule T, (3), (4) and I_{10}.
$\{6\}$	(6) $q \rightarrow r$	Rule P
$\{1,2,4,6\}$	(7) r	T, (5), (6) and I_{11}.
$\{1,2,4,6\}$	(8) $r \wedge (p \vee q)$	T, (4), (7) and I_9.

Example 1.54

Show I_{12} : $\overline{q}$, $p \rightarrow q \Rightarrow \overline{p}$

Solution:

$\{1\}$	(1) $p \rightarrow q$	Rule P
$\{1\}$	(2) $\overline{q} \rightarrow \overline{p}$	Rule T, (1) and E_{18}.

{3}	(3) $\overline{q}$	Rule P
{1,3}	(4) $\overline{p}$	Rule T, (2), (3) and I_{11}.

Example 1.55

Show that the conclusion C: ~ P follows from the premises

H_1: ~ P ∨ Q, H_2: ~ (Q ∧ ~ R) and H_3: ~ R.

Solution: We get

	(1) ~ R	Rule P (assumed premise)
	(2) ~ (Q ∧ ~ R)	Rule P
{2}	(3) ~ Q ∨ R	Rule T
{3}	(4) R ∧ ~ Q	Rule T
{4}	(5) ~ R → ~ Q	Rule T
{1, 5}	(6) ~ Q	Rule T
	(7) ~ P ∨ Q	Rule P
{7}	(8) ~ Q → ~ P	Rule T
{6, 8}	(9) ~ P	Rule T

So C logically follows from H_1, H_2 and H_3.

Example 1.56

Show that S ∨ R is tautologically implied by (P ∨ Q) ∧ (P → R) ∧ (Q → S).

Solution: Note that P ∨ Q, P → R, Q → S are three premises.

	(1) P ∨ Q	Rule P
{1}	(2) ~ P → Q	Rule T
	(3) Q → S	Rule P
{2, 3}	(4) ~ P → S	Rule T
	(5) ~ S → P	Rule T (as P → Q ⇔ ~ Q → ~ P)
	(6) P → R	Rule P
{5, 6}	(7) ~ S → R	Rule T
{7}	(8) S ∨ R	Rule T

Example 1.57

Show that R → S can be derived from the premises P → (Q → S), ~ R ∨ P and Q.

Solution: We get

	(1) R	Rule P
	(2) ~ R ∨ P	Rule P
{2}	(3) R → S	Rule T

{1, 3}	(4) P	Rule T
	(5) P → (Q → S)	Rule P
{4, 5}	(6) Q → S	Rule T
	(7) Q	Rule P
{7, 6}	(8) S	Rule T
	(9) R → S	Rule CP

Example 1.58

Prove or disprove the conclusion given below from the following axioms.

"If Socrates is a man, Socrates is mortal. Socrates is a man." Therefore Socrates is mortal.

[JNTUH, Nov 2010, Set No.3]

Solution: The argument is valid because it follows the pattern of Modus ponens.

Consider the argument

p: Socrates is a man.

q: Socrates is mortal.

p → q: If Socrates is a man, then Socrates is mortal.

The modus ponens is

$$p \rightarrow q$$
$$\underline{p}$$
$$\therefore q$$

Therefore, Socrates is mortal is true.

1.8   Consistency of Premises and Indirect Method of Proof

A set of m formulas H_1, H_2, ...H_m is said to be consistent if their conjunction $(H_1 \wedge, H_2 \wedge ... \wedge H_m)$ has truth value "T" for some assignment of the truth values to the atomic variables appearing in the statement formulas H_1, H_2, ...H_m.

If $H_1 \wedge$, $H_2 \wedge ... \wedge H_m$ is false for every assignment of the truth values of the automic variables appearing in the statement formulas H_1, H_2, ...H_m then we say that H_1, H_2, ...H_m are inconsistent. In other words, a set of formulas H_1, H_2, ...H_m are inconsistent if their conjunction $(H_1 \wedge, H_2 \wedge ... \wedge H_m)$ implies a contradiction, that is,

$$H_1 \wedge, H_2 \wedge ... \wedge H_m \Rightarrow R \wedge \bar{R}$$

where R is any statement formula. Note that $R \wedge \bar{R}$ is a contradiction for any formula R].

This concept will be used in a procedure called "proof by contradiction" (or indirect method of proof).

1.8.1   Indirect Method of Proof

In order to show that a conclusion C follows logically from the premises H_1, H_2, ..., H_m we assume that C is FALSE and consider ~C as an additional premise. If $H_1 \wedge H_2 \wedge ... \wedge H_m \wedge {\sim}C$ is a contradiction, then C follows logically from H_1, H_2, ..., H_m.

Example 1.59

Show that $\sim (P \wedge Q)$ follows from $\sim P \wedge \sim Q$.

Solution: Assume $\sim (\sim (P \wedge Q))$ as an additional premise. Then

	$(1) \sim (\sim (P \wedge Q))$	Rule P
{1}	$(2) P \wedge Q$	Rule T
	$(3) P$	Rule T
	$(4) \sim P \wedge \sim Q$	Rule P
{4}	$(5) \sim P$	Rule T
{3, 5}	$(6) P \wedge \sim P$	Rule T

Therefore $P \wedge \sim P$ is a contradiction. Hence by the indirect method of proof $\sim(P \wedge Q)$ follows from $\sim P \wedge \sim Q$.

Example 1.60

"If there was a party then catching the train was difficult. If they arrived on time then catching the train was not difficult. They arrived on time. Therefore there was no party." Show that the statement constitutes a valid argument.

Solution:

Let   p: There was a party

q: Catching the train was difficult.

r: They arrived on time.

We have to prove $\overline{P}$ follows from the premises $p \rightarrow q$, $r \rightarrow \overline{q}$ and r.

	$(1) r$	Rule P
	$(2) r \rightarrow \overline{q}$	Rule P
{1, 2}	$(3) \overline{q}$	Rule T
	$(4) p \rightarrow q$	Rule P
{4}	$(5) \overline{q} \rightarrow \overline{p}$	Rule T
{3, 5}	$(6) \overline{p}$	Rule T

Example 1.61

How does an indirect proof technique differ from a direct proof.

[JNTUH, June 2010, Set No.2]

[JNTUH, Nov 2008, Set No.3]

Solution: We know that a set of formulas H_1, H_2, ..., H_m is said to be consistent if their conjunction has the truth value T for some assignment of the truth value to the atomic variables appearing in H_1, H_2, ..., H_m.

If for every assignment of the truth values to the atomic variables, their conjunction $H_1 \wedge H_2 \wedge ... \wedge H_m$ is identically false, then the formulas H_1, H_2, ..., H_m are called inconsistent.

In other words, a set of formulas H_1, H_2, ..., H_m is inconsistent if their conjunction $H_1 \wedge H_2 \wedge ... \wedge H_m$ implies a contradiction, that is,

$$H_1 \wedge H_2 \wedge \ ... \wedge H_m \Rightarrow R \wedge \daleth R$$

where R is any formula.

Note that $R \wedge \daleth R$ is a contradiction, for any formula R.

In "Direct Method of Proof" we prove directly the conclusion by using the set of premises and inference rules. But the notion of inconsistency is used in a procedure called "proof by contradiction" (or "reduction and absurdum") or "indirect method of proof".

In this indirect method of proof, to show that a conclusion 'C' follows logically from the premises H_1, H_2, ..., H_m, we assume that 'C' is false and consider $\daleth C$ as an additional premise. If the new set of premises H_1, H_2, ..., H_m, $\daleth C$ are inconsistent, then the assumption that "$\daleth C$ is true" does not hold simultaneously with $H_1 \wedge H_2 \wedge \ ... \wedge H_m$ being true.

There C is true whenever $H_1 \wedge H_2 \wedge \ ... \wedge H_m$ is true. This concept is used in Indirect Method of Proof.

Proof by Indirect Method is sometimes convenient. However it can always be eliminated and replaced by Conditional Proof (CP).

Consider the following.

$$P \to (Q \wedge \daleth Q) \Rightarrow \daleth P \qquad \qquad .....(1.14)$$

In the proof by indirect method, we show that

$$H_1, H_2, ..., H_m \Rightarrow C$$

by proving that

$$H_1, H_2, ..., H_m, \daleth C \Rightarrow R \wedge \daleth R \qquad \qquad .....(1.15)$$

for same formula R.

Now (ii) can be written (by using rule C.P) as follows

$$H_1, H_2, ..., H_m, \Rightarrow \rceil C$$

$$\rightarrow (R \wedge \rceil R) \hspace{4cm} .....(1.16)$$

From (iii) and (i) and "$\rceil \rceil P \Leftrightarrow P$" we get that

$$H_1, H_2, ..., H_m, \Rightarrow C$$

which is the required derivation. Hence in some cases, the indirect method of proof is more convenient.

Example 1.62

Show that the following set of premises are inconsistent using indirect method of proof.

$$p \rightarrow q, q \rightarrow r, \sim (p \wedge r), p \vee r \Rightarrow r$$

[JNTUH, Nov 2008, Set No.2]

Solution: We wish to state that the given set of premises is not inconsistent.

Reason: when $(p, q, r) = (F, T, T)$, that is, the truth values of p, q, r are equal to F, T, T respectively, then the truth values of $p \rightarrow q$, $q \rightarrow r$, $\sim (p \wedge r)$ and $p \vee r \rightarrow r$ are all equal to T.

Hence $p \rightarrow q$, $q \rightarrow r$, $\sim (p \wedge r)$, $p \vee r \rightarrow r$ will not form a set of inconsistent formula.

Exercises

Statements and Notations

1. Write down the negation of the statement "No real number is greater than its square."

Ans: At least one real number is greater than its square.

2. Write down the following statement in symbolic form, and find its negation:

   "If all triangles are right-angled, then no triangle is equiangular"

Ans: $\{\forall x \in T, p(x)\} \rightarrow \{\forall x \in T, \sim q(x)\}$

   The negation is

   $\{\forall x \in T, p(x)\} \wedge \{\exists x \in T, q(x)\}$

3.   Prove that for any three statements p, q, r

Ans: $[(p \vee q) \rightarrow r] \Leftrightarrow \{(p \rightarrow r) \wedge (q \rightarrow r)\}$

4.   Let p and q be primitive statements for which the implication p → q is false. Determine the truth value of (i) p ∧ q (ii) (~p) ∨ q (iii) q → p (iv) (~q) → (~p)

Ans: (i) 0 (ii) 0 (iii) 1 (iv) 0

Connectives and Truth Tables

1.   Using the following statements

p: Raju is rich

q: Raju is happy

Write the following statements in symbolic form

(i)   Raju is rich or unhappy

(ii)   Raju is neither rich nor happy

(iii) Raju is poor, but happy

Ans:

(i) $(p \vee \sim q)$

(ii) $(\sim p \wedge \sim q)$

(iii) $(\sim p \wedge q)$

2.   Construct the truth table for the following

(i) $[(p \wedge q) \vee (\sim r)] \rightleftarrows q$

(ii) $(p \wedge q) \vee (\sim p \wedge \sim r)$

(iii) $[(p \vee q) \wedge (\sim p \vee r)] \wedge (q \vee r)$

Ans:

p	q	r	~r	p ∧ q	[(p ∧ q) ∨ ~r]	[(p ∧ q) ∨ ~r] ⇄ q
T	T	T	F	T	T	T
T	T	F	T	T	T	T
T	F	T	F	F	F	T
T	F	F	T	F	T	F
F	T	T	F	F	F	T
F	T	F	T	F	T	T
F	F	T	F	F	F	F
F	F	F	T	F	T	T

3. Given that p is true and q is false, find the truth values of the following

    (i) $(\sim p \vee q) \wedge (\sim q \rightleftarrows p)$

    (ii) $(p \vee q) \rightarrow (p \wedge q)$

    (iii) $(p \rightarrow \sim q) \wedge (q \rightarrow \sim p)$

Ans:

    (i)   False

    (ii) False

    (iii) True

Tautology and Contradiction

1. Prove the following are tautology

    (i) $\left[\left[p \rightarrow (q \vee r) \right] \wedge \sim q \right] \rightarrow (p \rightarrow r)$

    (ii) $\left[(p \wedge \sim q) \rightarrow r \right] \rightarrow \left[p \rightarrow (q \vee r) \right]$

    (iii) $\sim (p \vee q) \vee \left[\sim p \wedge q \right] \vee p$

2. Prove the following contradiction

    (i) $(p \vee q) \wedge (p \rightleftarrows q)$ (ii) $p \wedge \sim p$

3. Show that the statement $(p \wedge q) \Rightarrow p$ is a tautology

Ans:

p	q	p ∧ q	(p ∧ q) ⇒ p
T	T	T	T
T	F	F	T
F	T	F	T
F	F	F	T

Truth table for $(p \wedge q) \Rightarrow p$ which is a tautology

Equivalence of Statements or Formulas

1. Prove the following logical equivalence

    (i) $\left[p \rightarrow (q \rightarrow r) \right] \Leftrightarrow \left[p \rightarrow (\sim q \vee r) \right] \Leftrightarrow \left[(p \wedge q) \rightarrow r \right] \Leftrightarrow (p \rightarrow r) \vee (q \rightarrow r)$

    (ii) $\left[p \rightarrow (q \vee r) \right] \Leftrightarrow \left[(p \rightarrow q) \vee (p \rightarrow r) \right] \Rightarrow \left[\sim r \rightarrow (p \rightarrow q) \right] \Rightarrow \left[(p \wedge \sim q) \rightarrow r \right]$

    (iii) $\left[(p \rightarrow q) \wedge (r \rightarrow q) \right] \Leftrightarrow \left[(p \vee r) \rightarrow q \right]$

2. Show the following equivalence

    (i) $(\sim p \wedge \sim q) \Leftrightarrow \sim (p \vee q)$

    (ii) $\sim (p \rightleftarrows q) \Leftrightarrow (p \wedge \sim q) \vee (\sim p \wedge q)$

3. Prove that

    $$\left[(\sim p \vee q) \wedge \left(p \wedge (p \wedge q)\right)\right] \Leftrightarrow p \wedge q$$

    Hence deduce that

    $$\left[(\sim p \wedge q) \vee \left(p \vee (p \vee q)\right)\right] \Leftrightarrow p \vee q$$

Duality Law and Tautological Implication

1. Prove the following implications

    (i) $\left[p \rightarrow (q \rightarrow r)\right] \Rightarrow (p \rightarrow q) \rightarrow (p \rightarrow r)$

    (ii) $p \Rightarrow (q \rightarrow p)$

2. Prove that $\left[(\sim p \vee q) \wedge p \wedge (p \wedge q)\right] \Leftrightarrow p \wedge q$ by duality principle

3. Verify the principle duality for the logical equivalence

    $$(p \vee q) \wedge \left(\sim p \wedge (\sim p \wedge q)\right) \Leftrightarrow (\sim p \wedge q)$$

Normal Forms

1. Obtain the principal disjunctive Normal Form for

    (i) $(p \wedge q) \vee (\sim p \wedge r) \vee (q \wedge r)$

    (ii) $p \rightarrow \left[(p \rightarrow q) \wedge \sim (\sim q \vee \sim p)\right]$

Ans: $(p \wedge q \wedge r) \vee (p \wedge q \wedge \sim r) \vee (\sim p \wedge q \wedge r) \vee (\sim p \wedge \sim q \wedge \sim r)$

2. Find the DNF and CNF of the following formulas

    (i) $(\sim p \rightarrow r) \wedge (q \rightleftarrows p)$

    (ii) $p \rightarrow \left[p \wedge (q \rightarrow p)\right]$

Ans: ???

3. Find the principal disjunctive normal form of the following

    (i) $\sim (p \wedge q)$

    (ii) $p \rightleftarrows q$

Ans:

    (i) $(p \wedge \sim q) \vee (\sim p \wedge q) \vee (\sim p \wedge \sim q)$

    (ii) $(p \wedge q) \vee (\sim p \wedge \sim q)$

4. Find the principal conjunctive normal Form of the following

    (i) $(\sim p) \wedge q$

    (ii) $\sim (p \vee q)$

Ans:

    (i) $(\sim p \vee q) \wedge (\sim p \vee q) \wedge (p \vee q)$

    (ii) $(p \vee \sim q) \wedge (\sim p \vee q) \wedge (\sim p \vee \sim q)$

The Theory of Inference for Statement Calculus

1. Prove the argument $(p \rightarrow q) \wedge (r \rightarrow s), (p \vee r) \wedge (q \vee r) / -q \vee s$ is valid without using truth table

2. Show that $\sim (p \wedge q)$ follows from $\sim p \wedge \sim q$ without using truth table

3. Prove the following using the rule CP if necessary

    (i) $p \rightarrow q \Rightarrow p \rightarrow (p \wedge q)$

    (ii) $p, p \rightarrow [q \rightarrow (r \wedge s)] \Rightarrow q \rightarrow s$

Consistency of Premises and Indirect Method of Proof

1. Prove that "the square of an even integer is an even integer" by the method of contradiction.

2. Provide an indirect proof of the following statement For all positive real numbers x and y, if the product xy exceeds 25, then x > 5 or y > 5.

3. Show that the following sets of premises are inconsistent.

$$p \rightarrow (q \rightarrow r), s \rightarrow (q \wedge \sim r), p \wedge s$$

4. Show the following (use indirect method if need)

    (i) $(r \rightarrow \sim q), r \vee s, s \rightarrow \sim q, p \rightarrow q \Rightarrow \sim p$

    (ii) $\sim (p \rightarrow q) \rightarrow \sim (r \vee s), [(q \rightarrow p) \vee \sim r], R \Rightarrow P \rightleftarrows q$

CHAPTER - 2

Predicate Calculus

In Chapter-1, we considered the statements and statement formulas. In the inference theory studied, all the premises and conclusions are all statements. If any two statements have common feature, then we cannot express this situation. In order to study this common feature or property of atomic statements, we study the concept "predicate" in a primary statement. The logic based on the analysis of predicate in any statement is called as predicate logic.

2.1   Predicate Logic

For example, let us consider the following two statements:

1. Satya is beautiful.
2. Lakshmi is beautiful.

If we express these statements by symbols, we require two different symbols to denote them. These symbols do not reveal the features of these two statements.

If we introduce some symbol to denote "is beautiful" and a method to join it with symbols denoting the names of individuals, then we will have a suitable symbolism to denote the statements about any individuals, being "beautiful". Now we introduce "predicate".

2.1.1   Predicate

The part "is beautiful" is called a predicate.

Consider the following argument:

All human beings are mortal

Satya is a human being.

Therefore, Satya is a mortal.

This type of argument also lead to some similar device to denote. Intuitively, this argument seems to be valied. This argument is of the form $\dfrac{\begin{array}{c} p \\ q \end{array}}{\text{Therefore } r}$ where p, q and r are statements.

Let us discuss about this argument. According to statement logic, this is not a valid argument. But we understand that this argument is a valid one. This new argument validity not depend on the form but it is depended on the content of the statement.

We shall symbolize a predicate by a capital letter and individuals (or objects) in general by small letters. We shall soon see letters to symbolize statements as well as predicates with no confusion. Every predicate describes something about one or more objects.

Example 2.1

Consider the statements:

1. Satya is beautiful.
2. Lakshmi is beautiful.

We denote the predicate "is beautiful" by the capital letter B (which is the first letter of the word beautiful); "Satya" by "s" and "Lakshmi" by "l".

Now the statement (1) and (2) can be written as B (s) and B (l).

In general any statement of the form "p is Q" where Q is a predicate and p is noun (or subject is denoted by Q (p).

So   B(s) means "Satya is beautiful".

B(l) means "Lakshmi is beautiful".

Example 2.2

Consider the statements:

1. Mallikarjun is a student.
2. Gnyana is a student.

We denote the predicate "is a student" by the capital letter "S". The nouns (or subjects) Mallikarjun by small letter "m" and Gnyana by the small letter "g".

Now S (m) means "m is S" (i.e., Mallikarjun is a student) S (g) means "Gnyana is a student".

2.1.2   2–place Predicate

In the above example, we considered the statement "Mallikarjun is a student". Here the predicate "is a student" have one noun (or subject) namely "Mallikarjun". So it is called as 1–place predicate.

Consider the statement "Mallikarjun is taller than Gnyana". Here "is taller than" is a predicate and it deals with two names (or nouns). Such predicates are called as 2–place predicates.

2.1.3   m–place Predicate

A predicate requiring m names (where m > 0) is called an m–place predicate. In order to extent this definition to m = 0, we shall call a statement a O–place predicate if no names are associated with a statement.

Example 2.3

Consider the following statement:

Andhra Pradesh is to the north of Tamilnadu.

Here "is to the north of" is the predicate denoted by the capital letter "N".

Andhra Pradesh is denoted by "a".

Tamilnadu is denoted by "t".

So N (a, t) means "Andhra Pradesh is to the north of Tamilnadu".

Thus this predicate is a 2–place predicate.

Example 2.4

Consider the following statement:

"Satya sits between Mallikarjun and Gnyana". Here "Sits between" is the predicate, let us denote it by B. Let us denote Satya, Mallikarjun and Gnyana by s, m, g respectively. Then B(s, m, g) represents the given statement. It is a 3–place predicate.

2.1.4   Connectives

The connectives $(\wedge, \vee, \rceil)$ used earlier in statement logic can now used to form compound statements such as

"Satya is beautiful" and "Lakshmi is beautiful".

"Satya is beautiful" or "Lakshmi is beautiful".

These sentences is symbolic form were represented by

B(s) ∧ B(l)

B(s) ∨ B(l)

"The painting is Red" is denoted by R(p).

"The painting is not Red" is denoted by $\rceil$ R(p) or ~R(p) or $\overline{R(p)}$

Example 2.5

Represent the statement "Rama is handsome and Sita is beautiful" by predicate logic.

Solution: Let us use "H" for the predicate "is handsome"; "B" for "is beautiful"; "r" for Rama; "s" for Sita.

Then the given statement is denoted by

$H(r) \wedge B(s)$

2.2 Statement Functions, Variables and Quantifiers

A simple statement function of one variable is defined to be an expression consisting of a predicate symbol and an individual variable. Such a statement function becomes a statement when the variable is replaced by the name of any object.

For example, consider the predicate B ("is beautiful") and "s" (the name Satya). Then B(s) means "s is beautiful" (that is, Satya is beautiful).

In place of Satya we use a variable x. We write B(x) for "x is beautiful". In place of x we may substitute "Lakshmi" (or l). Then B(l) means "Lakshmi is beautiful". Now we consider B(x) as a statement and x is a place holder (or variable) B(x) is a statement function.

Example 2.6

In the statement function T(x, y) (means x is taller than y) T denotes the predicate "is taller than" x and y are place holders in the 2–place predicate T(x, y).

If we x by Mallikarjun, and y by Gnyana then T (m, g) denotes the statement.

Mallikarjun is taller than Gnyana.

T(x, y) is a statement function and x, y are variables.

Note:

Let B be the predicate "is beautiful".

B(s): Satya is beautiful.

B(g): Gnyana is beautiful.

B(l): Lakshmi is beautiful.

Here B(s), B(g), B(l) all denote statements, but they have common form.

If we write B(x) for "x is beautiful" then B(s), B(g), B(l) and others with same form can be obtained from B(x) by replacing x by the suitable name s, g, l, ... we note that B(x) is not a statement but it result in many statements when we replace x by appropriate names (or subjects or nouns).

2.2.1 Combined Statements (or Connectives)

Consider the statement functions

B(x): x is beautiful.

M(x): x is mortal.

Now

B(x) ∧ M(x) denotes "x is beautiful and x is mortal".

B(x) ∨ M(x) denotes "x is beautiful or x is mortal".

⏋B(x) (or $\overline{B(x)}$) denotes "x" is not beautiful".

If T(x, y) denotes "x" is taller than y" then ⏋T(x, y) (or $\overline{T(x,y)}$) denotes that "x" is not taller than y".

Example 2.7

Construct statement function in predicate calculus for the statement "x is rich and y is tall".

Solution: "x is rich" is denoted by R(x); and "y is tall" is denoted by T(y).

Therefore "R(x) ∧ T(y)" denotes the statement "x is rich and y is tall".

Note that, here, R and T are predicates; and x and y are variables.

2.2.2 Quantifiers

There is a way for obtaining statements with common predicate. Now we introduce the notion quantifiers such as "all" and "same". This provides some extension of the notations used earlier. The word "all" is called as universal quantifier; and the word "same" is called as existential quantifier.

2.2.3 Universal Quantifier

The word "all" (the universal quantifier) is denoted by (x) or ∀x.

We place this symbol before the statement function. For example, consider the statement functions:

B(x): x is beautiful

W(x): x is a woman

Then (x) (w(x) → B(x)) denotes "for all x, if x is a women then x is beautiful"(that is all women are beautiful).

Here we note that (x) (w(x) → B(x); and (y) (w(y) → B(y)) are equivalent (because x, y are variables) (x) (w(x) → B(x)) may be denoted by ∀x (w(x) → B(x))

Consider the 2–place predicate:

R(x, y): x is richer than y

P(y, x): y is poorer than x

⌐P(x, y): x is not poorer than y

"(x)(y) (R(x, y) → ⌐P(x, y))"

means that

"For any x and for any y, if x is richer than y then x is not poorer than y". This statement may be denote as follows:

$$\forall x\ \forall y\ (R(x, y) \rightarrow \urcorner P(x, y))$$

2.2.4   Existential Quantifier

The word "same" (the existential quantifier) is denoted by "∃x". This also means "for some" or "there exists atleast one".

The symbol ∃! (or ∃) is used for "there is a unique x.

We place this symbol before the statement functions.

Example 2.8

Consider the following statement functions:

M(x): x is a man

C(x): x is clever

I(x): x is an integer

E(x): x is even

P(x): x is prime.

Then

∃ x M(x) symbolizes "There exists a man"

∃ x (M(x) ∧ C(x)) symbolizes "There are some men who are clever"

∃ x (I(x) ∧ E(x)) symbolizes "Some integers are even" or "There are some integers which are even"

∃! x (E(x) ∧ P(x)) symbolizes "There exists unique even prime"

2.2.5   The Universe of Discourse

Variables which are quantified stand for only those objects which are members of a particular set or class. Such a set is called the universe of discourse or the domain or simply universe.

The universe may be, the class of human beings, or numbers (real, complex, and rational) or some other objects. The truth value of a statement function depends upon the universe.

Example 2.9

Consider the predicate $Q(x)$: x is less than 10 and the statements (x) $Q(x)$ and $\exists$ x $Q(x)$.

Also consider the following universes U_1, U_2 and U_3:

U_1: {−1, 0, 1, 2, 4, 6, 8};

U_2: {3, −2, 12, 14, 10} and

U_3: {10, 20, 30, 40}

(i) The statement (x) $Q(x)$ is true in U_1 because for every x in U_1 the statement function (x) $Q(x)$ [that is, x < 10] is true.

(ii) The statement (x) $Q(x)$ is not true in U_2 [because there is the element 12 in U_2 such that 12 is not less than 10].

(iii) The statement (x) $Q(x)$ is not true in U_3.

(iv) The statement "$\exists$ x $Q(x)$" is true in U_1 and U_2 [because there exist atleast one element in U_1 (also in U_2) which is less than 10].

(v) The statement "$\exists$ x $Q(x)$" is not true (that is, false) in U_3 [because there is no element in U_3 which is less than 10].

Example 2.10

Let the universe of discourse be the set of integers. Determine the truth values of the following sentences:

1. (x) $(x^2 \geq 0)$
2. (x) $(x^2 - 5x + 6 = 0)$
3. $\exists(x)$ $(x^2 - 5x + 6 = 0)$
4. (y) $(\exists$ x $(x^2 = y))$

Solution: 1. True, 2. False, 3. True, 4. False.

Example 2.11

Using proof by contradiction show that $\sqrt{2}$ is not a rational number.

[JNTUH, Nov 2010, Set 3]

Solution: Write $p(x)$: x is a rational number.

We have to prove that $\neg p(\sqrt{2})$

Contrary way suppose $\neg\,(\neg p\sqrt{2}\,)$

(That is, suppose p($\sqrt{2}$))

Since p($\sqrt{2}$) we have that $\sqrt{2} = \dfrac{a}{b}$ where a, b are two integers, gcd (a, b) = 1 and b ≠ 0.

Now $\left(\sqrt{2}\right)^2 = \dfrac{a^2}{b^2}$

$\Rightarrow 2 = \dfrac{a^2}{b^2}$

$\Rightarrow 2b^2 = a^2$ ⠀⠀⠀⠀⠀⠀⠀⠀⠀⠀⠀⠀⠀⠀⠀⠀.....(2.1)

$\Rightarrow 2$ divides a^2

$\Rightarrow 2$ divides a ⠀⠀⠀⠀⠀⠀⠀⠀⠀⠀⠀⠀⠀⠀.....(2.2)

$\Rightarrow a = 2k$ for some integer k

$\Rightarrow a^2 = 4k^2$ ⠀⠀⠀⠀⠀⠀⠀⠀⠀⠀⠀⠀⠀.....(2.3)

By (2.1) & (2.3) we get $2b^2 = 4k^2$

$\Rightarrow 2b^2 = 4k^2$

$\Rightarrow b^2 = 2k^2$

$\Rightarrow 2$ divides b ⠀⠀⠀⠀⠀⠀⠀⠀⠀⠀⠀⠀⠀.....(2.4)

Now 2 divides 'a' and 'b' [from (2.2), (2.4)]

This is a contradiction to the fact gcd (a, b) = 1.

Thus $p\left(\sqrt{2}\right)$ is not true

That is $\sqrt{2}$ is not a rational number.

Example 2.12

Write the quantifiers of the following statements where predicate symbols denote.

⠀⠀K(x): x is two-wheeler

⠀⠀L(x): x is a scooter

⠀⠀M(x): x is manufactured by Bajaj

(a)⠀Every two wheeler is a scooter.

(b)⠀There is a two wheeler that is not manufactured by Bajaj.

(c) There is no two wheeler manufactured by Bajaj that is not a scooter.

(d) Every two wheeler that is a scooter is manufactured by Bajaj.

[JNTUH, June 2010, Set No.1]

Solution: Given that

K(x): x is a two-wheeler

L(x): x is a scooter

M(x): x is manufactured by Bajaj

(a) We have to write quantifier for the statement:

"Every two wheeler is a scooter"

(x) (K (x) → L(x))

is the required expression.

(b) We have to write quantifier for the statement:

"There is a two wheeler that is not manufactured by Bajaj". "$\exists x$ (K(x) ∧ ⌐M(x))" is the required expression.

(c) We have to write quantifier for the statement

"There is no two wheeler manufactured by Bajaj that is not a scooter".

"K(x) ∧ M(x) ∧ ⌐L (x)" is equivalent to "two wheeler manufactured by Bajaj that is not a scooter".

Therefore the required quantifier expression is ⌐ ($\exists x$ (K(x) ∧ M(x) ∧ ⌐ L(x))

(d) We have to write quantifier for the statement "Every two wheeler that is a scooter is manufactured by Bajaj". "K(x) ∧ L(x)" represents "two wheeler that is a scooter". So the required expression is

(x) ((K(x) ∧ L(x) → M(x)).

2.3   Free and Bound Variable

Suppose that (x) p(x) or $\exists$ x p(x) is a part of a formula. Such part of the form (x) p(x) or $\exists$ x p(x) is called as x–bound part of that formula. The formula p(x) either in "(x) p(x)" or "$\exists$ x p(x)" is called the scope of the quantifier.

Example 2.13

Suppose the universe of discourse is the set of integers

p(x) be the statement that $x^2 \geq 0$

We know that $x^2 \geq 0$ for all integers

So p(x) is true for all x.

We write this fact as (x) p(x).

This (x) p(x) is x–bound part.

Example 2.14

If y is a complex number then there exist x (a complex number) such that $x^2 = y$

This statement can be written as

$$(y) \, (\exists \, x \, (x^2 = y))$$

In this formula, $\exists \, x \, (x^2 = y)$ is x–bound part.

Definition

(i) Any occurrence of in an x–bound part of a formula is called as bound occurrence of x.

(ii) Any occurrence of x (or a variable) which is not a bound occurrence is called a free occurrence.

Example 2.15

Consider the formula

$$\exists \, x \, (p(x) \wedge q(x))$$

Here the scope of $(\exists \, x)$ is $p(x) \wedge q(x)$

Hence in "$\exists \, x \, (p(x) \wedge q(x))$". So all the occurrences of x are bound occurrences.

Example 2.16

Consider the formula

$$(\exists \, x \, (p(x) \wedge q(x))$$

Here the scope of $(\exists \, x)$ is p(x). So the occurrence of x in p(x) is a bound occurrence. But the occurrence of x in q(x) is a free occurrence.

Example 2.17

Consider the statement:

Satya is beautiful.

The symbolic representation is B(s). We know the statement formula B(x) means x is beautiful. Here x is a variable. In "B(x)" there is no quantifier. Hence the occurrence of variable x in "B(x)" is a free occurrence.

Example 2.18

Let us consider the statement:

"All birds can fly"

Now we symbolize this statement. Write

B(x): x is a bird

F(x): x can fly

Now "(x) (B(x) → F(x))" denotes the statement "All birds can fly".

In this "(x) (B(x) → F(x))", all occurrences of x are bound occurrences.

Example 2.19

Symbolize "All Dogs are Animals"

Solution: D(x): x is a dog

A(x): x is an animal

D(x) → A(x): The dog 'x' is an animal.

(x) (D(x) → A(x)): All dogs are animals.

Example 2.20

Symbolize "Some horses are black"

Solution: H(x): x is a horse

B(x): x is black

H(x) → B(x): The horse 'x' is black

∃ x (H(x) → B(x)): There exists a horse which is black (or) some horses are black.

Example 2.21

Symbolize "All the people respects selfless leaders".

Solution: We write

P(x): x is a person

S(x): x is a selfless leader

R(x, y): x respects y

Now the required symbol

S(y) → R(x, y): If y is a selfless leader then x respects y.

(y) (S(y) → R(x, y)): x respects every selfless leader y.

p(x) → (y) (S(y) → R(x, y)): person x respects every selfless leader y.

(x) [p(x) → (y) (S(y) → R(x, y)): All the people respects selfless leaders.

Example 2.22

The following table presents the negation of some statement functions

Statement function	Negation
∃ x F(x)	(x) (~F(x))
(x) F(x)	∃ x (~F(x))
∃ x (~ F(x))	(x) F(x)
(x) (~F(x))	∃ x F(x)

Example 2.23

Write the negation of "(x) (E(x) → S(x))"

Solution: Suppose F(x) be "E(x) → S(x)" then the given formula is of the form "(x) F(x)".

From the above table we conclude that the negation is ∃ x (~F(x)).

Now ~F(x) is the negation of E(x) → S(x).

The negation of S(x) → S(x) is "E(x) → ~S(x)"

Hence ∃ x (~F(x)) is equivalent to ∃ x (E(x) → ~S(x)).

Hence ∃ x (E(x) → ~S(x)) is the negation of (x) (E(x) → S(x))

Example 2.24

Write the quantifiers of the following statements where predicate symbols denotes,

F(x): x is fruit

V(x): x is vegetable and

S(x, y): x is sweeter than y

(a)   Some vegetables are sweeter than all fruits

(b)   Every fruit is sweeter than all vegetables

(c)   Every fruit is sweeter than some vegetables

(d)   Only fruits are sweeter than vegetables

[JNTUH, Nov 2008, Set No.1]

Solution:

(a)   We have to symbolize the statement "Some vegetables are sweeter than all fruits".

(y) (F(y) → S(x, y)): (means) x is sweeter than all fruits y.

So ∃ x [V(x) → ((y) (F(y) → S(x, y))] is the required predicate formula.

(b)   We have to symbolize the statement "Every fruit is sweeter than all vegetables".

(y) (V(y) → S(x, y) : means x is sweeter than all vegetables y. Therefore the required predicate formula is (x) [F(x) → (y) (V(y) → S(x, y))].

(c) We have to symbolize the statement "Every fruit is sweeter than some vegetables". ∃ y (V(y) ∧ F(x, y)): (means) there exists a vegetable y such that x in sweeter than y. The required predicate formula is (x) [F(x) → ∃ y (V(y) ∧ F(x, y))].

(d) We have to symbolize the statement "only fruits are sweeter than vegetables". This statement is equivalent to say that "if x is sweeter than all vegetables, then x is a fruit". Hence the required predicate formula is (y) (V(y) → S(x, y) → F(x).

[Note that (y) (V(y) → S(x, y)) represents the statement "x is sweeter than all vegetables"].

2.4 Inference Theory for the Predicate Calculus

The method of derivation involving predicate formulas uses the rules of inference provided for the statement calculus. In addition to the rules of inference provided for the statement calculus, it also uses certain additional rules or principles that are mentioned below:

(i) Universal Specification (US)
(ii) Universal Generalization (UG)
(iii) Existential Specification (ES)
(iv) Existential Generalization (EG)

2.4.1 Universal Specification (US)

If (x) p(x) is assumed to be true then the universal quantifier can be dropped to obtain p(c) is true, where c is an arbitrary object in the universe.

Example 2.25

All women are mortal.

Sita is a woman.

Here the universe is the set of all women.

M(x): x is mortal

(x) M(x): all women are mortal (as x is in the universe).

Since "Sita is a woman", the "Sita" is in universe. So we replace x by "Sita". Then Sita is mortal. This was done by US.

2.4.2 Universal Generalization (UG)

If P(c) is true for all c in the universe then the universal quantifier may be prefixed to obtain (x) P(x).

Example 2.26

Let U = {1, 2, 3, 4} be the universe.

Let P(x) be the statement that "$x^2 \le 100$".

For every x $\in$ U we know that $x^2 \leq 100$.

If x = 1 then $x^2 = 1 \leq 100$. So P(1) is true

If x = 2 then $x^2 = 4 \leq 100$. So P(2) is true

If x = 3 then $x^2 = 9 \leq 100$. So P(3) is true

If x = 4 then $x^2 = 16 \leq 100$. So P(4) is true

Now we verified that P(c) is true for all c in the universe u.

Hence, by UG, we write (x) P(x).

2.4.3 Existential Specification (ES)

If $\exists$ x P(x) is assumed to be true then P(c) is true for some element c in the universe.

2.4.4 Existential Generalization (EG)

If P(c) is true for some element c in the universe then $\exists$ x P(x) is true.

Example 2.27

Prove that (x) (P(x) $\rightarrow$ Q(x)) $\wedge$ (x) (Q(x) $\rightarrow$ R(x))

$\Rightarrow$ (x) (P(x) $\rightarrow$ R(x))

Solution: Given that

(x) (P(x) $\rightarrow$ Q(x))   Premise–1

and

(x) (Q(x) $\rightarrow$ R(x))   Premise–2

Assuming premise–1 and premise–2 we have to obtain the conclusion "(x) (P(x) $\rightarrow$ R(x))".

Derivation:

1. (x) (P(x) $\rightarrow$ Q(x))   P (Premise–1)
2. P(c) $\rightarrow$ Q(c)   US and (1)
3. (x) (Q(x) $\rightarrow$ R(x))   P (Premise–2)
4. Q(c) $\rightarrow$ R(c)   US and (3)
5. P(c) $\rightarrow$ R(c)   [(2), (4) and Inference Rule (Hypothetical Syllogism)]
6. (x) (P(x) $\rightarrow$ R(x))   UG and (5)

Example 2.28

Consider the following statements.

All women are selfish.   (Premise–1)

All queens are women.   (Premise–2)

Prove that all queens are selfish.

Solution: Let

      W(x): x is woman.

      Q(x): x is queen.

      S(x): x is selfish.

(x) (W(x) → S(x)) is Premise–1; and (x) (Q(x) → W(x)) is Premise–2. Now above arguments are symbolized as follows.

1.	(x) (W(x) → S(x))	P (Premise–1)
2.	W(c) → S(c)	US, (1)
3.	(x) (Q(x) → W(x))	P (Premise–2)
4.	Q(c) → W(c)	US, (3)
5.	Q(c) → S(c)	[(2), (4) and Inference Rule hypothetical syllogism]
6.	(x) (Q(x) → S(x))	UG and (5)

Therefore all queens are selfish.

Example 2.29

Prove or disprove the validity of the following statements using the rules of inferences.

      All men are fallible.   (Premise–1)

      All Kings are men.   (Premise–2)

      Therefore All Kings are fallible.

[JNTUH, Nov 2010, Set No 2]

Solution: Let

      M(x): x is a man.

      K(x): x is a king

      F(x): x is fallible

(x) (M(x) → F(x)) (Premise–1), (x) (K(x) → M(x)) (Premise–2). Now arguments are symbolized as follows:

1.	(x) (M(x) → F(x))	P (Premise–1)
2.	M(c) → F(c)	US and (1)
3.	(x) (K(x) → M(x))	P (Premise–2)
4.	K(c) → M(c)	US and (3)
5.	K(c) → F(c)	[(2), (4), and Inference rule: hypothetic syllogism]
6.	(x) (K(x) → F(x)	UG and (5)

Example 2.30

Show that $\exists\, x\, (p(x) \wedge q(x)) \Rightarrow (\exists\, x\, p(x)) \wedge (\exists\, x\, q(x))$

[JNTUH Nov 2010, Set No. 2]

Solution: The given statement can be symbolized as follows:

1. $\exists\, x\, (p(x) \wedge q(x))$ P(Premise)
2. $p\,(y) \wedge q\,(y)$ ES and (1)
3. $p\,(y)$ [(2) and Inference Rule (Simplification)]
4. $q\,(y)$ [(2) and Inference Rule (Simplification)]
5. $\exists\, x\, p\,(x)$ EG and (3)
6. $\exists\, x\, q\,(x)$ EG and (4)
7. $\exists\, x\, p\,(x) \wedge \exists\, x\, q\,(x)$ [(5), (6), and Inference Rule:I_9]

Example 2.31

Show that $\exists\, x\, (M(x))$ follows logically from the premises.

$(x)\,(H(x) \rightarrow M(x))$ and $\exists\, x\, H(x)$

[JNTUH, Nov 2010, Set No. 4]

Solution: The given premises are

$(x)\,(H(x) \rightarrow M(x))$ Premise–1

$\exists\, x\, H(x)$ Premise–2

We have to derive the conclusion: $\exists\, x\, (M(x))$.

Derivation:

1. $\exists\, x\, H(x)$ P (Premise–2)
2. $H\,(c)$ ES and (1)
3. $(x)\,(H(x) \rightarrow M(x))$ P (Premise–1)
4. $H(c) \rightarrow M(c)$ US and (3)
5. $M(c)$ [(2), (4) and Inference rule (Modus Ponens)]
6. $\exists\, x\, M(x)$ EG and (5)

Example 2.32

Explain with an example of the following:

$(x)\,[A(x) \wedge B(x)]$ need not be a conclusion form $\exists\, x\, A(x)$ and $\exists x\, B(x)$

[JNTUH, Nov 2010, Set No. 4]

Solution: Suppose $U = \{1, 2\}$ is the universe.

Let us consider the statements:

      A(x): x is even

      B(x): x is odd

Since 1 is an element of U such that 1 is odd, we have that $\exists$ x B(x) (by EG).

Since 2 is an element of U such that 2 is even, we have that $\exists$ x A(x) (by EG).

A(x) $\wedge$ B(x): means x is both even and odd.

If x = 1 then x is not both even and odd.

If x = 2 then x is not both even and odd.

So there is no element in the universe which is both even and odd.

So A(x) $\wedge$ B(x) is not true for any x in the universe.

Hence (x) (A(x) $\wedge$ B(x)) is not true. So we conclude that (x) (A(x) $\wedge$ B(x)) need not be a conclusion from $\exists$ x A(x) and $\exists$ x B(x).

Example 2.33

Using predicate logic, prove that validity of the following statement.

      Every husband argues with his wife. x is a husband. Therefore, x argues with his wife.

[JNTUH, Nov 2008, Set No. 3]

[JNTUH, June 2010, Set No. 2]

Solution: Write H is a husband and 'w' for wife.

    H(x): x is a husband.

    A(x, w): x argues with his wife.

(x) (H(x) $\rightarrow$ A(x, w)): Every husband x argues with his wife w.

(x) (H(x) $\rightarrow$ A(x, w))	Premise–1
H(x): x is husband	Premise–2
1. H (x)	P (Premise–2)
2. (x) (H(x) $\rightarrow$ A(x, w))	P (Premise–1)
3. H(x) $\rightarrow$ A(x, w)	US and (2)
4. A (x, w)	[(1), (3) and modus ponens]

Therefore, A(x, w): x argues with his wife.

Example 2.34

Prove or disprove the conclusion of the below from the following axioms:

    "All men are mortal. Mahatma Gandhi is a man. Every mortal lives less than 100 years. Mahatma Gandhi was born in 1869. Now, it is 2008. Therefore, Is Mahatma Gandhi alive now?
[JNTUH, Nov 2008, Set No. 4]

Solution: First we symbolise the given statements.

       $A(x)$: x is a man

       $M(x)$: x is mortal

We use symbol 'g' for Mahatma Gandhi.

       $C(x)$: (Birth year of x) + 100 ≤ 2008

       $L(x)$: x is alive in 2008

Premise-1: All men are mortal, that is, $(x) (A(x) \rightarrow M(x))$

Premise-2: Mahatma Gandhi is a man, that is $A(g)$

Premise-3: Every mortal lives less than 100 years. So if (birth year of x) + 100 ≥ 2008 then x may alive.

In other words, if (birth year of x) + 100 ≤ 2008) then x is not alive in 2008, that is

$(x) (\rceil C(x) \rightarrow \rceil L(x))$.

So the given sentence with respect to 2008 is

$(x) (M(x) \rightarrow (\rceil C(x) \rightarrow \rceil L(x))$

Premise-4: Mahatma Gandhi was born in 1869 and 1869 + 100 = 1969 < 2008. So C (g) is not true.

Hence $\rceil C(g)$:

Derivation:

1.	$(x) (A(x) \rightarrow M(x))$	[P (Premise–1)]
2.	$A(g) \rightarrow M(g)$	[(1) and US]
3.	$A(g)$	[P(Premise–2)]
4.	$M(g)$	[(2), (4) and Modus Ponens]
5.	$(x) (M(x) \rightarrow (\rceil C(x) \rightarrow \rceil L(x))$	[P(Premise–3)]
6.	$M(g) \rightarrow (\rceil C(g) \rightarrow \rceil L(g))$	[(5) and US]
7.	$\rceil C(g) \rightarrow \rceil L(g)$	[(4), (6) and Modus Ponens]
8.	$\rceil C(g)$	[P (Premise–4)]
9.	$\rceil L(g)$	[(7), (8) and Modus Ponens]

$\rceil L (g)$: Mahatma Gandhi is not alive in 2008. Hence it is proved that Mahatma Gandhi is not alive in 2008.

2.4.5 Formulas with More than One Quantifier

We may consider the formulas with more than one quantifier.

For example, if we consider $P(x, y)$, a 2–place predicate formula then the follows cases exists.

(x) (y) P(x, y)

(x) (∃y) P(x, y)

(∃x) (y) P(x, y)

(∃x) (∃y) P(x, y)

(y) (x) P(x, y)

(y) (∃x) P(x, y)

(∃y) (x) P(x, y)

(∃y) (∃x) P(x, y)

The following diagram shows the logical relation of the above mentioned predicate formulas:

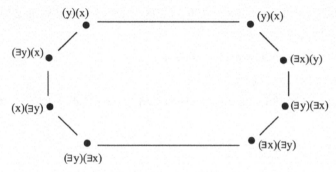

Exercises

Predicate Logic

1. Let p (x): $x^2 - 8x + 15 = 0$, q (x): x is odd, r (x): x > 0 with the set of all integers as the universe. Determine the truth or falsity of each of the following statements:

    (a)   ∀ x, [p(x) → q(x)]   (Ans: True)

    (b)   ∀ x, [q(x) → p(x)]   (Ans: False, For example x = 7)

    (c)   ∃ x, [q(x) → p(x)]   (Ans: True)

2. Translate the following statements into symbols by using predicate logic:

    (a)   All birds can fly

    (b)   Some men are giants

    (c)   There is a student who likes English but not Mathematics.

Ans:

    (a)   ∀ x, [p(x) → q(x)]   (b)   ∃ x, [p(x) ∧ q(x)]

    (c)   ∃ x, [p(x) ∧ q(x) ∧ ~ r(x)]

Show that the following statements constitute a valid argument "if A works hard then either B or C enjoys himself. If B enjoys himself then A will not work hard. If D enjoys himself then C will not. Therefore if A works hard D will not enjoy himself".

Statement Functions, Variables and Quantifiers

1. Write the following arguments using quantifiers, variables and predicate symbols:
    (i)   There is a student who likes mathematics, but not history.
    (ii) Some people are genius.
    (iii) Some numbers are not rationals.
    (iv) Not all men are good.
    (v)   All birds can fly.

Free and Bounded Variables

1. Indicate the variables that are free and bound.
    (i) $(x) [p(x) \wedge r(x)] \rightarrow (x) p(x) \wedge q(x)$
    (ii) $(x) [p(x) \wedge (\exists x) q(x)] \vee [(x) p(x) \rightarrow q(x)]$
2. For the universe of all integers, let
    $p(x)$: $x > 0$, $q(x)$: x is even
    $r(x)$: x is divisible by 3.
    Then write down the following statements in symbolic form
    (i)   At least one integer is even.
    (ii) There exists a positive integer that is even.
    (iii) Some even integers are divisible by 3.

Ans:
    (i) $\exists x, q(x)$
    (ii) $\exists x, [p(x) \wedge q(x)]$
    (iii) $\exists x, [q(x) \wedge r(x)]$

Inference Theory for the Predicate Calculus

1. Show that $\exists x [p(x) \wedge q(x)] \Rightarrow [\exists x\, p(x)] \wedge [\exists x\, q(x)]$
2. Show that $p(x) \wedge (x)\, q(x) \Rightarrow (\exists x) [p(x) \wedge q(x)]$

3. Test the validity of the argument

    (i)   If a person is poor, he is unhappy.

          If a person is unhappy, he dies young.

          ∴ poor persons die young

Ans: Valid

    (ii) If there is a strike by students, the examinations will be postponed.

          There was no strike by students

          ∴ The examination was not postponed.

Ans: Not Valid

CHAPTER - 3

Number Theory

The first scientific approach to the study of integers (that is, the true origin of the theory of numbers) is generally attributed to the Greeks.

Around 600 B.C., Pythagoras and his disciples made rather thorough study of integers. They were the first to classify integers in various ways, such as even numbers, odd numbers, prime numbers, composite numbers etc.

After Euclid in 300 B.C., no significant advances were made in number theory until about A.D. 250.

In the 17th century, the subject was revived in Western Europe, largely through the efforts of a remarkable French Mathematician Pierre de Fermat (1601-1665), who is generally acknowledged to be the father of Modern Number Theory. He was the first to discover really deep properties of the integers.

After Fermat's time, the names Euler (1707-1783), Lagrange (1736-1833), Legendre (1752-1833), Gauss (1777-1855) and Dirichlet (1805-1859) became prominent in the further development of the subject.

3.1 Properties of Integers

We know that the set $\{0, \pm1, \pm2, \ldots\}$ is called the set of integers. It is denoted by Z. The elements of Z are called Integers.

3.1.1 Properties of Integers

We consider two binary operations '+' and '0' (called usual addition and usual multiplication respectively) on the set Z of integers, with the following properties.

(i) $p, q \in z \Rightarrow p + q \in z$ (closure property)

$p, q, r \in z \Rightarrow (p + q) + r = p + (q + r)$ (associate property)

(ii) $p, q \in z \Rightarrow p + q = q + p$ (Ablian property)

(iii) $p \in z \Rightarrow p + 0 = 0 + p = p$ (Identity property)

(iv) $p \in z \Rightarrow \exists - p \in z$ such that $p + (- p) = (-p) + p = 0$ (Inverse property)

(v) $p, q \in z \Rightarrow p, q \in z$ (Closure property)

(vi) $p, q, r \in z \Rightarrow (pq) r = p(qr)$ (Associate property)

(vii) $p, q \in z \Rightarrow pq = qp$ (Commutative property)

(viii) $p \in z \Rightarrow p.1 = 1, p = p$ (Identity property)

(ix) $p, q, r \in z \Rightarrow p (q + r) = pq + pr$ (Distributive property)

3.1.2   Z (the set of Integers)

Note: There is an order relation denoted by '$<$' with the following properties:

(i) $p \in z \Rightarrow$ one and only one of the conditions

$p \in n, p = 0, - p \in n$ holds equivalently,

$p > 0, p = 0, p < 0$

(ii) $p, q, r \in z, p < q, q < r \Rightarrow p < r$ (transitive property)

(iii) $p, q, r \in z, p < q \Rightarrow p + r < q + r$

(iv) $p, q, r \in z, p < q, 0 < r \Rightarrow pr < qr$

(v) $p, q, r \in z, p < q, r < 0 \Rightarrow qr < pr$

3.2   Division Algorithm (or Division Theorem)

Let m and n be two integers where 'n' is positive, then there exist two unique integers q and r such that $m = nq + r$, where $0 \leq r \leq n$. The integers 'q' and 'r' are called the quotient and the remainder, respectively.

Here we may note that: 'm' is divisible by 'n' if and only if the remainder 'r' is equal to zero.

3.3   Greatest Common Divisor (GCD)

Definitions

Let m and n be two integers.

(i)   We say that "m divides n" if there exists an integer z such that $m.z = n$

(ii)   The greatest common divisor (gcd) of two positive integers a and b is the largest positive integer that divides both a and b.

(iii)   The notation gcd (a,b) denotes the greatest common divisor of a and b.

We say that two integers a and b are relatively prime if their greatest common divisor is 1.

Example 3.1

Find the gcd of 237 and 204.

Solution: Given integers are 237 and 204

$$
\begin{array}{r}
204)\ 237\ (1 \\
\underline{204} \\
33)204(6 \\
\underline{198} \\
6)33(5 \\
\underline{30} \\
3)6(2 \\
\underline{6} \\
0
\end{array}
$$

Therefore gcd (237, 204) = 3

Note:

Let m and n be positive integers and r the smallest positive integer for which there exist integers p and q such that r = p.m + q.n. Then we have that gcd (m, n) = r.

3.4   Euclidian Algorithm

Theorem 1

If m, n ∈ N then by repeated application of the division algorithm.

$$m = n\,p_1 + r_1,\ 0 < r_1 < c$$
$$n = r_1.p_2 + r_2,\ 0 < r_2 < r_1$$
$$r_1 = r_2 p_3 + r_3,\ 0 < r_3 < r_2$$
$$r_{j-z} = r_{j-1}\,p_j + r_j,\ 0 < r_j < r_{j-1}$$
$$r_j - 1 = r_j\,q_j + 1$$

The greatest common divisor (m, n) of m and n is r_j, the last non zero reminder in this division process. Values of x_0 and y_0 in (m, n) = $mx_0 + ny_0$ could be obtained by writing each r_i as a linear combination of two numbers m and n.

Proof: The chain of equations is obtained by dividing n_1 into m; r_1 into n; r_2 into r_1;…r_j into r_{j+1}.

This process stops if the division is exact. That is, when the remainder is zero.

So in the application of division algorithm, we have written the inequalities for the remainder without the equality sign.

So for example, $0 < r_1 < n$ in place of $0 \le r_1 \le n$ because if r_1 is equal to zero, the chain would stop at the first equation $m = cp_1$. In this case the gcd (m, n) would be n.

$$(m, n) = (m - np_1, n) = (r_1, n)$$
$$= (r_1, n - r_1 p_2)$$
$$= (r_1, r_2)$$
$$= (r_1 - r_2 p_3, r_2)$$
$$= (r_3, r_2)$$

Continuing this, by mathematical induction, we get that $(m, n) = (r_{j-1}, r_j) = (r_j, 0) = r_j$

Now
$$r_1 = m - np_1$$
$$r_2 = n - r_1 p_2$$
$$= n - (m - np_1)\, p_2$$
$$= n\,(1 + p_1 p_2) - mp_2$$

and
$$r_3 = r_1 - r_2 p_3$$
$$= m - np_1 - [n\,(1 + p_1 p_2) - mp_2]p_3$$
$$= m\,(1 + p_2 p_3) - n\,(p_1 + p_3 + p_1 p_2 p_3)$$

Continuing this, by mathematical induction, we get that each r_i is a linear combination of m and n.

3.5   Least Common Multiple

Definition

Let 'm' and 'n' be two non-zero integers. Then the least common multiple (LCM) of m and n, is the positive integer 'p' such that

(i) $m|p$, $n|p$

(ii)   If $m|q$, $n|q$ with $q > 0$, then $p|q$

The condition (ii) is equivalent to "if $m|q$ and $n|q$ then $|q| \ge p$" we write LCM (a, b) as $[a, b]$ also.

Example 3.2

Find the L.C.M of 12 and 15.

Solution: The integers given as 12 and 15

$$12 = 2.6 = 2.2.3 = 2^2.3$$
$$15 = 3.5$$
$$\text{LCM } (12, 15) = 2^2.3.5 = 4.3.5 = 60$$

Note:

Let m and n be any two integers then gcd (m, n). lcm (m, n) = mn.

3.6 Testing for Prime Numbers

Definition

An integer p > 1 is called a prime number if its only divisors are 1 and p.

Example 3.3

3, 5, 7, 13 are prime numbers (since their divisors are 1 and themselves).

Note:

Let p be a prime number and 'm' be any integer. Then either p divides m or gcd (p, m) = 1

3.7 Fundamental Theorem of Arithmetic (or Unique Factorization Theorem)

Theorem 2

Every positive integer can be factored into product of some prime numbers.

Proof: We prove this by mathematical induction p(n) be the statement that n can be written as product of some prime numbers. If n = 2 or n = 3 then n is a prime number. So p(2) and p(3) are true.

Let K be a positive integer and assume the induction hypothesis that $\phi(n)$ is true for all $n \leq k$.

Now we prove p (K + 1).

If K + 1 is a prime number then p (K + 1) is true. If K + 1 is not a prime number, then (K + 1) = a.b for some positive integers a and b with $1 < a \leq k$ and $1 < b \leq k$.

By induction hypothesis p(a) and p(b) are true. So a is a product of some prime numbers and b is a product of some prime numbers. Therefore (K + 1) = a.b is a product of some prime numbers.

Hence p (K + 1) is true.

Thus by mathematical induction, we conclude that p(n) is true for all positive integers n.

Theorem 3

Every positive integer n > 1 can be expressed as a product of primes uniquely apart from the order of the prime factors.

Proof: We know that (by above theorem) "Every integer $n \geq 1$ can be expressed as a product of primes".

Suppose if possible "there exists a positive integer n with different factorings".

Dividing out any primes common to the two representations, we would have an equality of the form

$$r_1 r_2 ... r_k = s_1 s_2 ... s_r$$

where the factors r_i and q_j are primes, not necessarily all distinct, but where no prime on the LHS occurs on the RHS.

Now $r_1 | r_1 r_2 ... r_k \Rightarrow r_1 | s_1 s_2 ... s_r \Rightarrow r_1 | s_j$ for $j = 1, 2, ..., n$

$\Rightarrow$ $r_1 = s_j$ for some $j = 1, 2, ..., n$.

This is a contradiction.

Thus every positive integer $n > 1$ can be expressed as a product of primes uniquely apart from the order of the prime factors.

Definition

Every positive integer $m > 1$ can be written uniquely in the form

$$m = q_1^{\beta_1} q_2^{\beta_2} ... q_k^{\beta_k}$$

where $q_1, q_2, ..., q_k$ are primes such that $1 < q_1 < q_2 < ... < q_k$ and $\beta_1, \beta_2, ..., \beta_k$ are positive integers.

This representation of 'm' is called prime factorization of 'm' in canonical form or prime power factorization of m.

Note 1:

If $c = q_1^{\alpha_1} q_2^{\alpha_2} ... q_m^{\alpha_m}$, $d = r_1^{\beta_1} r_2^{\beta_2} ... r_n^{\beta_n}$ are prime factorizations of 'c' and 'd' in canonical form, then $c = d$ if and only if $m = n$, $q_i = r_i$ and $\alpha_i = \beta_i$ for $i = 1, 2, ..., n$.

Note 2:

If 'h' is the gcd of m and n then $h = p_1^{n_1} p_2^{n_2} ... p_r^{n_r}$, where $p_1, p_2, ..., p_r$ are the common prime factors of m and n and for each $i = 1, 2, ..., r$, m_i is the minimum of the exponents of p_i in 'm' and n.

Note 3:

If T is the LCM of m and n, then $T = q_1^{l_1} q_2^{l_2} ... q_s^{l_s}$, where $q_1, q_2, ..., q_s$ are all the prime factors of both m and n and for each $i = 1, 2, ..., s$, m_i is the maximum of the exponents of q_i in m and n.

Example 3.4

Find the gcd and the lcm of m = 5040, n = 14850 by writing each of the numbers m and n into prime factorization (canonical form).

Solution: Given integers are

$$m = 5040$$

$$n = 14850$$

Now $$m = 5040 = 2^4 \times 3^2 \times 5 \times 7$$

$$n = 14850 = 2 \times 3^3 \times 5^2 \times 11$$

Therefore gcd (5040, 14850) $= 2 \times 3^2 \times 5 = 90$

$$\text{LCM } [5040, 14850] = 2^4 \times 3^3 \times 5^2 \times 7 \times 11$$

$$= 831600$$

3.8   Modular Arithmatic

"Equivalence relation" is a tool for abstraction (which is useful to ignore irrelevant details). Real applications are to algebric structures that are rather complex.

Such applications include the study of groups rings, including symmetries, and finite state machines. One application, particularly, useful in computer science, when we are dealing with familiar structures

Due to the finite storage limitations, and finite accuracy limitations of hardware arithmetic operations on computers, there is a frequent need for counting "modulo" some given number m.

Definition

Let 'p' be any positive integer. The relation congruence modulo p, (written as = (mod p)) on the integers by q = r (mod m) if and only if q = r + a.p for some integer 'a'. (Equivalently p divides q – r).

Note 1:

For any positive integer 'p', the relation $\equiv$ (mod p) is an equivalence relation on the integers, and partitions the integers into 'p' distinct classes (called as equivalence classes):

$$[0], [1], \ldots, [p-1].$$

Note 2:

The operations +, and '.' defined as follows on Z_m are well defined functions.

$$[a] + [b] = [a + b], [a].[b] = [a.b].$$

Note 3:

The notation x (mod p) is ordinarily used to denote the smallest non negative integer y such that

$$x \equiv y \pmod{p}$$

3.9   Fermats Theorem

Theorem 4

If 'p' is a prime number and 'm' is any integer such that 'p' is not a divisor of 'm' (so that (m, p) = 1), then $m^{p-1} \equiv 1 \pmod{p}$.

Proof: Set p is a prime number and 'm' be any integer such that

$$\gcd(p, m) = 1$$

If the numbers m, 2m, 3m, ..., (p – 1) m when divided by p, the remainders are 1, 2, ..., p – 1 (not necessarily in this order).

Let $r_1, r_2, ..., r_{p-1}$ be the remainders obtained on dividing m, 2m, ..., (p – 1) m respectively by p. Then as mentioned above $r_1, r_2, ..., r_{p-1}$ are precisely 1, 2, ..., p – 1 placed in some order so that the product $r_1, r_2, ..., r_{p-1}$ = the product 1, 2, ... (p – 1)

Now $m \equiv r_1 \pmod{p}$

$$2m \equiv r_2 \pmod{p}$$

$$(p - 1) \, m \equiv r_{p-1} \pmod{p}$$

Multiplying the congruence relations on both sides, we have

$$m, 2m, ... (p - 1) \, m \equiv r_1, r_2, ... r_{p-1} \pmod{p}$$

$$\Rightarrow 1, 2, 3, ... (p - 1) \, m^{p-1} \equiv 1, 2, 3, ... (p - 1) \pmod{p}$$

$$\Rightarrow (p - 1)! \, m^{p-1} \equiv (p - 1)! \pmod{p}$$

Since p is prime, we have that

$$\gcd(p, 1) = 1, (p, 2) = 1, ..., (p, p - 1) = 1$$

and hence

$$\gcd(p, p - 1) = 1$$

Therefore by cancelling (p – 1)! on both sides of (2) we set that

$$m^{p-1} \equiv 1 \pmod{p}$$

3.9.1 Corollary

If p is a prime number and 'm' is any integer, the $m^p \equiv m \pmod{p}$.

Definition (Euler's Function)

The Euler's ϕ – function is the function $\phi: N \to N$ defined as follows:

(i) $\phi(1) = 1$ and

(ii) For $n \in N$ and $n > 1$, we define $\phi(n)$ = the number of positive integers less than n and relatively prime to n. [that is $\phi(n) = |\{k$ is an integer $| 1 < k < n$ and $(k, n) = 1\}|$

Example 3.5

(i) $\phi(6) = |\{1, 5\}| = 2$ (Since 1 and 5 are relatively prime to 6)

(ii) $\phi(10) = |\{1, 3, 7, 9\}| = 4$ (Since 1, 3, 7, 9 are relatively prime to 10)

Note

If m and n are relatively prime positive integers, then $\phi(mn) = \phi(m)\,\phi(n)$.

Theorem (Euler) 5:

If m is a positive integer and n is any integer such that gcd (n, m) = 1, then $n^{\phi(m)} \equiv 1 \pmod{m}$.

Proof: Let $\{r_1, r_2, ..., r_{\phi(m)}\}$ be a reduced set of residues module m.

Since gcd (n, m) = 1 , we have that $\{nr_1, nr_2, ..., nr_{\phi(m)}\}$ is also a reduced set of residue module m, and consequently each nr_1 is congruent module m, to one and only one r_1.

Let $\qquad nr_1 \equiv y_1 \pmod{m}$

$\qquad\qquad\quad nr_2 \equiv y_2 \pmod{m}$

$\qquad\qquad\quad nr_{\phi(m)} \equiv y_{\phi(m)}^{(mod\ m)}$

then $y_1, y_2, ..., y_{\phi(m)}$ are precisely $r_1, r_2, ..., r_{\phi(m)}$ placed in some order so that the product $y_1, y_2, ..., y_{\phi(m)}$ = the product $r_1, r_2, ..., r_{\phi(m)}$ .....(3.1)

multiplying the above congruences, we have $nr_1, nr_2, ..., nr_{\phi(m)} \equiv y_1, y_2, ... y_{\phi(m)} \pmod{m}$

$\qquad \Rightarrow \qquad r_1 r_2 ... r_{\phi(m)}\, n^{\phi(m)} \equiv r_1 r_2 ... r_{\phi(m)} \pmod{m}$ .....(3.2)

Since each r_i, (i = 1, 2, ..., $\phi(m)$) is relatively prime to m, we have that the product $r_1 r_2 ... r_{\phi(m)}$ is also relatively prime to m. So cancelling $r_1 r_2 ... r_{\phi(m)}$ on both sides of (3.2), we get that

$\qquad\qquad n^{\phi(m)} \equiv 1 \pmod{m}$

The proof is complete.

Note:

In the statement of "Euler's Theorem", if we take m = p where p is a prime number, then ϕ (m) = ϕ (p) = p – 1. Therefore the result $n^{\phi(m)} \equiv 1$ (mod m) takes the form $n^{p-1} \equiv 1$ (mod p), which is the "Fermat's Theorem".

Example 3.6

If m = 12, then ϕ(m) = ϕ(12) = 4.

Since 1, 5, 7 and 11 are only positive numbers that are less than 12 and relatively prime to 12.

Therefore by Euler's theorem we have that

$1 \equiv 1^4 \equiv 5^4 \equiv 7^4 \equiv 11^4$ (mod 12).

Exercises

Properties of Integers

　　1.　List of all positive divisors of 32.

Ans:　　1, 2, 4, 8, 16, 32

　　2.　List of all positive divisors of 105

Ans:　　1, 3, 5, 7, 15, 105

　　3.　List of all positive integers divides 26.

Ans:　　1, 2, 13, 26

Division Theorem

　　1.　If a, b, c $\in$ z, such that a|b and a|c, then show that a| mb + nc for m, n $\in$ z.

　　2.　If a, b, c are positive integers; a|b; and a|c then show that a^2|bc.

　　3.　Suppose a, b, c are three positive integers. Suppose that

　　　　(i)　p is a number such that p^2 divides 'a'.

　　　　(ii)　pq divides b

　　　　(iii)　$2p^2$ divides c

Then show that $p^2 \mid a + 3b + c$

The Greatest Common Divisor

　　1.　Show that 403 and 517 are co-prime. (Hint: To show this it is enough to show that GCD of 403 and 517 is 1).

2. Find the Greatest Common Divisor of the numbers 2923, 3239

Ans: 79

3. Find the Greatest Common Divisor of 561 and 748.

Ans: 187

Euclidian Algorithm

1. Using Euclidian Algorithm find the gcd (1806, 3174).

Ans: 6

2. Using Euclidian Algorithm find gcd (1254, 779)

Ans: 19

3. Using Euclidian Algorithm find the gcd (2, 3)

Ans: 1

The Least Common Multiple

1. Find the L.C.M. of 8, 12, 27, 40

Ans: 1080

2. Find the L.C.M. of 7, 17, 51, 63

Ans: 1071

3. The L.C.M. of the two numbers 3855 and their G.C.D is 257. If one of the number is 1285. Then find the other.

Ans: 771

Testing of Prime Numbers

1. Is log a prime number (or) not?

Ans: Yes

2. Is 437 a prime number or not?

Ans: Not a prime

3. Show that if a positive integer 'n' is composite, then it has a prime factor 'p' such that $p \leq \sqrt{n}$.

The Fundamental Theorem of Arithmetic

1. Find the prime factorization of 744.

Ans: $2^3.3.31$

2. Find the prime factorization of 1862

Ans: $2.7^2.19$

3. Express 3000 as a product of prime powers

Ans: $2^3.5^3.3^1$

Modular Arithmatics

1. Construct the addition table for $Z_5 = \{0, 1, 2, 3, 4\}$, the set of integers module 5.

Ans:

Z_5	0	1	2	3	4
0	0	1	2	3	4
1	1	2	3	4	0
2	2	3	4	0	1
3	3	4	0	2	3
4	4	0	1	2	3

2. Construct multiplication table for $Z_5^* = \{1, 2, 3, 4\}$ (mod 5). Find an element $x \in Z_5^*$ such that $4.x = 1$

Ans:

Z_5	1	2	3	4
1	1	2	3	4
2	2	4	1	3
3	3	1	4	2
4	4	3	2	1

3. Write down the addition table for Z_6. (The set of integers module 6). Find an element $x \in Z_6$ such that $z + x = 0$

Z_6	0	1	2	3	4	5
0	0	1	2	3	4	5
1	1	2	3	4	5	0
2	2	3	4	5	0	1
3	3	4	5	0	1	2
4	4	5	0	1	2	3
5	5	0	1	2	3	4

4.   Find elements $0 \neq a$, $0 \neq b$ in Z_6 such that $a.b = 0$

Ans: The element s a, b $\in Z_6$ satisfying the given property are mentioned in the following:

$(a, b) = (2, 3)$, $(a, b) = (3, 2)$

$(a, b) = (4, 3)$, $(a, b) = (3, 4)$

Fermat's Theorem and Eulers Theorem

1.   If x and y are both prime to 'n', where n is a prime, then show that

$x^{n-1} - y^{n-1}$ is divisible by n.

2.   Show that the 8th power of any number is of the form 17n (or) $17n \neq 1$.

3.   Find (i) $\phi (35)$; (ii) $\phi (360)$

Ans: (i) 24 ; (ii) 96

CHAPTER - 4

Mathematical Induction

In Mathematics, as in Science, there are two main aspects of inquiry whereby we can discover new results: deductive and inductive.

As we have said, the deductive aspect involves accepting certain statements as premises and axioms and then deducting other statements on the basis of valid inferences.

The inductive aspect, on the other hand, is concerned with the search for facts by observation and experimentation – we arrive at a conjecture for a general rule by inductive reasoning.

Frequently we may arrive at a conjecture that we believe to be true for all positive integers 'n'. But then before we can put any confidence in our conjecture we need to verify the truth of the conjecture.

There is a proof technique that is useful to verifying such conjectures. Let us describe this technique now.

LEARNING OBJECTIVES

♦ *to understand the Principle of Mathematical Induction*
♦ *in solving the problems related to induction method*

4.1 The Principle of Mathematical Induction

4.1.1 The Principle

Let $S(n)$ be a statement which for each integer n, may be either true or false. To prove $S(n)$ is true for all integers $n \geq 1$, it is suffices to prove the following two conditions:

(i) $S(1)$ is true

(ii) For all $m \geq 1$, $S(m)$ implies $S(m + 1)$.

Note:

If one replaces (i) and (ii) by (i)^1 $S(n_o)$ is true, and (ii)^1 for all $m \geq n_o$, $S(m)$ implies $S(m + 1)$, then we can conclude that $S(n)$ is true for all $n \geq n_0$ and the starting point n_0, or basis of induction, may be any integer – positive, negative, or zero.

Normally we expect to prove $S(m) \rightarrow S(m + 1)$ directly. So there are 3-steps to a proof by using the "principle of mathematical induction":

  (i)   Show $S(n_0)$ is true. (Basis of induction)

  (ii) Assume $S(m)$ is true for $m \geq n_0$ (inductive hypothesis)

  (iii) Show that $S(m + 1)$ is true on the basis of the inductive hypothesis.

Example 4.1

Using principle of mathematical induction show that $1 + 2 + \ldots + n = \dfrac{n(n+1)}{2}$ for all $n \in N$.

Solution: Given statement is

$$S(n): 1 + 2 + \ldots + n = \dfrac{n(n+1)}{2} \qquad\qquad \ldots\ldots(4.1)$$

  (i)   *Basis of induction:* $n = 1$

  Now $S(1) = 1 = \dfrac{1(1+1)}{2}$, and so the formula is true for $n = 1$

  (ii) *Inductive hypothesis:* for $n = m$

  Assume that the statement $S(n)$ is true for $n = m$ that is,

  $S(m)$ is true. So $1 + 2 + \ldots + m = \dfrac{m(m+1)}{2}$ $\qquad\qquad \ldots\ldots(4.2)$

  (iii) *Inductive step:* Take $n = m + 1$

  Consider $S(m + 1) = 1 + 2 + \ldots + m + (m + 1)$

$$= S(m) + (m + 1) \quad \text{(by (4.2))}$$
$$= \dfrac{m(m+1)}{2} + m + 1$$
$$= \dfrac{m(m+1) + 2(m+1)}{2}$$
$$= \dfrac{(m+1)\,(m+2)}{2}$$

Therefore the statement holds for $n = m + 1$. So $S(n)$ is true for all n.

Therefore $1 + 2 + \ldots + n = \dfrac{n(n+1)}{2}$ is true for all positive integers n.

Example 4.2

Show that $1.4 + 2.5 + 3.6 + \ldots + n\,(n + 3) = \dfrac{n(n+1)\,(n+5)}{3}$ for all $n \in N$, by using principle of mathematical induction.

Solution: Suppose the statement S(n) denotes

$$1.4 + 2.5 + 3.6 + \ldots + n(n + 3) = \frac{n(n+1)\,(n+5)}{3} \text{ for all } n \in N. \qquad \ldots\ldots(4.3)$$

$$\text{if } n = 1 \text{ then LHS} = 1.4 = 4 = \frac{1(1+1)\,(1+5)}{3}$$

$$= \frac{12}{3} = 4 = \text{RHS}$$

Therefore S(1) is true.

Assume that the statement is true for $n = m$ that is, S(m) is true.

That is $1.4 + 2.5 + \ldots + m(m + 3) = \dfrac{m(m+1)\,(m+5)}{3}$

$$\text{put } n = m + 1$$

$$S(m + 1) = 1.4 + 2.5 + \ldots + m\,(m + 3) + (m + 1)\,(m + 4)$$

$$= S(m) + (m + 1)\,(m + 4)$$

$$= \frac{m(m+1)\,(m+5)}{3} + (m+1)\,(m+4)$$

$$= \frac{(m)\,(m+1)\,(m+5) + 3(m+1)\,(m+4)}{3}$$

$$= \frac{(m+1)\,[m^2 + 5m + 3m + 12]}{3}$$

$$= \frac{(m+1)\,[m^2 + 8m + 12]}{3}$$

$$= \frac{(m+1)\,(m+2)\,(m+6)}{3}$$

$$= S(m + 1)$$

Therefore S(m + 1) is true.

Hence by principle of Mathematical Induction S(n) is true for all n.

Example 4.3

Show that $2.7^n + 3.5^n$ is divisible by 24 by using principle of mathematical induction.

Solution: Suppose S(n) is "$2.7^n + 3.5^n - 5$ is divisible by 24."

1. If n = 1 then S(1) = 2.7 + 3.5 − 5
$$= 14 + 15 - 5 = 29 - 5 = 24$$

   So S(1) is divisible by 24.

2. Assume S(n) is true

   That is, $2.7^n + 3.5^n - 5$ is divisible by 24

3. Put n = m + 1

$$2.7^{m+1} + 3.5^{m+1} - 5$$
$$= 2.7.7^m + 3.5.5^m - 5$$
$$= 7(2.7^m) + 15.5^m - 5$$
$$= 7(24p - 3.5^m + 5) + 15.5^m - 5 \quad (\because \text{ where } 2.7^m + 3.5^m - 5 = 24\ p)$$
$$= 168p - 21.5^m + 35 + 15.5^m - 5$$
$$= 168p - 6.5^m + 30$$
$$= 168p - 30(5^{m-1} - 1)$$
$$= 168p - 30(5 - 1)(5^{m-2} + 5^{m-3} + \dots + 1)$$
$$= 24[7p - 5(5^{m-2} + 5^{m-3} + \dots + 1)]$$

   is divisible by 24

Therefore S(m + 1) is true.

Hence S(n) is true for all n ∈ N.

Example 4.4

Show that $B \cup \left(\bigcap_{i=1}^{n} A_i \right) = \bigcap_{i=1}^{n} (B \cup A_i)$ by using Mathematical Induction

Solution: Let S(n) be the statement $B \cup \left(\bigcap_{i=1}^{n} A_i \right) = \bigcap_{i=1}^{n} (B \cup A_i)$

For n = 1 it is obviously true.

For n = 2, $B \cup (A_1 \cap A_2) = (B \cup A_1) \cap (B \cup A_2)$ is the distributive law of union and intersection.

Note that here we are trying to prove that the distributive law holds over the intersection (for any number of sets). Now assume that the statement (1) is true for n = k, that is S(k) is true. So $B \cup \left(\bigcap_{i=1}^{k} A_i \right) = \bigcap_{i=1}^{k} (B \cup A_i)$,

Now for n = k + 1

Consider $B \cup \left(\bigcap_{i=1}^{k+1} A_i \right) = B \cup \left(\bigcap_{i=1}^{k} A_i \cap A_{k+1} \right)$

$$= \left(B \cup \left(\bigcap_{i=1}^{k} A_i \right) \right) = \cap (B \cup A_{k+1})$$

$$= \bigcap_{i=1}^{k} (B \cup A_i) \cap (B \cup A_{k+1})$$

$$= \bigcap_{i=1}^{k+1} (B \cup A_i)$$

Here we have used the fact that S(2) and S(k) holds, and we have proved that S(k+1) holds.

Hence the given statement S(n) is true for all positive integers n.

Exercises

1. Show that 2n < n! for n ≥ 4 by using mathematical induction.
2. Show that $n^3 + 2n$ is divisible by 3, by using mathematical induction.
3. Prove that
    $$= \frac{1}{1.2} + \frac{1}{2.3} + ... + \frac{1}{n(n+1)} = \frac{n}{n+1}$$ by using mathematical induction.
4. Show that $2 + 2^2 + 2^3 + ... + 2^n = 2^{n+1} - 2$ by using mathematical induction.
5. Show that $1^2 + 2^2 + ... + n^2 = \frac{n(n+1)(2n+1)}{6}$
6. Show that $2 + 6 + 12 + ... + (n^2 - n) = \frac{n(n^2 - 1)}{3}$
7. Show that $1^2 - 2^2 + 3^2 - ... + (-1)^{n-1} n^2 = \frac{(-1)^{n-1} n(n+1)}{2}$
8. Prove that 4^{2n-1} is divisible by 15 for n ≥ 1.
9. Prove that 43 divides $6^{n+2} + 7^{2n+1}$ for n ≥ 1.
10. Show that $n! < n^n$ for n ≥ 2.

CHAPTER - 5

Set Theory

The concept "Set" have fundamental importance in Modern Mathematics. The idea of a set has been intuitively used in Mathematics since ancient time of Greeks. For all the modern algebra branches like: Group, Theory, Ring theory the fundamental base is the concept 'set'.

The systematic development of set theory is attributed to the German mathematician George Cantor (1845s–1918).

Some elementary definitions of set theory have been studied by students in the high school standard. In this chapter we briefly give some preliminaries of set theory and discuss the relations and functions.

5.1 Sets

5.1.1 Set

A set is a collection of objects in which we can say whether a given object is in the collection. Sets will be denoted by capital letters.

The fact that a is a member of a set A is denoted by $a \in A$ and we call it as 'a belongs to A'. The members of a set are called elements.

If x is not an element of A then we write $x \notin A$.

Note:

There are five ways used to describe a set:

(i) Describe a set by describing the properties of the members of the set.

(ii) Describe a set by listing its elements.

(iii) Describe a set A by its characteristic function, defined as

$$\mu_A(x) = 1 \quad \text{if } x \in A$$
$$= 0 \text{ if } x \notin A$$

for all x in $\cup$, where $\cup$ is the universal set, sometimes called the "universe of discourse," or just universe.

(iv) Describe a set by recursive formula. This is to give one or more elements of the set and a rule by which the rest of the elements of the set may be generated.

(v) Describe a set by an operation (say union, intersection, complement etc.,) on some other sets.

Example 5.1

Describe the set containing all the non-negative integers less than or equal to 4.

Let X denote the set. Then X can be described in the following ways:

(i) $X = \{x \mid x$ is a nonnegative integer less than or equal to 4$\}$ and
$N = \{x \mid x$ is a natural number$\} = \{1, 2, 3, 4, \ldots\}$ is an infinite set.

(ii) $X = \{0, 1, 2, 3, 4\}$.

(iii) $\mu_A(x) = 1 \quad$ for $x = 0, 1, 2, 3, 4$
$$= 0 \quad \text{if } x \notin A.$$

(iv) $X = \{x_{i+1} = x_i + 1, \quad i = 0, 1, 2, 3$ where $x_0 = 0\}$.

5.1.2 Sub Set

Suppose A and B are two sets. Then we say that A is a subset of B (written as $A \subseteq B$) if every element of A is also an element of B. A set A is said to be a proper subset of B if there exists an element of B which is not an element of A. That A is a proper subset of B if $A \subset B$ and $A \neq B$.

Note:

The containment of sets has the following properties: Let X, Y and Z be sets.

(i) $X \subseteq X$

(ii) If $X \subseteq Y$ and $Y \subseteq Z$, then $X \subseteq Z$

(iii) If $X \subset Y$ and $Y \subset Z$, then $X \subset Z$

Example 5.2

(i) Let N, Z, Q, R denote the set of natural numbers; the set of integers; the set of rational numbers; the set of real numbers respectively. Then
$$N \subset Z \subset Q \subset R.$$

(ii) If $A = \{1, 3, 5\}$, $B = \{1, 3, 5, 7\}$ then A is a proper subset of B.

5.1.3   Equity of Sets

Two sets A and B are said to be equal (denoted by A = B) if A is a subset of B, and B is a subset of A.

To show two sets A and B are equal, we must show that

(i)   Each element of A is also an element of B, and

(ii)   Each element of B is also an element of A

Definitions

(i)   The set that contains no members is called the empty set and it is denoted by ϕ.

(ii)   A set which contains a single element then it is called singleton set.

X = {2} is a singleton set.

5.1.4   Powerset

Let A be a set. Define $\wp(A)$ = the set of all subsets of A. This $\wp(A)$ is called the power set of A. It is also denoted by P(A).

Example 5.3

Let A = {1, 3, 5}. Then

$\wp(A)$ = {{ϕ}, {1}, {3}, {5}, {1, 3}, {3, 5}, {5, 1},{1, 3, 5}}.

Note that there are $2^3 = 8$ elements in $\wp(A)$.

Example 5.4

If the set A has n elements, formulate a conjecture about the number of elements in $\wp(A)$.

Solution: Suppose A has n elements. Let m be an integer such that $0 \le m \le n$. We can select m elements from the given set A in ^nC_m ways. So A contains ^nC_m distinct subsets containing m elements.

Therefore the number of elements in P(A)

= number of subsets containing 0 number of elements

+ number of subsets containing only 1 element

+ ...

+ number of subsets containing n elements

= $^nC_0 + {}^nC_1 + {}^nC_2 + ... + {}^nC_n = 2^n$.

Now we can verify the following statements from the definitions of union, intersection and complement.

5.1.5   Notations

Some of the more important set notations are given below:

N: The set of all natural numbers = {n | n is a natural number} = {1, 2, 3, ...};

Z: The set of all integers = {x | x is an integer} = {..., –1, 0, 1, 2, ...};

Q: The set of all rational numbers = {p/q | p, q $\in$ z where q $\neq$ 0};

R: The set of all real numbers = {x | x is a real number};

C: The set of all complex numbers + {z | z is a complex number}.

5.2   Operations on Sets

5.2.1   Union and Intersection

(i)   If A and B are two sets, then the set {x | x $\in$ A or x $\in$ B} is denoted by A $\cup$ B and we call it as the union of A and B.

(ii)   The set {x | x $\in$ A and x $\in$ B} is denoted by A $\cap$ B and we call it as the intersection of A and B.

5.2.2   Properties of Union and Intersection

Some Properties Satisfied by Union and Intersection

S. No	Property	Union	Intersection
1	Idempotent	A $\cup$ A = A	A $\cap$ A = A
2	Commutative	A $\cup$ B = B $\cup$ A	A $\cap$ B = B $\cap$ A
3	Associative	A$\cup$(B$\cup$C) =(A$\cup$B)$\cup$C	A$\cap$ (B$\cap$C) = (A$\cap$B) $\cap$ C

5.2.3   Symmetric Difference

Let X and Y be two sets. The symmetrical difference of X and Y is defined as {x | x $\in$ X, or x $\in$ Y, but not both}. This is denoted as X Δ Y .

So X Δ Y = {x | x $\in$ X, or x $\in$ Y, but not both}. It is also called Boolean sum of two sets.

5.2.4   Set Complement

If A and B are two sets, then the set {x $\in$ B | x $\neq$ A} is denoted by B – A (or B \ A) and it is called as the complement A in B.

Example 5.5

Let A = {1, 2, 3, 4, 5}, B = {a, b, c, d}, C = {2, b, d}, D = {3, a, c} and E = {x | is an integer and 1 < x < 2}.

  (i) $A \cup B = \{x \mid x \in A \text{ or } x \in B\} = \{1, 2, 3, 4, 5, a, b, c, d\}$.

  (ii) $A \cap B = \{x \mid x \in A \text{ } x \in B\}$

           = the set of all elements that are both in A and B

           $= \phi$ (the empty set).

 (iii) $A \cap C = \{x \mid x \in A \text{ and } x \in C\} = \{2\}$

 (iv) $A \cap D = \{x \mid x \in A \text{ and } x \in D\} = \{3\}$

  (v) $B \cup C = \{x \mid x \in B \text{ and } x \in C\} = \{2, a, b, c, d\}$

Here we may note that, in roster form, there is no necessity of writing the same element in second time. So we avoid writing the same element second time in roster form.

 (vi) $B \cap C = \{a, b, c, d\} \cap \{2, b, d\} = \{b, d\}$

 (vii) $E = \{x \mid \text{is an integer and } 1 < x < 2\} = \phi$

         (Since there is no integer strictly lies between 1 and 2)

(viii) $A \cup \phi = \{x \mid x \in A \text{ and } x \in \phi\} = A$

 (ix) $A \cap \phi = \{x \mid x \in A \text{ and } x \in \phi\} = \phi$

5.2.5   Cartesian Product

  (i)   If S and T are two sets, then the set $\{(s, t) \mid s \in S \text{ and } 1 \in T\}$ is called the Cartesian product of S and T (here $(a, b) = (s, t) \Leftrightarrow a = s \text{ and } b = t$). The cartesian product of S and T is denoted by $S \times T$. Thus

         $S \times T = \{(s, t) \mid s \in S \text{ and } t \in T\}$

Note that if S and T are two sets, then $S \times T$ and $T \times S$ may not be equal (an example is given under 5.6).

  (ii)   If $S_1, S_2, \ldots, S_n$ are n sets, then the cartesian product is defined as

         $S_1 \times S_2 \times \ldots \times S_n = \{(s_1, s_2, \ldots, s_n) \mid s_i \in S_i \text{ for } 1 \leq i \leq n\}$.

         Here the elements of $S_1 \times S_2 \times \ldots \times S_n$ are called ordered n-tuples: For any two n-tuples, we have $(s_1, s_2, \ldots, s_n) = (t_1, t_2, \ldots, t_n) \Leftrightarrow s_i = t_i, 1 \leq i \leq n$.

Example 5.6

  (i)   {Rama, Sita, Lakshmana} is a set.

  (ii)   Suppose $A = \{a, b, c\}$ and $B = \{a, b, c, 3, 4\}$

         Then A is a subset of B.

         If $X = \{c, b, a\}$, then $A = X$

         If $D = \{a, b, 2, 4\}$, then $A \cap D = \{a, b\}$ and $A \cup D = \{a, b, c, 2, 4\}$

 (iii)   Empty set is a subset of every set. For any set Y we have $\phi \subseteq Y$.

(iv)   If $X = \{a, b\}$ and $Y = \{x, y\}$, then

$X \times Y = \{(a, x), (a, y), (b, x), (b, y)\}$ and

$Y \times X = \{(x, a), (x, b), (y, a), (y, b)\}$

It is clear that $X \times Y \neq Y \times X$

(v)   If $A = \{a, b\}$, $B = \{2\}$, $C = \{x\}$, then $A \times B \times C = \{(a, 2, x), (b, 2, x)\}$

Example 5.7

List the elements of the set $\{\frac{a}{b} \,/\, a$ and b are prime integers with $1 < a < 10$ and $3 < b < 9\}$.

Solution: We know that the prime numbers that are greater than 1 and less than 10 are 2, 3, 5, 7. Therefore 'a' may be 2 or 3 or 5 or 7 and 'b' may be 5 or 7.

Therefore the set is: $\{2/5, 2/7, 3/5, 3/7, 5/5, 5/7, 7/5, 7/7\} = \{2/5, 3/5, 1, 7/5, 2/7, 3/7, 5/7\}$.

5.2.6   Disjoint Sets

(i)   Two sets A and B are said to be disjoint if $A \cap B = \phi$.

(ii)   A collection $\{A_i\}_{i \in I}$ of sets is said to be mutually disjoint if $A_i \cap A_j = \phi$ for all $i \in I, j \in I$ such that $i \neq j$.

Example 5.8

(i)   The sets $A = \{1, 2\}$ and $B = \{a, b\}$ are disjoint sets.

(ii)   The sets $A = \{1, 2\}$, $B = \{a, b\}$ and $C = \{x, y, z\}$ are mutually disjoint.

(iii)   If $B_i = \{2i, 2i + 1\}$ for all $i \in N$, then $\{B_i\}_{i \in N}$ is a collection of mutually disjoint sets.

5.2.7   Distributive Laws

Let A, B and C be three sets. Then

(i) $A \cap (B \cup C) = (A \cap B) \cup (A \cap C)$

(ii) $A \cup (B \cap C) = (A \cup B) \cap (A \cup C)$

5.2.8   Properties Related to Set Operations

Now list the properties related to the set operations.

(i) $A \cup B = B \cup A$; $A \cap B = B \cap A$ (commutative properties)

(ii) $A \cup (B \cup C) = (A \cup B) \cup C$; $A \cap (B \cap C) = (A \cap B) \cap C$ (Associative)

(iii) $A \cup (B \cap C) = (A \cup B) \cap (A \cup C)$; $A \cap (B \cup C) = (A \cap B) \cup (A \cap C)$ (Distributive)

(iv) $A \cup A = A$; $A \cap A = A$ (Idempotent)

(v) $(A')' = A$

(vi) $A \cup A' = \cup$

(vii) $A \cap A = \Phi$

(viii) $\Phi' = \cup$

(ix) $U' = \Phi$

(x) $(A \cup B)' = A' \cap B'; \ (A \cap B)' = A' \cup B'$ (D' Morgan laws)

(xi) $A \cup \Phi = A; A \cap \Phi = \Phi; A \cup U = U; A \cap U = A$ (Universal)

5.2.9 Venn Diagrams

Introduction of the universal set permits the use of a pictorial device to study the connections between the subsets of a universal set and their intersection, union, difference and other operations. The diagrams used here are called Venn diagrams.

A Venn diagram is a pictorial representation of a set by a set of points.

The universal set $\cup$ is represented by a set of points inside a rectangle and the subset say X, of $\cup$ is represented by the interior of a circle or some other simple closed curve inside the rectangle that represents the universal set.

Example 5.9

In the following the shaded area represents the sets. The related set was indicated below the diagram.

(i)

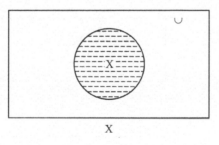

X

(ii)

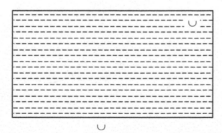

$\cup$

(iii)

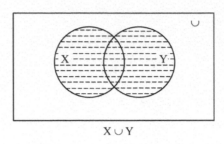

$$X \cup Y$$

(iv)

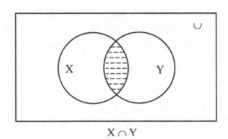

$$X \cap Y$$

(v)

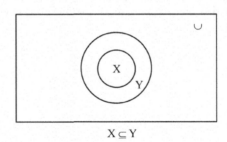

$$X \subseteq Y$$

(vi)

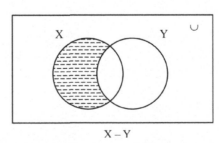

$$X - Y$$

(vii)

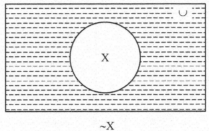

~X

(viii)

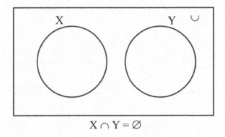

$X \cap Y = \emptyset$

Note 1:

It is easy to verify that the following through Venn diagrams

    (a) $X \cup Y = Y \cup X$ (Commutative law)

    (b) $X \cap Y = Y \cap X$ (Commutative law)

    (c) $\sim (\sim X) = X$

    (d) If $X \subseteq Y$ then

$$X - Y = \emptyset$$

$$X \cap Y = X$$

$$X \cup Y = Y$$

Note 2:

It should be emphasized that the above relations (given in above note) between the subsets can be verified by the Venn diagrams.

    Venn diagrams do not provide proofs, but these are helpful to understand the relations between the sets involved.

Example 5.10

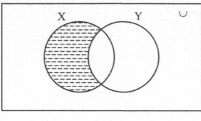 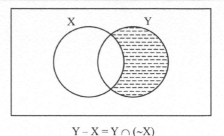

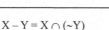

$$X - Y = X \cap (\sim Y) \qquad\qquad Y - X = Y \cap (\sim X)$$

Consider the Venn diagram given above. From these two diagrams, we may understand that

$$X \cup Y = [X \cap (\sim Y)] \cup [Y \cap (\sim X)] \cup (X \cap Y) \qquad\qquad .....(5.1)$$

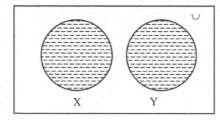

$$X \cup Y$$

From the above diagram we can understand that

$$X \cup Y = [X \cap (\sim Y)] \cup [Y \cap (\sim X)]$$

This equality is not true in general, although it happens to be true for the two disjoints sets X and Y.

Note 1:

A formal proof of equation (5.1) of 5.10 is given as follows:

For any x,

$$x \in X \cup Y \Leftrightarrow x \in \{x \mid x \in X \cup x \in Y\}$$
$$x \in (X \cap \sim Y) \cup (Y \cap \sim X) \cup (X \cap Y)$$
$$\Leftrightarrow x \in \{x \mid x \in (X \cap \sim Y) \cup x \in (Y \cap \sim X) \cup x \in (X \cap Y)\}$$
$$\Leftrightarrow x \in (X \cap \sim Y) \cup x \in (Y \cap \sim X) \cup x \in (X \cap Y)$$
$$\Leftrightarrow (x \in X \cap x \in \sim Y) \cup (x \in Y \cap x \in \sim X) \cup (x \in X \cap x \in Y)$$
$$\Leftrightarrow (x \in X \cap (x \in \sim Y \cup x \in Y) \cup (x \in X \cap (x \in Y \cap x \in \sim X)$$
$$\Leftrightarrow (x \in X \cup x \in Y) \cap (x \in X \cup x \in \sim X)$$

$$\Leftrightarrow (x \in X \cup x \in Y)$$
$$\Leftrightarrow x \in \{x \mid x \in X \cup x \in Y\}$$
$$\Leftrightarrow x \in \{X \cup Y\}$$

Note 2:

Consider the following Venn – diagrams:

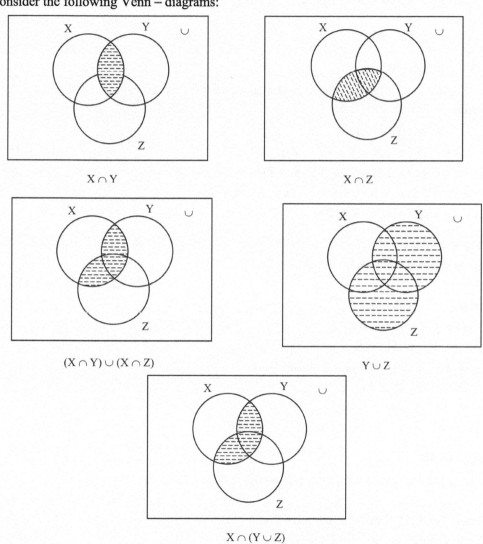

(i) From the 3rd and 5th Venn diagrams, we can understand that

$$X \cap (Y \cup Z) = (X \cap Y) \cup (X \cap Z) \qquad \qquad .....(5.2)$$

(ii) Similarly, one can verify through Venn diagrams that

$$X \cup (Y \cap Z) = (X \cup Y) \cap (X \cup Z) \qquad \qquad .....(5.3)$$

These two equations (5.2) and (5.3) are known as the distributive laws of union and intersection.

Example 5.11

Verify the distributive law.

$X \cup (Y \cap Z) = (X \cup Y) \cap (X \cup Z)$ without using Venn diagrams

Solution:

$x \in X \cup (Y \cap Z)$

$\Leftrightarrow x \in \{x \mid x \in X \cup x \in (Y \cap Z)\}$

$\Leftrightarrow x \in \{x \mid x \in X \cup (x \in Y \cap x \in Z)\}$

$\Leftrightarrow x \in \{x \mid (x \in X \cup x \in Y) \cap (x \in X \cup x \in Z)\}$

$\Leftrightarrow x \in (X \cup Y) \cap (X \cup Z)$

Example 5.12

Prove that $A - (B \cap C) = (A - B) \cup (A - C)$.

[JNTU, Nov 2010, Set No.1]

Solution: Let $x \in A - (B \cap C)$

$\Leftrightarrow x \in A$ and $x \notin B \cap C$

$\Leftrightarrow x \in A$ and $(x \notin B$ or $x \notin C)$

$\Leftrightarrow x \in A$ and $x \notin B$ or $x \in A$ and $x \notin C$

$\Leftrightarrow x \in A - B$ or $x \in A - C$

$\Leftrightarrow x \in (A - B) \cup (A - C)$

Therefore $A - (B \cap C) = (A - B) \cup (A - C)$

Example 5.13

If A, B, C are three sets such that $A \subseteq B$. Show that $(A \times C) \subseteq (B \times C)$.

[JNTUH, Nov 2010, Set No 4]

Solution: Let A, B, C be three sets such that $A \subseteq B$.

We have to show that $(A \times C) \subseteq (B \times C)$

Let $x \in A \times C$

Then x = (a, c) for some a ∈ A and c ∈ C.

  ⇒ x = (a, c) for some a ∈ B and c ∈ C [Since A ⊆ B]

  ⇒ x = B × C this is true for all x ∈ A × C.

Therefore A × C ⊆ B × C.

Example 5.14

If A = {1, 2, 3}, B = {4, 5} find

  (i) A × B   (ii) B × A

[JNTUH, Nov 2010, Set No 4]

Solution:

  Let A = {1, 2, 3}

      B = {4, 5}

  Now A × B = {(a, b) | a ∈ A and b ∈ B}

          = {(1, 4), (1, 5), (2, 4), (2, 5), (3, 4), (3, 5)}

    B × A = {(b, a) | b ∈ B and a ∈ A}

          = {(4, 1), (4, 2), (4, 3), (5, 1), (5, 2), (5, 3)}

Example 5.15

How many relations are there on a set with 'n' elements? Of a set A has 'm' elements and a set B has 'n' elements, how many relations are there been A to B? If a set A = {1, 2}, determine all relations from A to A.

[JNTUH, Nov 2008, Set No 2]

Solution:

  (i) Since we have that a relation from A to A is a subset of A × A, the set of all relations from A to A which are the set of all subsets of A × A.

    So, the number of relations from A to A is equal to the number of subsets of A × A.

    In the given problem, |A| = n, so we have |A × A| = n.n = n^2.

    Therefore A × A has 2^{n^2} number of subsets.

    ⇒ there are 2^{n^2} relations from A to A.

  (ii) Since |A| = m and |B| = n, we have |A×B| = mn.

    Therefore A × B has 2^{mn} subsets, thus there are 2^{mn} relations from A to B.

  (iii) Given that

          A = {1, 2}

    A × A = {(1, 1), (1, 2), (2, 1), (2, 2) }

$R_1 = \{(1, 1)\}$

$R_2 = \{(1, 2)\}$

$R_3 = \{(2, 1)\}$

$R_4 = \{(2, 2)\}$

$R_5 = \{(1, 1), (1, 2)\}$

$R_6 = \{(1, 1), (2, 1)\}$

$R_7 = \{(1, 1), (2, 2)\}$

$R_8 = \{(1, 2), (2, 1)\}$

$R_9 = \{(1, 2), (2, 2)\}$

$R_{10} = \{(2, 1), (2, 2)\}$

$R_{11} = \{(1, 1), (1, 2), (2, 1)\}$

$R_{12} = \{(1, 1), (1, 2), (2, 2)\}$

$R_{13} = \{(1, 2), (2, 1), (2, 2)\}$

$R_{14} = \{(1, 1), (2, 1), (2, 2)\}$

$R_{15} = A \times A$

5.3   Principle of Inclusion and Exclusion

5.3.1   Principle

Let A and B be any finite sets:

Then $n(A \cup B) = n(A) + n(B) - n(A \cap B)$

Clearly, to find the number $n(A \cup B)$ of elements in the union $A \cup B$, we add $n(A)$ and $n(B)$ and then subtract $n(A \cap B)$, i.e., we include $n(A)$ and $n(B)$ and exclude $n(A \cap B)$.

This principle holds for any number of sets.

For any three sets A, B and C, we have

$$n(A \cup B \cup C) = n(A) + n(B) + n(C) - n(A \cap B) - n(B \cap C) - n(C \cap A) + n(A \cap B \cap C)$$

Example 5.16

If A and B be two sets containing 3 and 6 elements respectively, what can be the minimum number of elements in $A \cup B$? Find also, the maximum number of elements in $A \cup B$.

Solution: By inclusion-exclusion principle

$$n(A \cup B) = n(A) + n(B) - n(A \cap B)$$

Now, $n(A \cup B)$ is minimum or maximum according as $n(A \cap B)$ is maximum or minimum respectively.

Case I: If $n(A \cap B)$ is minimum, then $n(A \cap B) = 0$

(For example $A = \{a, b, c, d, e, f\}$ and $B = \{g, h, i\}$)

Then $[n(A \cup B)]_{max.} = n(A) + n(B) = 6 + 3 = 9$

Case II: If $n(A \cap B)$ is maximum, i.e., then $n(A \cap B) = 3$

(For example $A = \{a, b, c, d, e, f\}$ and $B = \{b, d, e\}$

Then $[n(A \cup B)]_{min.} = n(A) + n(B) - n(A \cap B) = 6 + 3 - 3 = 6$

Example 5.17

In a group of 1000 people, there are 750 who can speak Hindi and 400 who can speak Punjabi. How many can speak both Hindi and Punjabi?

Solution: Let H and P be the set of those people who can speak Hindi and Punjabi respectively, then according to the problem, we have

$$n(H \cup P) = 1000, n(H) = 750, n(P) = 400$$

By inclusion-exclusion principle, we have

$$n(H \cup P) = n(H) + n(P) - n(H \cap P)$$

$$\Rightarrow 1000 = 750 + 400 - n(H \cap P)$$

$$\therefore \ n(H \cap P) = 150$$

$\therefore$ Number of people speaking both Hindi and Punjabi is 150.

Example 5.18

A survey of 500 television watchers produced the following information: 285 watch football, 195 watch hockey, 115 watch basketball, 45 watch football and basket ball, 70 watch football and hockey. 50 watch hockey and basketball, 50 do not watch any of three games. How many watch all the three games?

Solution: Let F, H and B be the sets of television watchers who watch Football, Hockey and Basketball respectively. Then according to the problems, we have

$n(\cup) = 500$, $n(F) = 285$, $n(H) = 195$, $n(B) = 115$, $n(F \cap B) = 45$, $n(F \cap H) = 70$, $n(H \cap B) = 50$ and $n(F' \cup H' \cup B') = 50$, where $\cup$ is the set of all the television watchers.

Since $n(F' \cup H' \cup B') = n(\cup) - n(F \cup H \cup B)$

$$\Rightarrow 50 = 500 - n(F \cup H \cup B)$$

$$\Rightarrow n(F \cup H \cup B) = 450$$

By inclusion-exclusion principle,

$$n(F \cup H \cup B) = n(F) + n(H) + n(B) - n(F \cap H) - n(H \cap B) - n(B \cap F) + n(F \cap H \cap B)$$

$$\Rightarrow 450 = 285 + 195 + 115 - 70 - 50 - 45 + n(F \cap H \cap B)$$

$$\therefore \ n(F \cap H \cap B) = 20$$

which is the number of those who watch all the three games.

Example 5.19

There are 20 teachers who teach mathematics and physics in a school. Of these 12 teach mathematics and 4 teach both physics and mathematics. How many teach physics?

*Solution:*Let M be the set of teachers who teach mathematics.

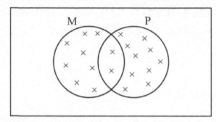

Let P be the set of teachers who teach physics.

Number of teachers who teach mathematics and physics is n(M $\cup$ P) = 20.

Number of teachers who teach mathematics only is n (M) = 12.

Number of teachers who teach both mathematics and physics is n(M $\cap$ P) = 4.

We have to find number of teachers who teach physics only.

We have that

$$n (M \cup P) = n(M) + n(P) - n(M \cap P)$$
$$20 = 12 + n(P) - 4$$
$$\Rightarrow n(P) = 20 - 12 + 4 = 12$$

Therefore, 12 teachers teach physics only.

Example 5.20

An elocution competition was held in Telugu and Hindi. Out of 80 students 45 students took part in Telugu, 35 in Hindi, 15 both in Telugu and Hindi. Then find the number of students.

 (i) Who took part in Telugu but not Hindi

 (ii) Who took part in Hindi but not Telugu

 (iii) Who took part in either Telugu or Hindi

 (iv) Who took part in neither

Solution: Suppose T is the set of students who took part in Telugu, H is the set of students who took part in Hindi. Then T $\cap$ H gives the set of students who took part in both Telugu and Hindi.

 (i) The number of students who took part in Telugu but not in Hindi.

$$= n (T) - n (T \cap H)$$
$$= 45 - 15$$
$$= 30$$

(ii)   The number of students who took part in Hindi but not in Telugu

   $= n(H) - n(T \cap H)$

   $= 35 - 15$

   $= 20$

(iii)   The number of students who took part either in Telugu or in Hindi is $n(T \cup H)$

   So $n(T \cap H) = n(T) + n(H) - n(T \cap H)$

   $= 45 + 35 - 15 = 65$

(iv)   The number of students who took part neither in Telugu nor in Hindi

   $= n(\mu) - n(T \cup H)$

   $= 80 - 5$

   $= 15$

We can solve the problem by using Venn diagram easily.

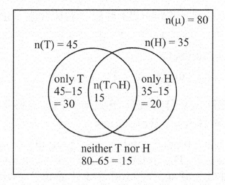

Example 5.21

Out of 100 students who appeared for 9th class examination from a school, 15 students in English; 12 students in Mathematics; 8 students in Science; 7 students in Mathematics and Science; 4 students in English and Science; 6 students in English and Mathematics; 4 students in all the three subjects could get first class marks. How many of them have got first class marks?

(i)   Only in Mathematics

(ii)   Only in Science

(iii)   Only in English

(iv)   Exactly in two subjects;

(v)   In more than one subject

Solution: Observe the diagram to understand the solution.

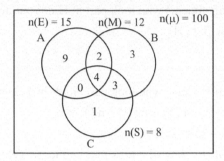

The number of students who have got first class marks

(i) Only in Mathematics $= n(M) - n(E \cap M) - n(M \cap S) + n(E \cap M \cap S)$
$$= 12 - 6 - 7 + 4 = 3$$

(ii) Only in English $= n(E) - n(E \cap M) - n(E \cap S) + n(E \cap M \cap S)$
$$= 15 - 6 - 4 + 4 = 9$$

(iii) Only in Science $= n(S) - n(M \cap S) - n(S \cap E) + n(M \cap S \cap E)$
$$= 8 - 4 - 7 + 4 = 1$$

(iv) Exactly in two subject $= 2 + 0 + 3 = 5$

(v) In more than one subject $= 2 + 4 + 0 + 3 = 9$

5.4   Relations, Binary Operations on Relations and Properties of Binary Relations

The concept 'Relation' is a fundamental concept used in our daily life. We use it, in the same meaning, in Mathematics. The relations used in Mathematics also include the natural relations such as "father and son", "brother and sister", "wife and husband", "mother and daughter", "greater than", "less than", etc.

We define the relation in the following way.

Definition

A relation S between the sets A_1, A_2, ..., A_n is a subset of $A_1 \times A_2 \times ... \times A_n$ $= \{(x_1, x_2, ..., x_n) \mid x_i \in A_i \text{ for } 1 \leq i \leq n\}$. The relation S is called n-ary relation (two – ary is called binary, three – ary is called ternary).

Example 5.22

A relation of mother to his child can be described by a set. Say M, of ordered pairs in which the first member is the name of mother and the second name of his child.

That is M = {(x, y) / x is the mother of y}   ..... (5.4)

5.4.1   Domain and Range

Let M be a binary relation. The set D(M) of all objects a such that (a, b) ∈ M for some b, is called the domain of M.

That is D (M) = {a | (a, b) ∈ M for some b}

The set R(M) of all objects b such that (a, b) ∈ M for some a, is called the range of M.

That is R(M) = {b | (a, b) ∈ M for some a}

Let A and B be any two sets. A subset of the cartesian product A × B defines a relation say C. For any such relation C, we have D(c) ⊆ A and R(c) ⊆ B, we say that C is a relation from A to B.

If A = B, then C is said to be a relation from A to A.

In such a case, C is also called as a relation in 'A' (or on A).

Thus any relation in A is a sub set of A × A.

The set A × A itself defines a relation in A and it is called a universal relation on A.

The empty set which is also a subset of A × A is called a void relation (or null relation or empty relation) on A.

5.4.2   Notation

If R is a binary relation from A to B then R ⊆ A × B. So R contains some ordered pairs of the form (a, b) where a ∈ A and b ∈ B.

If (a, b) ∈ R then we write a R b.

If ~ is a binary relation and (a, b) ∈ ~ then we write a ~ b.

Let us consider example 5.22 in which

M = {(x, y) | x is the mother of y}

If x is the mother of y then (x, y) ∈ M and so we may write x M y.

Hence x M y means "x is the mother of y". The set of all mothers is the domain of M, and the set of all the human beings is the range of M.

Example 5.23

A familiar example is the relation "less than" for real numbers. The relation is denoted by "<".

For any a, b real numbers such that a < b, then we write that (a, b) ∈ < or a < b.

So the relation '<' is given by

R = {(a, b) | a, b are real numbers such that a < b}

This is a binary relation on R the set of all real numbers.

Since 2 < 3 we may write (2, 3 ∈ <)

Since 2.01 < 2.02 we may write (2.01, 2.02) ∈ <

The set of all real numbers in the domain of the binary relation '< '.

Also the set of all real numbers is the range of the binary relation '< '.

5.4.3   Some Binary Operations on Relations

Let B and M denotes two relations.

(i)   The intersection of B and M, denoted by B ∩ M, defined by a relation such that

x (B ∩ M) y if and only if x B y and x M y.

(ii)   The union of B and M, denoted by B ∪ M, defined by a relation such that

x (B ∪ M) y if and only if x B y or x M y

Example 5.24

Let A = {1, 2, 3}, B = {3, 4, 5}, C = {4, 6}, and D = {3, 4, 5, 6}

then A × B = {(1, 3) (1, 4) (1, 5), (2, 3) (2, 4) (2, 5), (3, 3) (3, 4) (3, 5)} is a relation from A to D.

Also A × C = {(1, 4) (1, 6) (2, 4) (2, 6) (3, 4) (3,6)} is also a relation from A to D.

Now we have two relations A × B and A × C from A to D. The intersection of these two relations is

(A × B) ∩ (A × C) = {(1, 4) (2, 4) (3, 4)}

Example 5.25

Let A = {a, b}, B = {x , y}, and C = {a, c, y} and D = {a, c, x, y}

Then A × B = {(a, x) (a, y) (b, x) (b, y)} is a relation from A to D

A × C = {(a, a), (a, c) (a, y), (b, a), (b, c), (b, y)} is a relation from A to D.

Now the union of these two relations is

(A × B) ∪ (A × C) = {(a, x) (a, a) (a, y), (a, c) (b, x) (b, y) (b, a) (b, c)}

5.4.4   Properties of Relations

In general, a relation means binary relation on a set S (means a subset of S × S).

A relation R on S is said to be

(i)   Transitive if (a, b) ∈ R, (b, c) ∈ R implies (a, c) ∈ R

(ii)   Reflexive if (a, a) ∈ R for all a ∈ S

(iii)   Anti-symmetric if (a, b) ∈ R and (b, a) ∈ R ⇒ a = b

(iv)   Symmetric if $(a, b) \in R$ implies $(b, a) \in R$

(v)   Equivalence relation if it is reflexive, symmetric and transitive

(vi)   Irreflexive of $(x, x) \notin R$ for all $x \in S$

If (a, b) is an element of the equivalence relation, then we write $a \sim b$ and we say that a and b are equivalent.

Example 5.26

(i)   The relation "is equal to" is a reflexive, symmetric and transition relation on any set of numbers.

(ii)   Consider the relation:

$x \ B \ y \Leftrightarrow x$ is a brother of y

If $x \ B \ y$, $y \ B \ z$ then x is a brother of y and y is a brother of z. In this case, x is also a brother of z. Hence $x \ B \ z$. This shows that $x \ B \ y$, $y \ B \ z \Rightarrow x \ B \ z$. i.e., the relation 'B' is a transitive relation.

5.4.5   Composition of Relations

Let $R \subseteq A \times B$ and $S \subseteq B \times C$.

i.e., R is a relation from A to B, and S is a relation from B to C.

We define a relation (RoS) from A to C by

$RoS = \{(a, c) \mid$ there exists $b \in B$ such that $(a, b) \in R$ and $(b, c) \in S\}$

RoS is called as composition of R and S.

Example 5.27

Write $A = \{1, 2, 3\}$, $B = \{a, c\}$, and $C = \{x, y, z\}$

$R = \{(1, a), (2, c), (3, c)\}$

$S = \{(a, x), (a, y), (c, z)\}$

Since $(1, a) \in R$ and $(a, x) \in S$ we have $(1, x) \in RoS$

Since $(1, a) \in R$ and $(a, y) \in S$ we have $(1, y) \in RoS$

Since $(2, c) \in R$ and $(c, z) \in S$ we have $(2, z) \in RoS$

Since $(3, c) \in R$ and $(c, z) \in S$ we have $(3, z) \in RoS$

Finally we get that

$RoS = \{(1, x), (1, y), (2, z), (3, z)\}$

5.4.6   Notation

Let R be a binary relation on a set A. The relation RoR is denoted by R^2. The relation R^2oR is denoted by R^3.

In general $R^n = (R^{n-1})oR$.

Example 5.28

Write A = {1, 2, 3, 4, 5}

$\qquad$ R = {(1, 1), (1, 2), (1, 3) (2, 3) (2, 4) (2, 5)}

Now R^2 = RoR = {(1, 1) (1, 2) (1, 3) (1, 4) (1, 5)}

$\qquad$ $R^3 = R^2$oR = {(1, 1) (1, 2) (1, 3) (1, 4) (1, 5)}

Note:

The composition SoR of relations may also be defined as follows:

$\qquad$ Let R be a relation from A to B, S be a relation from B to C. Then the relation SoR from A to C is defined by a (SoR) c $\Leftrightarrow$ a R b and b S c for some b $\in$ B for all a $\in$ A and c $\in$ C.

Example 5.29

Take A = {1, 2, 3, 4}

Define R = {(1, 1), (1, 2) (2, 3), (2, 4), (3, 4), (4, 1), (4, 2)} and

$\qquad$ S = {(3, 1), (4, 4), (2, 3), (2, 4), (1, 1), (1, 4)}

Since (1, 1) $\in$ R, (1, 1) $\in$ S we have (1, 1) $\in$ SoR

Since (1, 2) $\in$ R, (2, 3) $\in$ S we have (1, 3) $\in$ SoR

Since (2, 3) $\in$ R, (3, 1) $\in$ S we have (2, 1) $\in$ SoR

Continuing this way, we get that

SoR = {(1, 1), (1, 4), (1, 3), (2, 1), (2, 4), (3, 4), (4, 1), (4, 4), (4, 3)}

Similarly,

RoR = {(1, 1), (1, 2), (1, 3), (1,4), (2,4), (2,1), (2,2), (3,1), (3,2), (4,1), (4,2), (4,3), (4,4)}

Note:

(a) If R_1 and R_2 are relations from A to B; R_3 and R_4 are relations from B to C, then if $R_1 \subseteq R_2$ and $R_3 \subseteq R_4$, then R_1 o $R_3 \subseteq R_2$ o R_4.

(b) (Associative law): If R is a relation from A to B, S is a relation form B to C, and T is a relation from C to D, then (RoS)o T = Ro(SoT).

Example 5.30

If R is a relation from A to B and S is a relation from B to C, then $(RoS)^{-1} = S^{-1}$ o R^{-1}

Solution: Since R is a relation from A to B, we have R^{-1} is a relation from B to A. Similar way, S^{-1} is a relation from C to B. Therefore S^{-1} or R^{-1} is a relation from C to B.

If (x, y) $\in$ R, (y, z) $\in$ S, then (x, z) $\in$ R o S $\Rightarrow$ (z, x) $\in$ $(R o S)^{-1}$

$\qquad$ Since (z, y) $\in$ S^{-1} and (y, x) $\in$ R^{-1}, we have (z, x) $\in$ S^{-1} o R^{-1}

$\qquad$ This is true for any x $\in$ A and z $\in$ C. Hence $(RoS)^{-1} = S^{-1}$ o R^{-1}

Example 5.31

If R is a relation on a set A, then R is transitive $\Leftrightarrow R^2 \subseteq R$.

Solution: Suppose R is transitive, take $(x, y) \in R^2 \Rightarrow \exists \, z \in A$ such that $(x, z) \in R$, $(z, y) \in R$. Since R is transitive, we have $(x, y) \in R$. Thus $R^2 \subseteq R$

Converse: Suppose $R^2 \subseteq R$. Take $(x, y) \in R$, and $(y, z) \in R$. Then $(x, z) \in RoR = R^2 \subseteq R$. Thus R is transitive.

5.5   Relation Matrix and the Graph of a Relation

A relation S from a finite set A to a finite set B can also be represented by a matrix called the relation matrix of S. In this section we considered binary relations only.

Definition

Let $A = \{a_1, a_2, ..., a_m\}$ and $B = \{b_1, b_2, ..., b_n\}$ and S be a relation from A to B. The relation matrix of S can be obtained by first constructing a table whose columns are preceded by a column consisting of successive elements of A and whose rows are headed by a row consisting of the successive elements of B.

If $(a_i, b_j) \in S$, then we enter 1 in the ith row and j^{th} column. If $(a_i, b_j) \notin S$, then we enter 0 in the ith row and j^{th} column. Note that each entry of a relation matrix is either 0 or 1.

Example 5.32

Consider $A = \{a_1, a_2, a_3\}$ and $B = \{b_1, b_2\}$ and the relation $S = \{(a_1,b_1) \, (a_2,b_1) \, (a_3,b_2) \, (a_2, b_2)\}$.

Then the table for relation S is

	b_1	b_2
a_1	1	0
a_2	1	1
a_3	0	1

Hence, the relation S can be represented by the following matrix

$$\begin{bmatrix} 1 & 0 \\ 1 & 1 \\ 0 & 1 \end{bmatrix}$$ which is a relation matrix of S.

Note:

(i)   If set A has m elements and B has n elements then the relation matrix is an m × n matrix.

(ii)   One can write a relation matrix when a relation S is given. Also we can obtain a relation from a relation matrix.

5.5.1   Some Properties of Relation Matrices

(i)   If a relation is reflexive, then all the diagonal entries must be 1.

(ii)   If a relation is symmetric, then the relation matrix is symmetric.

(iii) If a relation is anti symmetric, then there exist i and j matrix is with $i \neq j$ such that $a_{ij} \neq a_{ji}$.

5.5.2   Pictorial Representation

A relation can also be represented pictorially by drawing its graph. We use graphs only as a tool to represent the relations.

Let S be a relation on a set $A = \{a_1, a_2, ..., a_m\}$ denote the elements of A with points or small circles and call them as nodes. (Nodes may also be called as Vertices).

Label the nodes with corresponding elements a_i. If $(a_i, a_j) \in S$, then draw an arc (edge) from a_i to a_j and put an arrow on the arc in the direction from a_i to a_j. Then we get a graph (directed graph) of the relation S.

If $(a_i, a_j) \in S$ and $(a_j, a_i) \in S$, then we draw two arcs between a_i and a_j (one from a_i to a_j, and the other from a_j to a_i).

For the sake of simplicity, we may replace two arcs by one arc putting arrows on both directions. If $(a_i, a_j) \in S$, then we get an arc which starts from a_i and returns to a_i, such an arc is called a loop.

Example 5.33

Consider the set $A = \{2, 3, 4, 5\}$ and the relation $S = \{(a, b) \mid a < b\}$.

(i)   Construct the relation matrix.

(ii)   Draw the graph of R.

Solution: Given set $A = \{2, 3, 4, 5\}$ and relation $S = \{(a, b) \mid a < b\}$

That is $S = \{(2, 3) (2, 4) (2, 5) (3, 4) (3, 5) (4, 5)\}$ is the given relation.

(i)   First we write the relation matrix

First write the table of the given relation

	2	3	4	5
2	0	1	1	1
3	0	0	1	1
4	0	0	0	1
5	0	0	0	0

Now the matrix of the given relation

$$\begin{bmatrix} 0 & 1 & 1 & 1 \\ 0 & 0 & 1 & 1 \\ 0 & 0 & 0 & 1 \\ 0 & 0 & 0 & 0 \end{bmatrix}$$

(ii) Now we construct the graph of given relation.

   Take the elements of A = {2, 3, 4, 5} as nodes.

   We have the relation S = {(2,3), (2,4), (2,5), (3,4), (3,5), (4,5),}

   Then the corresponding graph as shown below

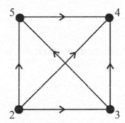

5.5.3 Properties of the Graph of a Relation

(i) If a relation is reflexive, then there must be a loop at each node.

   That is, if relation is irreflexive, then there is no loop at any node.

(ii) If a relation is symmetric, and one node is connected to another, then there must be a return arc from the second node to first node.

(iii) For anti symmetric relations, no such direct return path exists.

Example 5.34

(i) Observe the following graph of a relation.

   This graph denotes (a, b) ∈ S where S is a relation.

(ii)

This graph denotes (a, a) ∈ S

(iii)   Observe the following graph

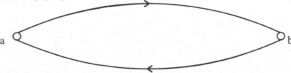

This graph says that (a, b) ∈ S and (b, a) ∈ S

(iv)   Consider the graph

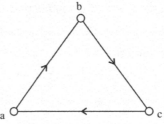

This graph corresponds the relation with (a, b) ∈ S, (b, c) ∈ S and (c, a) ∈ S

(v)   Consider the following graph

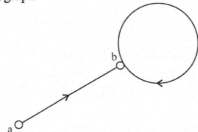

The corresponding relation is (a, b) ∈ S and (b, b) ∈ S

5.6   Partition, Covering of a Set and Equivalence Relations

Definition

Let B be a given set and M = {M₁, M₂, ..., Mₘ} where each M_i, i = 1, ..., m is a subset of B and $\bigcup_{i=1}^{m} M_i = B$. Then the set M is called a covering of B.

The sets $M_1, M_2, ..., M_m$ are said to cover B.

If the elements of the set M, which are sub-sets of B are mutually disjoint, then M is called a partition of B. In this case, the subsets $M_1, M_2, ..., M_m$ are called the blocks of the partition.

Example 5.35

Let B = {x, y, z}. Take M = {(x, y), (y, z)} S = {(x) (x, z)} T = {(x) (y, z)} P= {(x, y, z)}
Q = {(x), (y), (z)} R = {(x), (x, y), (x, z)}

   (i)   Clearly the sets M and R are coverings of B. The sets T, P and Q are partitions of B. The sets M and R are coverings but not partitions.

   (ii)   Since union of elements, of S is not equal to B, we have that S is not a partition of B.

   (iii) The partition P has only one block and Q has three blocks.

   (iv) In the case of the given set B, we cannot have more than three blocks in any partition, because it contains only three elements.

Note:

   (i)   For any finite set, the smallest partition consists of the set itself as a block.

   (ii)   The largest partition consists of blocks containing only single elements.

   (iii) Two partitions are said to be equal if they are equal as sets.

   (iv) For a finite set, every partition is a finite partition (that is, every partition contains only a finite number of blocks).

Example 5.36

Let A = {1, 2, 3, 4}

   (i)   Write x = {1, 2} and y = {3, 4}

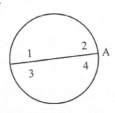

     A = X ∪ Y and X ∩ Y = φ

     So X and Y are disjoint sets whose union is A.

     Therefore {x, y} forms a partition for A.

(ii)   Write $A_1 = \{1\}$, $A_2 = \{2, 3\}$, $A_3 = \{4\}$

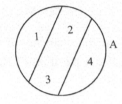

Now $A_1 \cap (A_2 \cup A_3) = \phi$

$A_2 \cap (A_1 \cup A_3) = \phi$

$A_3 \cap (A_1 \cup A_2) = \phi$

So A_1, A_2 and A_3 are mutually disjoint sets.

Also $A = A_1 \cup A_2 \cup A_3$

Hence $\{A_1, A_2, A_3\}$ forms a partition for A.

Example 5.37

Let A be a set and A_1, A_2 are two subsets of A.

Considers the subsets I_0, I_1, I_2, I_3 defined by

$$I_0 = \overline{A_1} \cap \overline{A_2}$$

$$I_1 = \overline{A_1} \cap A_2$$

$$I_2 = A_1 \cap \overline{A_2}$$

$$I_3 = A_1 \cap A_2$$

The sets I_0, I_1, I_2, I_3 (I_i for $a \le i \le (2^2 - 1)$) are mutually disjoint and $A = I_0 \cup I_1 \cup I_2 \cup I_3$
This is a partition for A (containing 2^n subsets I_0, I_1, I_2, I_3) obtained from the given n (= 2) subsets of A.

Observe the relation between the binary notation of indices and the intersections involved.

$$0 \equiv 00 \leftrightarrow I_0 = \overline{A_1} \cap \overline{A_2}$$

$$1 \equiv 01 \leftrightarrow I_1 = \overline{A_1} \cap A_2$$

$$2 \equiv 10 \leftrightarrow I_2 = A_1 \cap \overline{A_2}$$

$$3 \equiv 11 \leftrightarrow I_3 = A_1 \cap A_2$$

Example 5.38

We can obtain a partition for A (containing $8 = 2^3$ subsets) related to any given three subsets A_1, A_2, A_3 of A.

The method (is as it is in the above example).

Consider the subsets I_i ($0 \le i \le 2^3$) defined as follows:

$$I_0 = \overline{A_1} \cap \overline{A_2} \cap \overline{A_3}$$

$$I_1 = \overline{A_1} \cap \overline{A_2} \cap A_3$$

$$I_2 = \overline{A_1} \cap A_2 \cap \overline{A_3}$$

$$I_3 = \overline{A_1} \cap A_2 \cap A_3$$

$$I_4 = A_1 \cap \overline{A_2} \cap \overline{A_3}$$

$$I_5 = A_1 \cap \overline{A_2} \cap A_3$$

$$I_6 = A_1 \cap A_2 \cap \overline{A_3}$$

$\{I_i \mid 1 \le i \le 7\}$ forms a partition for A.

Observe the relation between the binary notation of indices and the intersections formed.

$$0 \equiv 000 \leftrightarrow I_0 = \overline{A_1} \cap \overline{A_2} \cap \overline{A_3}$$

$$1 \equiv 001 \leftrightarrow I_1 = \overline{A_1} \cap \overline{A_2} \cap A_3$$

$$2 \equiv 010 \leftrightarrow I_2 = \overline{A_1} \cap A_2 \cap \overline{A_3}$$

$$3 \equiv 011 \leftrightarrow I_3 = \overline{A_1} \cap A_2 \cap A_3$$

$$4 \equiv 100 \leftrightarrow I_4 = A_1 \cap \overline{A_2} \cap \overline{A_3}$$

$$5 \equiv 101 \leftrightarrow I_5 = A_1 \cap \overline{A_2} \cap A_3$$

$$6 \equiv 110 \leftrightarrow I_6 = A_1 \cap A_2 \cap \overline{A_3}$$

$$7 \equiv 111 \leftrightarrow I_7 = A_1 \cap A_2 \cap A_3$$

Note:

Observe the above two examples.

If A is a set and A_1, A_2, ..., A_n are given subsets of A, then we can define mutually disjoint subsets (2^n in number) I_0, I_1, I_2, ..., $I_{(2n-1)}$ such that $A = I_0 \cup I_2 \cup ... \cup I_{(2n-1)}$

This collection $\{I_i \mid 0 \le i \le (2^n-1)\}$ of subsets of A forms a partition for A.

Example 5.39

Suppose A = {1, 2, 3, 4, a, b} and A_1 = {1, 2, 3, 4} and A_2 = {3, 4, a, b}.

Write $\quad I_0 = \overline{A_1} \cap \overline{A_2}$ = {a, b} ∩ {1, 2} = ϕ

$\qquad I_1 = \overline{A_1} \cap A_2$ = {a, b} ∩ {3, 4, a, b} = {a, b}

$\qquad I_2 = A_1 \cap \overline{A_2}$ = {1, 2, 3, 4} ∩ {1, 2,} = {1, 2}

$\qquad I_3 = A_1 \cap A_2$ = {1, 2, 3, 4} ∩ {3, 4, a, b} = {3, 4}

The sets I_0, I_1, I_2, I_3 are mutually disjoint and $A = I_0 \cup I_1 \cup I_2 \cup I_3$

Hence $\{I_0, I_1, I_2, I_3\}$ forms a partition for A.

5.6.1 Equivalence Relations

Let us recall the definition.

A relation R in a set A is called an equivalence relation if it is reflexive, symmetric and transitive.

Let us consider an example:

Let $A = \{1, 2, ..., 7\}$ and $S = \{(a, b) \mid a - b$ is divisible by 3$\}$

Now we verify that S is an equivalence relation.

Verification:

Given the set $A = \{1, 2, ..., 7\}$ and relation is $S = \{(a, b) \mid a - b$ is divisible by 3$\}$

We have to show the relation S is an equivalence relation.

For any $a \in A$, $a - a$ is divisible by 3. So $(a, a) \in S$.

Hence the given relation is reflexive.

Let $a, b \in A$. If $a - b$ is divisible by 3, then $b - a$ is also divisible by 3.

That is., $(a, b) \in S \Rightarrow (b, a) \in S$.

This shows that the relation S is symmetric.

To verify the transitive property, let us take three elements $a, b, c \in A$ such that $(a, b) \in S$ and $(b, c) \in S$.

Then $a - b$ is divisible by 3, and

$\qquad b - c$ is divisible by 3.

Now $a - c = (a - b) + (b - c)$ is also divisible by 3.

Hence $(a, c) \in S$.

Therefore the given relation is transitive.

Hence the relation S is an equivalence relation.

5.6.2 Equivalence Class

Let S be an equivalence relation on set A. For any $a \in A$, the set $[a]_s$ (a sub set of A) defined by

$[a]_s = \{b \mid b \in A$ and $(a, b) \in S\}$ is called an S-equivalence class generated by $a \in A$.

Note:

Every equivalence relation on a set generates a unique partition of the set. The blocks of this partition are nothing but the S-equivalence classes.

Example 5.40

If A = {a, b, c}, then R = {(a, a), (b, b), (c, c), (a, b), (b, a)} is an equivalence relation on A. The equivalence classes are

$$[a] = \{a, b\}, [b] = \{a, b\}, \text{ and } [c] = \{c\}.$$

Note that   {[a], [b], [c] forms a partition for A = {a, b, c}.

Example 5.41

Let S be the set of all integers. Define a relation on S as follows: for any a, b $\in$ S,

a ~ b $\Leftrightarrow$ a – b is an even number. Then this relation is an equivalence relation.

Solution: Let   a $\in$ S.

Reflexive: a – a = 0 is an even number $\Rightarrow$ a ~ a.

Symmetric: Suppose a, b $\in$ S such that a ~ b.

Now a ~ b $\Rightarrow$ a – b is an even number

$$\Rightarrow b - a = -(a - b) \text{ is an even number}$$

$$\Rightarrow b \sim a.$$

Transitive: Suppose a, b, c $\in$ S such that a ~ b and b ~ c.

Now a ~ b, b ~ c $\Rightarrow$ a – b, b – c are even numbers

$$\Rightarrow a - c = (a - b) + (b - c) \text{ is an even number}$$

$$\Rightarrow a \sim c.$$

Hence the relation is an equivalence relation.

$$[a] = \{x \mid q \sim x\} \text{ is the equivalence class of } q \in S.$$

Example 5.42

Let Z be the set of all integers. Consider the set $X = \{\frac{a}{b} \mid a, b \in Z \text{ and } b \neq 0\}$.

Define $\frac{a}{c} \sim \frac{b}{d}$ (or $\frac{a}{c}$ is equal to $\frac{b}{d}$) if ad = bc.

Then this ~ is an equivalence relation on X.

An equivalence class of $\frac{a}{b}$ is called a rational number.

Example 5.43

Let $f : X \rightarrow Y$. Define a relation on X as follows:

$$x_1 \sim x_2 \Leftrightarrow f(x_1) = f(x_2).$$

This is an equivalence relation. What are the equivalence classes.

Solution: It is easy to verify that it is an equivalence relation. The set of all equivalence classes are $\{f^{-1}(y) \mid y \in f(x)\}$ where $f^{-1}(y) = \{x \in X \mid f(x) = y\}$.

Example 5.44

For two sets X and Y, we say that X is numerically equivalent to Y if there exist a bijection from X to Y. Take a set A and consider $P(A) = \{X \mid X \subseteq A\}$. Then the relation "numerically equivalent" is an equivalence relation on $P(A)$.

Example 5.45

Consider R, the set of all real numbers.

The collection $\{(a, b) \mid a, b \in Z \text{ and } b = a + 1\}$ of subsets of R, forms a partition for R.

Definition

(i) Let a, b are two integers. Then we say that 'b divides a' if there exists an integer c such that $a = bc$. If b divides a, then we say that b is a divisor of a.

(ii) Let $n > 0$ be a fixed integer. We define a relation namely "Congruence modulo n" on Z, the set of integers as:

$a \equiv b \pmod{n} \Leftrightarrow n$ divides $(a - b)$.

Some times we write $a \equiv b$ as $a \cong b$ and we read as "$a \cong b$ mod n" as 'a is congruent to b modulo n'.

Example 5.46

Show that the relation "$a \equiv b \pmod{n}$" defined above is an equivalence relation on Z, the set of integers.

Solution: Let $a \in Z$.

Since n divides $a - a = 0$, we have that $a \equiv a \pmod{n}$. Therefore the relation is reflexive. To show symmetric, let $a \equiv b \pmod{n}$

$\Rightarrow$ n divides $a - b$

$\Rightarrow$ n divides $-(a - b)$

$\Rightarrow$ n divides $b - a$

$\Rightarrow b \equiv a \pmod{n}$

To verify the transitivity, let a, b, c $\in$ Z such that $a \equiv b \pmod{n}$ and $b \equiv c \pmod{n}$

$\Rightarrow$ n divides a – b, and n divides b – c

$\Rightarrow$ n divides (a – b) + (b – c) = a – c

$\Rightarrow$ n divides a – c

$\Rightarrow$ a $\equiv$ c (mod n)

Hence the relation is an equivalence relation.

Example 5.47

Consider the above example. Take n = 5, then

$[0] = \{x \mid x \equiv 0 \ (mod \ 5)\} = \{x \mathbin{/} 5 \ divides \ x - 0 = x\} = \{..., -10, -5, 0, 5, 10, ...\}$

$[1] = \{x \mid x \equiv 1 \ (mod \ 5)\} = \{x \mathbin{/} 5 \ divides \ x - 1\} = \{..., -9, -4, 1, 6, ...\}$

$[2] = \{x \mid x \equiv 2 \ (mod \ 5)\} = \{x \mathbin{/} 5 \ divides \ x - 2\} = \{..., -8, -3, 2, 7, ...\}$

$[3] = \{x \mid x \equiv 3 \ (mod \ 5)\} = \{x \mathbin{/} 5 \ divides \ x - 3\} = \{..., -7, -2, 3, 8, 13, ...\}$

$[4] = \{x \mid x \equiv 4 \ (mod \ 5)\} = \{x \mathbin{/} 5 \ divides \ x - 4\} = \{..., -6, -1, 4, 9, 14, ...\}$

Also it is clear that

$[0] = [5] = [10] = ...$

$[1] = [6] = [11] = ...$

$[2] = [7] = [12] = ...$

$[3] = [8] = [13] = ...$

$[4] = [9] = [14] = ...$

Therefore the set of all equivalence classes is given by $\{[0], [1], [2], [3], [4]\}$.

Note that Z = $[0] \cup [1] \cup [2] \cup [3] \cup [4]$. Hence the set of equivalence classes $\{[0], [1], [2], [3], [4]\}$ forms a partition for Z.

Example 5.48

Every equivalence relation R on a set generates a unique partition of the set. The blocks of this partition correspond to the R-equivalence class.

[JNTUH, Nov 2010, Set No. 2]

Solution: Let R be an equivalence relation on a set B. Let b $\in$ B and M be a sub-set of B consisting of all those elements which are equivalent to b.

That is,

M = $\{x \mid x \in B \ and \ (x, b) \in R\}$

Clearly b $\in$ M [since b $\in$ B and (b, b) $\in$ R]

Now for any two elements x, y $\in$ M

$\Rightarrow$ x, y $\in$ B and (x, b) $\in$ R and (y, b) $\in$ R. [By definition of M]

$\Rightarrow$ (x, b) $\in$ R and (b, y) $\in$ R [Since R is symmetric]

$\Rightarrow$ (x, y) $\in$ R [Since R is transitive]

Thus M is an equivalence class.

Let $M_1 = \{x \mid x \in B$ and $(x, c) \in R\}$ where $c \neq b$. By a similar argument as above, we conclude that M_1 is the equivalence class containing e.

Moreover, M and M_1 are disjoint.

[*Verification:* Suppose $p \in M \cap M_1$ then $(p, b) \in R$ and $(p, c) \in R$. Since R is an equivalence relation, we have $(b, p) \in R$ and $(p, c) \in R \Rightarrow (b, c) \in R$ (since R is transitive) a contradiction to our hypothesis].

Thus the set B can be decomposed into equivalence class M, M_1, M_2, ... such that every element of B belongs to one and only one of these classes. Also these equivalence classes are mutually disjoint.

Thus we obtained the required partition of B.

Example 5.49

Let A = {0, 1, 2, 3, 4,}. Show that the relation R = {(0,0), (0,4), (1,1), (1,3), (2,2), (3,1), (3,3), (4,0), (4,4)} is an equivalence relation. Find the distinct equivalence classes of R.

[JNTUH, Nov 2010, Set No.3]

Solution: Let A = {0, 1, 2, 3, 4,}.

Given relation is R = {(0,0), (0,4), (1,1), (1,3), (2,2), (3,1), (3,3), (4,0), (4,4)}. Now we show that this relation is an equivalence relation.

Let a $\in$ A then (a, a) $\in$ R for every a $\in$ A.

So the given relation is reflexive.

Let a, b $\in$ A.

Suppose (a, b) $\in$ R. Observing the set R, we conclude that if (a, b) $\in$ R then (b, a) $\in$ R. So the given relation symmetric.

Let a, b, c $\in$ A. Suppose (a, b) $\in$ R and (b, c) $\in$ R.

Then observing the given relation conclude that (a, c) $\in$ R. Thus given relation is transitive.

Hence the given relation is an equivalence relation.

We know that equivalence class of 'a' is defined as $[a]_R = \{x \mid x \in A$ and $(x, a) \in R\}$

So $[0]_R = \{0, 4\} = [4]_R$

$[1]_R = \{1, 3\} = [3]_R$

$[2]_R = \{2\}$

are the equivalence classes of A on R.

Note:

(i) $[a]_S = \{x \mid x \in B$ and $(x, a) \in S\}$ where S is a relation on a set B denotes the equivalence class of a generated by 'a'.

(ii) $[a]_R$ also can be written as $a \mid R$

(iii) The family of equivalence classes of a relation R on A may be denoted as $A|R$ (or A modulo R, or simply A mod R).

(iv) $A|R$ is called quotient set of A by R

Example 5.50

Let Z be set of integers and R be a relation called "congruence modulo 3" defined by

$R = \{(x, y) \mid x \in z, y \in z$ and $(x - y)$ is divisible by 3$\}$.

Solution: The equivalence classes are

$[0]_R = \{\ldots, -6, -3, 0, 3, 6, \ldots\}$

$[1]_R = \{\ldots, -5, -2, 1, 4, 7, \ldots\}$

$[2]_R = \{\ldots, -4, -1, 2, 5, 8, \ldots\}$

$Z|R = \{[0]_R, [1]_R, [2]_R\}$

Example 5.51

Let $A = \{1, 2, 3, 4\}$ and R be the relation defined by the following M_R.

$$M_R = \begin{bmatrix} 1 & 0 & 0 & 0 \\ 0 & 1 & 1 & 1 \\ 0 & 1 & 1 & 1 \\ 0 & 1 & 1 & 1 \end{bmatrix}, \text{ compute } A|R$$

[JNTUH, June 2010, Set No. 3]

Solution: Let A $\{1, 2, 3, 4\}$

Given relation matrix is $M_R = \begin{bmatrix} 1 & 0 & 0 & 0 \\ 0 & 1 & 1 & 1 \\ 0 & 1 & 1 & 1 \\ 0 & 1 & 1 & 1 \end{bmatrix}$

First write the relation from the matrix.

$R = \{(1, 1)\ (2, 2)\ (2, 3)\ (2, 4)\ (3, 2)\ (3, 3)\ (3, 4)\ (4, 2)\ (4, 3)\ (4, 4)\}$.

First we verify that this relation is an equivalent relation. Since for any $a \in A$, $(a, a) \in R$, we have that R is a reflexive relation. For any $(a, b) \in R$, we have that $(b, a) \in R$. So R is a symmetric relation. Suppose $(a, b) \in R$ and $(b, c) \in R$. Then $(a, c) \in R$.

So R is a transitive relation.

Therefore, the relation R is an equivalence relation.

Now $[1]_R = \{1\}$

$[2]_R = \{2, 3, 4\}$

$[3]_R = \{2, 3, 4\}$

$[4]_R = \{2, 3, 4\}$ here $[2]_R = [3]_R = [4]_R$

Therefore $A|R = \{[1]_R, [2]_R\}$ is the family of distinct equivalence classes.

Example 5.52

Let $A = \{1, 2, 3, 4, 5\}$ and $P = \{A \times A\}$. R is a relational set such that $(x, y) \, R \, (x', y') \Leftrightarrow xy' = x'y$.

   (i)   Show that R is an equivalence relation

   (ii) Determine $A|R$.

[JNTUH, June 2010, Set No 4]

Solution: Let $A = \{1, 2, 3, 4, 5\}$

        Then $P = A \times A$

        $= \{(1, 1), (1, 2), (1, 3), (1, 4), (1, 5), (2, 1), (2, 2), (2, 3), (2, 4), (2, 5), (3, 1), (3, 2),$
        $(3, 3), (3, 4), (3, 5), (4, 1), (4, 2), (4, 3), (4, 4), (4, 5), (5, 1), (5, 2), (5, 3), (5, 4),$
        $(5, 5),\}$

Define a relation $(x, y) \, R \, (x', y') \Leftrightarrow xy' = x'y$.

That is $((x, y), (x', y')) \in R$. If and only if $xy' = x'y$. Now we have to show that this relation is an equivalence relation.

First we write the relation R.

   $R = \{[(1, 1) \, (1, 1)] \; [(1, 1) \, (2, 2)] \; [(1, 1) \, (3, 3)] \; [(1, 1) \, (4, 4)] \; [(1, 1) \, (5, 5)]$
      $[(2, 1) \, (2, 1)] \; [(2, 2) \, (2, 2)] \; [(2, 2) \, (3, 3)] \; [(2, 2) \, (4, 4)] \; [(2, 2) \, (5, 5)]$
      $[(2, 3) \, (2, 3)] \; [(2, 4) \, (2, 4)] \; [(2, 5) \, (2, 5)] \; [(3, 1) \, (3, 1)] \; [(3, 2) \, (3, 2)] \; [(3, 3) \, (5, 5)]$
      $[(3, 3) \, (3, 3)] \; [(3, 3) \, (4, 4)] \; [(3, 4) \, (3, 4)] \; [(3, 5) \, (3, 5)] \; [(4, 1) \, (4, 1)]$
      $[(4, 2) \, (4, 2)] \; [(4, 3) \, (4, 3)] \; [(4, 4) \, (4, 4)] \; [(4, 5) \, (5, 5)] \; [(4, 5) \, (4, 5)]$
      $[(5, 1) \, (5, 1)] \; [(5, 2) \, (5, 2)] \; [(5, 3) \, (5, 3)] \; [(5, 4) \, (5, 4)] \; [(5, 5) \, (5, 5)]\}$

    Since for any $(x, y) \in P$ we have that $xy = xy$ and so $(x, y) \, R \, (x, y)$.

    Therefore the given relation is a reflexive relation.

Let $(x_1, y_1), (x_2, y_2) \in P$ such that $[(x_1, y_1), (x_2, y_2)] \in R$.

Then $x_1 y_2 = x_2 y_1 \Rightarrow x_2 y_1 = x_1 y_2$

$$\Rightarrow [(x_2, y_2), (x_1, y_1)] \in R.$$

This shows that the given relation is a symmetric relation.

Let $(x_1, y_1) (x_2, y_2)$ and $(x_3, y_3) \in P$

Suppose that $[(x_1, y_1) (x_2, y_2)] \in R$ and $[(x_2, y_2) (x_3, y_3)] \in R$

Then $x_1 y_2 = x_2 y_1$ .....(5.5)

and $x_2 y_3 = x_3 y_2$ .....(5.6)

From this we have $y_2 = \dfrac{x_2}{x_1} y_1$ [From equation (5.5)]

Substitute y_2 in the equation (5.6), to get that

$$x_2 y_3 = x_3 y_2 \Rightarrow x_2 y_3 = x_3 . \dfrac{x_2}{x_1} y_1$$

$$\Rightarrow x_1 y_3 = x_3 y_1$$

$$\Rightarrow [(x_1, x_3), (y_1, y_2)] \in R$$

Thus given relation is an equivalence relation.

Now we find the quotient relation.

$[(1, 1)]_R = \{(1, 1), (2, 2), (3, 3), (4, 4), (5, 5)\}$

$[(2, 1)]_R = \{(2, 1)\}$

$[(2, 2)]_R = \{(1, 1), (2, 2), (3, 3), (4, 4), (5, 5)\}$

$[(2, 3)]_R = \{(2, 3)\}$

$[(2, 4)]_R = \{(2, 4)\}$

$[(2, 5)]_R = \{(2, 5)\}$

$[(3, 1)]_R = \{(3, 1)\}$

$[(3, 2)]_R = \{(3, 2)\}$

$[(3, 3)]_R = \{(1, 1), (2, 2), (3, 3), (4, 4), (5, 5)\}$

$[(3, 4)]_R = \{(3, 4)\}$

$[(3, 5)]_R = \{(3, 5)\}$

$[(4, 1)]_R = \{(4, 1)\}$

$[(4, 2)]_R = \{(4, 2)\}$

$[(4, 3)]_R = \{(4, 3)\}$

$[(4, 4)]_R = \{(1, 1), (2, 2), (3, 3), (4, 4), (5, 5)\}$

$[(4, 5)]_R = \{(4, 5)\}$

$[(5, 1)]_R = \{(5, 1)\}$

$[(5, 2)]_R = \{(5, 2)\}$

$[(5, 3)]_R = \{(5, 3)\}$

$[(5, 4)]_R = \{(5, 4)\}$

$[(5, 5)]_R = \{(1, 1), (2, 2), (3, 3), (4, 4), (5, 5)\}$

Therefore, the quotient relation is given by

$A|R = \{[(1,1)]_R, [(2,1)]_R, [(2,3)]_R, [(2,4)]_R, [(2,5)]_R$

$[(3,1)]_R, [(3,2)]_R, [(3,4)]_R, [(3,5)]_R,$

$[(4,1)]_R, [(4,2)]_R, [(4,3)]_R, [(4,5)]_R$

$[(5,1)]_R, [(5,2)]_R, [(5,3)]_R, [(5,4)]_R\}$

Example 5.53

A relation on a set A defined by "a is equals to b" is an equivalence relation.

Solution: Define a relation on a set A as "a is equals to b".

Now we show this is an equivalence relation.

Since "a is equals to a", we have that the given relation is a reflexive relation.

(i)   Let "a is equals to b", then b is equals to a

Therefore the given relation is a symmetric relation.

(ii)   Suppose "a is equals to b" and "b is equals to c". then "a is equals to c"

Therefore the given relation is a transitive relation.

Here the given relation satisfies the reflexive, symmetric and transitive properties. Hence the given relation is an equivalence relation.

5.7   Transitive Closure

Definition

(i)   Let R be a relation on a set A. R may or may not have some given property P (such as reflexivity, symmetry or transitivity). If there is a relation S with property P containing R such that S is a subset of every relation with P containing R, then S is called the closure of R with respect to the property 'P'.

(ii)   Let R be a relation on a set S. The reflexive closure of R is the smallest reflexive relation R_1 which contains R.

Note

$R_1 = R \cup \Delta$ where Δ is the diagonal relation on S, that is., $\Delta = \{(a, a) \mid a \in S\}$.

(iii)   The symmetric closure of R is the smallest symmetric relation containing R. That is $R \cup R^{-1}$ is symmetric closure R, where R^{-1} is the inverse of the relation R.

That is, $R^{-1} = \{(b, a) \mid (a, b) \in R\}$

(iv)   Transitive closure of a relation R is the smallest transitive relation containing R.

Example 5.54

Consider the set S = {1, 2, 3, 4}

   (i)   The relation R = {(1,2), (2,1), (1,1), (2,2)} is not reflexive, since (3, ,3) $\notin$ S.

       Consider Δ = {(1,1), (2,2), (3,3), (4,4)} = {(x, x) | x $\in$ S}.

       Now the reflexive closure R_1 = R $\cup$ Δ = {(1,2), (2,1), (1,1), (2,2), (3,3), (4,4)}.

       Observe that R_1 (the reflexive closure of R, sometimes we denote as $R^{(r)}$.

  (ii) Consider the relation K = {(1,2), (4,3), (2,2), (2,1), (3,1)}, which is not symmetric on S. Now K^{-1} = {(2,1), (3,4), (2,2), (1,2), (1,3)}. The symmetric closure $K^{(s)}$ of K is given by $K^{(s)}$ = K $\cup$ K^{-1} = {(1,2), (2,1), (4,3), (3,4), (3,1), (1,3)}.

Note:

Given a relation R on a set A. To make a relation R transitive, add all pairs of R^2, all pairs of R^3, ... all pairs of R^m (assume that |A| = m), unless these pairs are already in R. Note that R^2 = {(x, z) | there exists y such that (x, y), (y, z) $\in$ R}. Then the transitive closure of R, denoted by R^∞ or $R^{(T)}$

$$R^{(T)} = R \cup R^2 \cup ... \cup R^m, \text{ if } |A| = m$$

5.7.1   Properties of Transitive Closure

   (i) $R^{(T)}$ is transitive.

  (ii) $R \subseteq R^{(T)}$

 (iii) If S is any other transitive relation that contains R, then $R^{(T)} \subseteq$ S.

5.7.2   Representation of Closure Relations in the Form of Matrices

Let M be the relation matrix of the relation R. Then

   (i)   The symmetric closure of R, denoted by M_S, defined as M_S = M $\vee$ M' where M' is the transpose of M.

  (ii) The reflexive closure of R, denoted by M_R, defined as M_R = M $\vee$ I_n where n is the cardinality of the set on which the relation was defined, and I_n is the identity matrix of size n.

 (iii) The transitive closure of R, denoted by M_T or M_R^∞ defined as M_T = M $\vee$ M^2 $\vee$ M^3 $\vee$ ... $\vee$ M^n.

Example 5.55

Take A = {1, 2, 3} and R = {(1, 2), (2, 3), (3, 1)}. Find the reflexive, symmetric and transitive closures of R, using composition of matrix relation of R.

Solution: Let M be the relation matrix of R. Then $M = \begin{bmatrix} 0 & 1 & 0 \\ 0 & 0 & 1 \\ 1 & 0 & 0 \end{bmatrix}$

(i)   The Reflexive closure of R, $M_R = M \cup I_3 = \begin{bmatrix} 0 & 1 & 0 \\ 0 & 0 & 1 \\ 1 & 0 & 0 \end{bmatrix} \vee \begin{bmatrix} 1 & 0 & 0 \\ 0 & 1 & 0 \\ 0 & 0 & 1 \end{bmatrix} = \begin{bmatrix} 1 & 1 & 0 \\ 0 & 1 & 1 \\ 1 & 0 & 1 \end{bmatrix}$

One can write the reflexive closure $R^{(r)}$, using the above matrix as $R^{(r)} = \{(1,1),$ $(1,2), (2, 2), (2,3), (3,1), (3,3)\}$.

(ii)   The symmetric closure of R is $MS = M \, M' = \begin{bmatrix} 0 & 1 & 0 \\ 0 & 0 & 1 \\ 1 & 0 & 0 \end{bmatrix} \vee \begin{bmatrix} 0 & 0 & 1 \\ 1 & 0 & 0 \\ 0 & 1 & 0 \end{bmatrix} = \begin{bmatrix} 0 & 1 & 1 \\ 1 & 0 & 1 \\ 1 & 1 & 0 \end{bmatrix}$

One can write the symmetric closure $R^{(S)}$, using the above matrix as
$R^{(S)} = \{(1,2), (1,3), (2,1), (2,3), (3,1), (3,2)\}$

(iii)   To find the transitive closure of R. we first find M^2 and M^3 (since the cardinality of the set $A = 3$).

$$M^2 = \begin{bmatrix} 0 & 1 & 0 \\ 0 & 0 & 1 \\ 1 & 0 & 0 \end{bmatrix}\begin{bmatrix} 0 & 1 & 0 \\ 0 & 0 & 1 \\ 1 & 0 & 0 \end{bmatrix} = \begin{bmatrix} 0 & 0 & 1 \\ 1 & 0 & 0 \\ 0 & 1 & 0 \end{bmatrix} \text{ and}$$

$$M^2 = M^2.M = \begin{bmatrix} 0 & 0 & 1 \\ 1 & 0 & 0 \\ 0 & 1 & 0 \end{bmatrix}\begin{bmatrix} 0 & 1 & 0 \\ 0 & 0 & 1 \\ 1 & 0 & 0 \end{bmatrix} = \begin{bmatrix} 1 & 0 & 0 \\ 0 & 1 & 0 \\ 0 & 0 & 1 \end{bmatrix}$$

Therefore the transitive closure of R, $M_T = M \vee M^2 \vee M^3$

$$= \begin{bmatrix} 0 & 1 & 0 \\ 0 & 0 & 1 \\ 1 & 0 & 0 \end{bmatrix} \vee \begin{bmatrix} 0 & 0 & 1 \\ 1 & 0 & 0 \\ 0 & 1 & 0 \end{bmatrix} \vee \begin{bmatrix} 1 & 0 & 0 \\ 0 & 1 & 0 \\ 0 & 0 & 1 \end{bmatrix} = \begin{bmatrix} 1 & 1 & 1 \\ 1 & 1 & 1 \\ 1 & 1 & 1 \end{bmatrix}$$

One can write the transitive closure of R, $R^T = \{(1,1), (1,2), (1,3), (2,1), (2,2),$ $(2,3), (3,1), (3,2), (3,3)\}$.

Theorem 1

Let A, B, C, be finite sets. Let R be a relation from A to B and S be a relation from B to C. Then $M_{RoS} = M_R.M_S$ where M_R and M_S represents relation matrices of R and S respectively.

Proof: Let $A = \{a_1, a_2, \dots a_m\}$, $B = \{b_1, b_2, \dots, b_n\}$, and $C = \{c_1, c_2, \dots, c_p\}$

Suppose $M_R = [a_{ij}]$, $M_S = [b_{ij}]$, $M_{RoS} = [d_{ij}]$

Then $d_{ij} = 1 \Leftrightarrow (a_i, c_j) \in RoS$

$\Leftrightarrow (a_i, b_k) \in R$ and $(b_k, c_j) \in S$ for some $b_k \in B$

$\Leftrightarrow a_{ik} = 1 = b_{kj}$ for some k $1 \le k \le$ n.

If $d_{ij} = 0$, then $(a_i, a_k) \notin R$ or $(a_k, a_j) \notin S$. This condition is identical to the condition needed for M_R. M_S to have 1 or 0 in position (i, j) and thus $M_{RoS} = M_R \cdot M_S$.

Theorem 2

Let R be a relation on a set A. Then R^∞ is the transitive closure of R.

Proof: Clearly $R \subseteq R^\infty$. For a, b $\in$ A.

a R^∞b $\Leftrightarrow$ there is a path in R from a to b. Now a R^∞b and b R^∞c $\Rightarrow$ there is a path from a to b in R, and a path from b to c in R $\Rightarrow$ there is a path from a to c in R. This path is the composition of paths from a to b and b to c.

Next we verify that R^∞ is the smallest transitive relation containing R.

Let S be any transitive relation such that $R \subseteq S$, we have to show that $R^\infty \subseteq S$.

Since S is transitive, we have $S^n \subseteq S$ for all n (that is, if a and b are connected by a path of length n, then aSb) and so $S^\infty = \overset{\infty}{\underset{n=1}{U}} \subseteq S$

Since $R \subseteq S$, $R^\infty = \overset{\infty}{\underset{n=1}{U}} R_n \subseteq R \subseteq S$. Thus R^∞ is the transitive closure of R.

Let the relation R = {(1, 2), (2, 3), (3, 3)} on the set {1, 2, 3}.

[JNTUH, Nov 2008, Set 4]

Solution: Given set is

$\qquad$ A = {1, 2, 3} $\hspace{4cm}$ .....(5.7)

$\qquad$ Given relation is

$\qquad$ R = {(1, 2), (2, 3), (3, 3),} $\hspace{2.5cm}$ .....(5.8)

By the definition of transitive closure of a relation R, we have

$\qquad$ $R^+ = R \cup R^v \cup R^3 \cup ... \cup R^k$, k = 1,2, ...

Here $\;\;R^+ = \{(1, 2), (1, 3), (2, 3)\}$.

5.8   Compatibility Relations

Definition

A relation S on a set A is said to be a compatibility relation if it is reflexive and symmetric.

Example 5.56

Let A = {bat, chair, cat, jet, egg}

Write S = {(a, b) | a, b ∈ A a and b contain some common letter}

Since for any a ∈ A, (a, b) ∈ S implies that (b, a) ∈ S.

Therefore the relation is symmetric. Hence given relation is a compatibility relation.

Note:

    (i)   In a compatibility relation S, we say that a and b are compatible if (a, b) ∈ S.

    (ii)   Compatibility relation sometimes denoted by the symbol "≈".

    (iii)   The relation matrix of a compatibility relation is symmetric and has each of its diagonal elements equal to one.

5.8.1   Definition: (Maximal Compatibility Block)

Let A be a set and "≈" a compatibility relation on A. A sub set B ⊆ A is called a maximal compatibility block. If any element of B is compatible to every other element of B and no element of A − B is compatible to all the elements of B.

5.8.2   Composition of Binary Relation

Let R be a relation from A to B and S be a relation from B to C. Then the relation written as RoS (called as the composite relation of R and S) is defined as

RoS = {(a, c) | a ∈ A and c ∈ C and there exists b ∈ B such that (a, b) ∈ R and (b, c) ∈ S}.

The operation 'o' for obtaining RoS from R and S is called composition of the relations R and S.

Example 5.57

Let R = {(a, b), (c, d), (b, b)} and S = {(d, b), (b, e), (c, a), (a, c)}.

Find (i) RoS (ii) SoR (iii) Ro(SoR) (iv) (RoS)oR (v) RoR (vi) SoS and RoRoR

Proof: Given relations R = {(a, b), (c, d), (b, b)} and S = {(d, b), (b, e), (c, a), (a, c)}

Now

    (i)   RoS = {(a, e), (c, b), (b, e)}

    (ii)   SoR = {(d, b), (c, b), (a, d)}

            Clearly RoS ≠ SoR

    (iii)   Ro(SoR) = {(c, b)}

    (iv)   (RoS)oR = {(c, b)}

            Clearly Ro(SoR) = (RoS)oR

    (v)   RoR = {(a, b) (b, b)}

    (vi)   SoS = {(d, e), (c, c)}

            RoRoR = {(a, b), (b, b)}

Example 5.58

Let A = {1, 2, 3}, B = {a, b, c} and C = {x, y, z}.

The relation R from A to B is {(1, b), (2, a), (2, c)} and the relation S from B to C is {(a, y), (b, x), (c, y), (c, z)}.

Find the composition relation RoS.

[JNTUH, Nov 2008, Set No 1]

Solution: Given that

A = {1, 2, 3}

B = {a, b, c}

C = {x, y, z}

R = {(1, b), (2, a), (2, c)}

S = {(a, y), (b, x), (c, y), (c, z)}

RoS = {(1, x), (2, y), (2, y), (2, z)}

5.9 Partial Order Relations

There are various types of relations defined on a set. In this unit our interest is partially ordered relation which is defined on a set, referred as a partially ordered set. This would lead to the important concepts "lattices" and "Boolean algebras", which are used in computer science. We discuss different properties of partial order relations on a set, and representations of posets.

Definition

A relation R on a set A is called a partial order relation if R is reflexive, anti-symmetric and transitive. The set S with a partial order R is called a partially ordered set or Poset and it is denoted by (A, R). In general, a partial order R on a set is denoted by $\leq$.

Note that if (a, b) $\in$ R, then we write a $\geq$ b. If a $\geq$ b and a $\neq$ b, then we write a > b.

Example 5.59

Let A = Z^+ the set of all positive integers. Define R on A as aRb if and only if a $\leq$ b. Then (A, $\leq$) is a partially ordered set. It is clear that (A, <) is not a Poset, since it does not satisfy the reflexive property.

Example 5.60

(i) The relations '$\leq$' and '$\geq$' are the partial orderings on the set of real numbers.

(ii) Let X be the power set of the set A. Then define R on X as S_1RS_2 if and only if $S_1 \subseteq S_2$ for $S_1, S_2 \in X$. Then the relation inclusion '$\subseteq$' is a partial ordering on X.

Example 5.61

Let A be a non-empty set and S = P(A), the power set of A.

Define a relation R on S as R = {(X, Y) / X, Y are in P(A) such that X contains Y}. Now we verify that the relation is reflexive.

For this take X ∈ S. Then X is a subset of A. Since X contains X, we have (X, X) ∈ R. Therefore R is reflexive. To verify the anti-symmetric condition, let (X, Y), (Y, X) ∈ R. Then X contains Y, and Y contains X, which imply X = Y. Hence the relation is anti-symmetric. To verify the transitive condition, let (X, Y), (Y, Z) ∈ R. Then X contains Y, and Y contains Z. So X contains Z, which implies (X, Z) ∈ R. Hence R is transitive. Therefore S is a Poset.

Definition

Let (A, ≤) be any Poset. Two elements a and b of A are comparable if either a ≤ b or b ≤ a. If every pair of elements is comparable then it is called a linearly ordered set or a chain. The Poset (Z^+, R) where R is defined on A as aRb if and only if a ≤ b is a chain.

Example 5.62

The Poset (Z^+, R) where R is defined on Z^+ as aRb ⟺ a ≤ b, is a chain.

Definition

Let $P_1, P_2, ..., P_k$ be Posets. The lexicographic product of $P_1, P_2, ..., P_k$ is defined to be the Poset $P_1 \times P_2 \times ... \times P_k$ with $(a_1, a_2, ..., a_k) < (b_1, b_2, ..., b_k)$ if $a_1 < b_1$ or if $a_i = b_i$ for i = 1, ..., m and $a_{m+1} < b_{m+1}$ for some m < k.

Example 5.63

Prove that the lexicographic product of the Posets $P_1, ..., P_k$ is a partial order on $P_1 \times P_2 \times ... \times P_k$.

Proof:

  (i)   Now we verify the reflexive property. Let $(a_1, ..., a_k) \in P_1 \times ... \times P_k$.

        Since $a_i \leq a_i$, for all $1 \leq i \leq k$, we have that $(a_1, ..., a_k) \leq (a_1, ..., a_k)$.

  (ii) Now we verify the transitive property. Suppose $(a_1, ..., a_k) \leq (b_1, ..., b_k)$ and $(b_1, ..., b_k) \leq (c_1, ..., c_k)$.

        Case-i: Suppose $a_i = b_i$ for all $1 \leq i \leq k$. Then clearly $(a_1, ..., a_k) = (b_1, ..., b_k) \leq (c_1, ..., c_k)$. If $b_i = c_i$, for all $1 \leq i \leq k$, then we have that $(a_1, ..., a_k) \leq (b_1, ..., b_k) = (c_1, ..., c_k)$.

        Case-ii: Suppose $a_1 < b_1$. If $b_1 < c_1$, then $a_1 < c_1$ and hence $(a_1 ... a_k) \leq (c_1 ... c_k)$.

        Case-iii: Suppose $a_i = b_i$ for $1 \leq i \leq m$ and $a_{m+1} < b_{m+1}$ for m < k. If $b_1 < c_1$, then

$a_1 = b_1 < c_1$ and so $(a_1 \ldots a_k) \le (c_1 \ldots c_k)$. Otherwise, since $(b_1 \ldots b_k) \le (c_1 \ldots c_k)$, there exists m' such that $b_i = c_i$, for all $1 \le i \le m'$ and $b_{m'+1} < c_{m'+1}$ for $m' < k$.

If $m < m'$, then $a_i = b_i = c_i$ for all $1 \le i \le m$ and $a_{m+1} < b_{m+1} \le c_{m+1}$. If $m > m'$, then $a_i = b_i = c_i$ for all $1 \le i \le m'$ and $a_{m=1} = b_{m'1+1} < c_{m'1+1}$.

Hence $(a_1, \ldots, a_k) \le (c_1, \ldots, c_k)$.

(iii) Now we verify the anti-symmetric property. Suppose $(a_1, \ldots, a_k) \le (b_1, \ldots, b_k)$ and $(b_1, \ldots, b_k) \le (a_1, \ldots, a_k)$. If $(a_1, \ldots, a_k) \ne (b_1, \ldots, b_k)$, then there exists $m < k$ such that $a_i = b_i$ for $1 \le i \le m$ and $a_{m+1} \ne b_{m+1}$. If $(a_1, \ldots, a_k) < (b_1, \ldots, b_k)$, then $a_{m+1} < b_{m+1}$. If $(b_1, \ldots, b_k) < (a_1, \ldots, a_k)$, then $b_{m+1} < a_{m+1}$. This imply that $a_{m+1} < b_{m+1}$ and $b_{m+1} < a_{m+1}$, a contradiction (since P_{m+1} is a Poset and $a_{m+1}, b_{m+1} \in P_{m+1}$). This shows that the relation is anti-symmetric.

Definition

A finite Poset can be diagrammed on the plane. If S is a Poset and a, b are in S such that $a > b$ and there is no c in S such that $a > c$ and $c > b$, then we say that a covers b.

Example 5.64

If a covers b, then represent the point corresponding to a, above the point for b and join the points (This fact is illustrated in the following Fig 5.9.1a)

Now consider the Fig 5.9.1b. In this, we can observe that:

D covers E; B covers C; F covers C; A covers F.

Also note that B joined to E by a sequence of line segments all going downwards. So we have $B \ge E$.

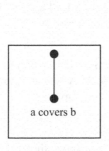

Fig 5.9.1a

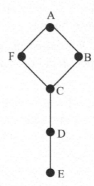

Fig 5.9.1b

Definition

(i)   An element x of a Poset S is said to be a minimal element if it satisfies the following condition: $y \in S$ and $x \geq y \Rightarrow y = x$.

(ii) An element 'a' of S is said to be a maximal element if it satisfies the following condition: $b \in S$ and $b \geq a \Rightarrow b = a$.

(iii) A Poset S said to be a totally ordered (or ordered) set if for a, b in S exactly one of the conditions: $a > b$, $a = b$, or $b > a$ holds.

Example 5.65

In a finite Poset S, show that there is always at least one maximal element and one minimal element.

Solution:

Part-I: For maximal element: In a contrary way, suppose S contains no maximal element. Let $x_1 \in S$. Since x_1 is not maximal, there exists x_2 in S such that $x_2 > x_1$. Since x_2 is not maximal, there exists x_3 in S such that $x_3 > x_2$. If we continue this process, we get an infinite sequence of distinct elements $x_1, x_2, x_3, \ldots$, such that $x_{i+1} > x_i$ for each i. This is a contradiction to the fact that S contains only a finite number of elements (since S is a finite Poset). Hence we conclude that S contains a maximal element

Part-II: This part of the proof is parallel to that of part-I.

Definition

(i)   A chain in a Poset is a sequence $a_0, a_1, \ldots, a_n$ of elements of the Poset such that $a_i > a_{i+1}$. The length of this chain is n.

(ii) Let $(P, \geq)$ be a Poset and $A \subseteq P$. An element $x \in P$ is called a lower bound for A if $a \geq x$, for all $a \in A$. A lower bound x of A is called a greatest lower bound of A if $x \geq y$ for all lower bounds y of A.

An element $x \in P$ is called an upper bound for A if $x \geq a$, for all $a \in A$. An upper bound x is called a least upper bound of A if $b \geq a$ for all upper bounds b of A.

Note:

Let R be the set of all real numbers, $\phi \neq A \subseteq R$. If A has a lower bound, then its greatest lower bound is called infimum and it is denoted by inf A. If A has an upper bound, then its least upper bound is called its supremum and it is denoted by sup A.

Example 5.66

For any subset A of R (the set of all real numbers), we have that inf A = min A and sup A = max A.

5.9.1   Zorn's Lemma

If P is a partially ordered set in which every chain has an upper bound, then P possesses a maximal element.

Definition

The covering matrix of a finite Poset $P = \{p_i \mid 1 \leq i \leq n\}$ is the matrix $(b_{ij})_{n \times n}$ where $b_{ij} = 1$ if p_i covers p_j or $i = j$; $= 0$ otherwise.

Example 5.67

The diagram of a Poset was given in Fig.5.9.2. The covering matrix of this Poset is given by

$$\begin{bmatrix} 1 & 1 & 0 \\ 0 & 1 & 1 \\ 0 & 0 & 1 \end{bmatrix}$$

Fig 5.9.2

Note 1:

(i)   The chain $p_0 > p_1 > p_2 > \ldots > p_k$ is said to have length k.

(ii)   An element p of a finite Poset is on level k if there exists a sequence $p_0 > p_1 > \ldots > p_k = p$ and any other such sequence has length less than or equal to k.

(iii)   Suppose p is on level k and $p_0 > p_1 > \ldots > p_k = p$. Then p_0 is a maximal element of the Poset. (if p_0 is not maximal then there exists p' such that $p' > p_0$. Then $p' > p_0 > p_1 > \ldots > p_k$ is of length (k + 1), a contradiction to the fact p is on level k).

(iv)   Fix J. An element p_j is maximal $\Leftrightarrow$ p_j has no cover $\Leftrightarrow$ $b_{ij} = 0$ for all $i \neq j$ and i = 1, 2, …, n. $\Leftrightarrow$ j^{th} column of (b_{ij}) contains 1 in the jth row and 0 elsewhere.

     $\Leftrightarrow$ The sum of the elements in the jth column is 1.

(v)   If the sum of the elements of the jth column of the covering matrix is "1", then the corresponding j^{th} element is a maximal element (that is, the element is of level 0).

Note 2:

A partial ordering $\leq$ on a poset, represented by a diagram called Hasse diagram. In a Hasse diagram, each element is represented by a small circle.

Example 5.68

Consider the Poset with the diagram (given in Fig. 5.9.3). Here a is of level 0; b is of level 1; c is of level 1; d is of level 2; e is of level 2; f is of level 3.

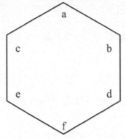

Fig. 5.9.3

Example 5.69

Let A = {a, b, c}. Then p(A) = {ϕ, {a}, {b}, {c}, {a, b}, {b, c}, {c, a}, {a, b, c}}. Consider the poset (p(A), ⊆). Then Hasse diagram is shown below.

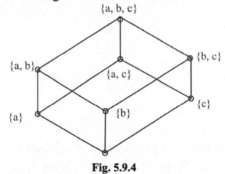

Fig. 5.9.4

Example 5.70

Let A = (2, 7, 14, 28, 56, 84) and a ≤ b if and only if a divides b. Then Hasse diagram for the poset (A, ≤) is given in Fig. 5.9.5.

Since 2 divides 14, we join 2 and 14 with a line segment; 7 divides 14 so we join 7 and 14 by a line segment; and so on.

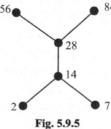

Fig. 5.9.5

Example 5.71

Let n be a positive integer and D_n denotes the set of all divisors of n. Consider the partial order 'divides' in D_n. The Hasse diagrams for D_6, D_{24} and D_{30} are given in the following figures.

$D_6 = \{1, 2, 3, 6\}$

$D_{24} = \{1, 2, 3, 4, 6, 8, 12, 24\}$

$D_{30} = \{1, 2, 3, 5, 6, 10, 15, 30\}$

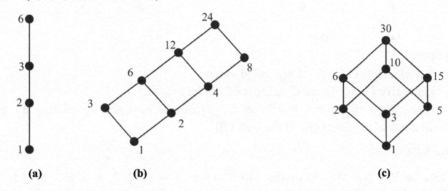

(a)   (b)   (c)

Example 5.72

Consider the Posets S and T represented in the following figures.

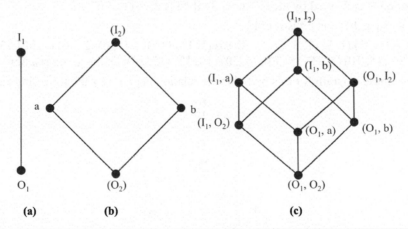

(a)   (b)   (c)

Example 5.73

Consider the Posets $(D_4, \leq)$ and $(D_9, \leq)$ given in (a) and (b). The Hasse diagram for $L = D_4 \times D_9$ under the partial order, is given by figure (c).

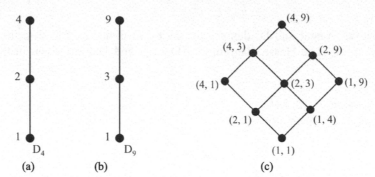

(a)   (b)   (c)

Observations:

   (i)   The elements in level – 1 are called atoms.

   (ii)   For a given Poset Hasse diagram need not be unique.

   (iii) Hasse diagram for the dual Poset (A, ≥) can be obtained by rotating the Hasse diagram of the Poset (A, ≤) through 180°.

Example 5.74

Let the relation R = {(a, b) | a is divisor of b} on a set S = {1, 2, 3, 5, 6, 10, 15, 30}. Draw Hasse Diagram.

Solution: Given Set S = {1, 2, 3, 5, 6, 10,15, 30}.

   The relation R is defined by aRb if and only if "a is divisor of b"

   That is R = {(a, b) | a is divisor of b}.

   Then R = {(1, 2) (1, 3) (1, 5) (1, 6) (1, 10) (1, 15) (1, 30), (2, 6) (2, 10) (2, 30), (3, 6) (3, 15) (3, 30), (5, 10) (5, 15) (5, 30), (6, 30) (10, 30) (15, 30)} is the given relation.

   Now the Hasse diagram for this partial order relation is given by the following figure.

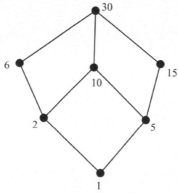

What is a poset? Draw the Hasse diagram of all the lattices with 5 elements.

[JNTUH, Nov 2008, Set No.1]

Solution: We know that a binary relation 'R' in a set P is called a partial order relation (or) a partially ordering in P if R is reflexive, antisymmetric and transitive. If " ≤ " is a partially ordering on P, then the ordered pair (P, ≤) is called a partially ordered set or a poset.

A lattice is a poset in which any two elements has their best upperbound and greatest lowerbound.

Lattices with 5 elements:

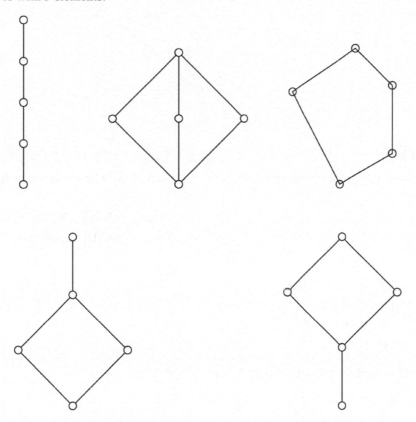

Example 5.75

Draw the Hasse diagram of {(P(s), ≤)} where P(s) is power set of the set S, S = {a, b, c}.

[JNTUH, Nov 2008, Set No. 2]

Solution: Given that

S = {a, b, c}

P(S) = {{a}, {b}, {c}, {a, b}, {b, c}, {a, c}, {a, b, c}, φ}.

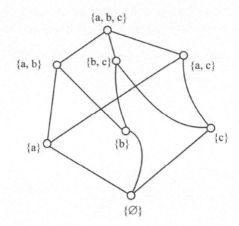

Example 5.76

Draw the Hasse diagram for the relation R on A = {1, 2, 3, 4, 5} whose relation matrix is given below.

[JNTUH, Nov 2008, Set No. 3]

[JNTUH, June 2010, Set No. 1]

$$M_R = \begin{bmatrix} 1 & 0 & 1 & 1 & 1 \\ 0 & 1 & 1 & 1 & 1 \\ 0 & 0 & 1 & 1 & 1 \\ 0 & 0 & 0 & 1 & 0 \\ 0 & 0 & 0 & 0 & 1 \end{bmatrix}$$

Solution: Given relation matrix of a set A = {1, 2, 3, 4, 5} is

$$\begin{bmatrix} 1 & 0 & 1 & 1 & 1 \\ 0 & 1 & 1 & 1 & 1 \\ 0 & 0 & 1 & 1 & 1 \\ 0 & 0 & 0 & 1 & 0 \\ 0 & 0 & 0 & 0 & 1 \end{bmatrix}$$

By examining the above relation matrix, we have the relation,

R = {(1,1), (1,3), (1,4), (1,5), (2,2), (2,3), (2,4), (2,5), (3,3), (3,4), (3,5), (4,4), (5,5)}

By using the above relation we can draw the Hasse diagram of R as follows:

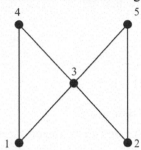

Exercises

Set

1. Describe the following sets in roster form:

    (i)   A = {x | x is an even number}

    (ii) B = {x ∈ R| $x^2 - 1 = 0$}

    (iii) C = {x | x is a positive integral divisor of 60}

Ans: (i) A = {2}; (ii) B = {–1, 1}; (iii) C = {1, 2, 3, 4, 5, 6, 10, 12, 15, 20, 30, 60}.

2. Describe the following in set builder form:

    (i)   A = {2, 4, 6, 8, 10}

    (ii) B = {–5, –4, –3, –2, –1, 0, 1, 2}

Ans:

    (i)   A = {x | x is even integer < 11}

    (ii) B = {x | – 5 ≤ x ≤ 2 and x ∈ z}

3. List the subsets of the set A = {1, 2, 3}

Ans: P (A) = {φ, {1}, {2},{3},{1,2}, {2,3}, {1,3}, A}

Operations on Sets

1. If P = {1, 2, 3}, Q = {a, b, c} and R = {5, 7, 9} then show that P, Q and R are mutually disjoint.

2. Show that for any two sets A and B
$$A - (A∩B) = A - B$$

3. Given A = {2, 5, 6}, B = {3, 4, 2}, C = {1, 3, 4}

Find A – B and B – A. Show that

(i)   A –B ≠ B – A; and

(ii) A – C = A

Ans: A – B = {5, 6}, B – A = {3, 4}

4.   Show that $A \subseteq B \Leftrightarrow A \cup B = B$

Principle of Inclusion and Exclusion

1.   Verify the principle of inclusion and exclusion for the sets A and B where
     A = {a, b, c, d, e} and B = {c, e, f, h, k}

2.   Set X, Y, Z are finite sets with
     $|X| = 6, |Y| = 8, |Z| = 6, |X \cup Y \cup Z| = 11, |X \cap Y| = 3, |X \cap Z| = 2$ and $|Y \cap Z| = 5$.
     Find $|X \cap Y \cap Z|$.

Ans: 1

3.   If A and B are finite sets, then prove that
     $$|A \cup B| = |A - B| + |B - A| + |A \cap B|$$

Relations

1.   Let A = {1, 2, 3, 4} of R = {(x,y) | x ∈ A ∧ y ∈ A ((x – y) is an integral non-zero multiple of 2)}
     ={(1,3), (3,1), (2,4), (4,2)}
     S = {(x,y) | x ∈ A ∧ y ∈ A ((x – y) is an integral non-zero multiple of 3)}
        = {(1,4), (4,1)}.
     Find (i) R ∪ S and (ii) R ∩ S

Ans: R ∪ S = {(1,3), (3,1), (2,4), (4,2), (1,4), (4,1)}
        R ∩ S = φ

Relation Matrix and Digraphs

1.   For A = {a, b, c, d, e, f}, the digraph is given below represents a relation R on A.
     Determine R as well as its relation matrix.

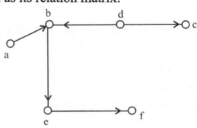

Ans: R = {(a,b), (b,c), (d,b), (d,c), (e,f)}

$$M(R) = \begin{bmatrix} 0 & 1 & 0 & 0 & 0 & 0 \\ 0 & 0 & 0 & 0 & 1 & 0 \\ 0 & 0 & 0 & 0 & 0 & 0 \\ 0 & 1 & 1 & 0 & 0 & 0 \\ 0 & 0 & 0 & 0 & 0 & 1 \\ 0 & 0 & 0 & 0 & 0 & 0 \end{bmatrix}$$

2. Write the relation matrix to the following digraph.

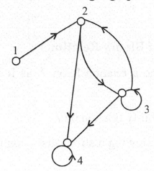

∴ $$M(R) = \begin{bmatrix} 0 & 1 & 0 & 0 \\ 0 & 0 & 1 & 1 \\ 0 & 1 & 1 & 1 \\ 0 & 0 & 0 & 1 \end{bmatrix}$$

Partitions and Covering of a Set, Equivalence Relations

1. If A is a non empty set then show that any equivalence relation R on A induces a partition of A.

2. Let A = {1, 2, 3} and R = {(1, 1), (2, 2), (3, 3)}. Verify that R is an equivalence relation.

3. Let A = {1, 2, 3, 4, 5, 6, 7, 8, 9, 10} and A_1 = {1, 2, 3, 4}, A_2 = {5, 6, 7}, A_3 = {4, 5, 7, 9}, A_4 = {4, 8, 10}, A_5 = {8, 9, 10}, A_6 = {1, 2, 3, 6, 8, 10}, which of the following are partitions of A?

   (i) {A_1, A_2, A_3} (ii) {A_1, A_3, A_5} (iii) {A_3, A_6}

   (i) Yes (ii) No (iii) Yes

Transitive Closure

1. Let A = {1, 2, 3}. Determine the nature of the relation R on A where

    R = {(1, 2), (2, 1), (1, 3), (3, 1)}

Ans: Symmetric and irreflexive, but neither reflexive nor transitive.

2. Show that the relation represented by the following matrix is transitive

$$\begin{bmatrix} 1 & 1 & 1 \\ 0 & 0 & 1 \\ 0 & 0 & 1 \end{bmatrix}$$

Compatibility Composition of Binary Relations

1. Let A = {1, 2, 3}. Define a relation R on A as R = {(1, 1), (2, 2), (3, 3), (1, 3), (3, 1)}.

    Show that R is a compatibility relation on A.

2. Verify that whether the following matrix satisfy compatibility relation or not.

$$\begin{bmatrix} 1 & 0 & 1 & 0 \\ 0 & 1 & 0 & 1 \\ 1 & 0 & 1 & 0 \\ 0 & 1 & 0 & 1 \end{bmatrix}$$

Ans: Yes

3. The matrix of a relation defined on a set s

    A = {a_1, a_2, a_3, a_4, a_5, a_6} is given below. Verify that R is a compatibility relation.

$$\begin{bmatrix} 1 & 1 & 1 & 1 & 0 & 0 \\ 1 & 1 & 1 & 1 & 1 & 0 \\ 1 & 1 & 1 & 1 & 0 & 1 \\ 1 & 1 & 1 & 1 & 0 & 0 \\ 0 & 1 & 0 & 0 & 1 & 1 \\ 0 & 0 & 1 & 0 & 1 & 1 \end{bmatrix}$$

Partially Ordered Relations and Hasse Diagrams

1. Prove that of R and S are equivalence relations on a Set A, then so is R ∩ S.

2. Check whether the relation

a R b if and only if a = 3b is partially ordered or not of all integers a, b.

Ans: No

3. Write down the Hasse diagram for the positive divisors of 45.

Ans:

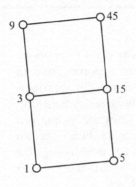

CHAPTER - 6

Functions

In this lesson, we study a particular class of relations called function. Functions play an important role in Mathematics, computer science and many applications. First we consider the discrete functions which transform a finite set into another finite set. Computer output can be considered as a function of the input. Functions can also be used for counting and for establishing the cardinality of sets. We also discuss the different types of functions and some of their applications.

LEARNING OBJECTIVES

♦ to know the Definitions and Examples of Functions and types of functions

♦ to find Composition of Functions and Inverse Functions

♦ to identify different types of Functions such as: Bijective functions, Permutation Functions, Recursive Functions

6.1 Definition and Examples of Functions

A function is a special case of relation. Let A, B be two non-empty sets and R be a relation from A to B, then R may not relate an element of A to an element of B or it may relate an element of A to more than one element of B. But a function relates each element of A to unique element of B.

Definition

Let S and T be sets. A function f from S to T is a subset f of $S \times T$ such that

  (i)   for $s \in S$, there exists $t \in T$ with $(s, t) \in f$,

  (ii) $(s, u) \in f$ and $(s, t) \in f \Rightarrow t = u$.

  If $(s, t) \in f$, then we write $(s, f(s))$ of $f(s) = t$.

  Here t is called the image of s; and s is called the preimage of t.

  The set S is called the domain of f and T is called the codomain.

  The set $\{f(s) \mid s \in S\}$ is a subset of T and it is called the image of S under f (or image of f). We denote the fact: 'f is a function from S to T' by "$f : S \to T$".

148

Example 6.1

Let X = {a, b, c} and Y = {0, 1}. Then observe the following.

(i)   f = {(a, 0), (b, 1), (c, 0)} is function. Hence f(a) = 0, f(b) = 1, f(c) = 0.

(ii) g = {(a, 0), (a, 1), (b, 0), (c, 1)} is not a function as a is related to 0 and 1.

(iii) h = {(a, 0), (b, 1)} is not a function as c is not related to any element in Y.

(iv) k = {(a, 0), (b, 0), (c, 0)} is a function. Domain of k is = {a, b, c} and the range of k is {0}.

We can write function by some rule.

Example 6.2

Let R be the set of real numbers. Define $f(x) = x^2$ for every $x \in R$. This represents a function
$$f = \{(x, x^2) \mid x \in R\}.$$

Example 6.3

Let f: N → N be a function such that

$$f(x) = \begin{cases} 1, & \text{if } x \text{ is odd} \\ 0, & \text{if } x \text{ is even} \end{cases}$$

Then the domain and the range of f respectively are N and {0, 1}.

Example 6.4

Let f: N → N be a function such that f(x) = x (mod 3). That is f(x) is the remainder obtained when x is divided by 3. Then the domain of f is N and the range of f is {0, 1, 2}.

Example 6.5

Let $f(x) = x^2 - 3x + 2$. Find

(i) $f(x^2)$;   (ii) f(x + 3)   [JNTU, Nov 2010, Set No.1]

Solution: Given function f(x) is defined as

$$f(x) = x^2 - 3x + 2 \qquad\qquad \text{….. (6.1)}$$

(i)   we have to find $f(x^2)$

Substitute 'x^2' in place of x in (6.1), we get that

$$f(x^2) = (x^2)^2 - 3(x^2) + 2 = x^4 - 3x^2 + 2$$

(ii)   substitute 'x + 3' in place of x in (6.1), we get that

$$f(x + 3) = (x + 3)^2 - 3(x + 3) + 2$$
$$= x^2 + 6x + 9 - 3x - 9 + 2$$
$$= x^2 + 3x + 2$$

6.2   Types of Functions (including Bijective Functions)

Definition

(i)   f: S → T is said to be one-one function (or injective function) if it satisfies the following condition:

$$f(s_1) = f(s_2) \Rightarrow s_1 = s_2.$$

(ii)   f: S → T is said to be onto function (or surjective function) if it satisfies the following condition:

$t \in T \Rightarrow$ there corresponds an element s in S such that $f(s) = t$.

(iii)   A function is said to be a bijection if it is both one-one and onto.

Example 6.6

(i)   f: R → R such that $f(x) = 3x + 2$ is an one-to-one and onto function.

(ii)   f: N → {0, 1} such that $f(x) = \begin{cases} 1, & \text{if x is odd} \\ 0, & \text{if x is even} \end{cases}$

is an onto function but not an one-one function.

(iii)   f: N → N defined by $f(x) = x^2 + 2$. It is an one-one function not an onto function, since there is no $x \in N$ such that $f(x) = 1$.

(iv)   f: R → R be such that $f(x) = |x|$ where $|x|$ is the absolute value of x. Then f is neither one-one nor onto.

Theorem 1

Let X and Y be two finite set with same number of elements. A function f: X → Y is one-to-one if and may is it is onto.

Proof: Let $X = \{x_1, x_2, ..., x_n\}$ and $Y = \{y_1, y_2, ..., y_n\}$. If f is one-to-one then $\{f(x_1), f(x_2), ..., f(x_n)\}$ is a set of n distinct elements of Y and hence f is onto.

If f is onto then $\{f(x_1), f(x_2), ..., f(x_n)\}$ form the entire set Y, so must all be different.

Hence f is one-to-one.

6.2.1 Observation

From the above theorem, we can understand that if we have a bijection between two finite sets, then the two sets must have same number of elements.

Example 6.7

The function $\sigma : Z^+ \rightarrow Z^+$ such that $\sigma(x) = x + 1$ is called the Peano's successor function. Here $\sigma(1) = 2$, $\sigma(2) = 3$, ... The range of σ is the set $\{2, 3, 4, ...\}$.

Definition

For any real number x, we define the floor of x as

$[x]$ = the greatest integer less than or equal to $x = \max\{n \mid n \leq x, n \text{ is an integer}\}$

Example 6.8

Take $x = 2.52$, then

$[x] = \max\{n \mid n \leq x, n \text{ is an integer}\} = \max\{1, 2\} = 2$.

Definition

For any real number x, we define the ceiling of x as

$[x]$ = the least integer greater than or equal to $x = \min\{n \mid n \geq x, n \text{ is an integer}\}$.

Example 6.9

Take $x = 3.732$, then

$[x] = \min\{n \mid n \geq x, n \text{ is an integer}\} = \min\{4, 5, 6, 7...\} = 4$.

6.2.2 Geometric Interpretation

Floor and Ceiling functions may be understood from their graphical (or geometrical) representation. Consider the line $f(x) = x$, the diagonal on I, III coordinates, take $x = e = 2.71828.....$ we describe floor and ceiling of e as follows:

From the graph

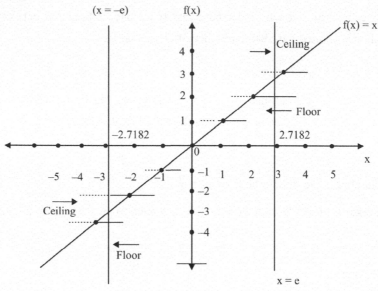

6.2.3   Properties

(i)   From the graph, it can be observed that, the two functions $\lceil x \rceil$ and $\lfloor x \rfloor$ are equal at integer points. That is, $\lceil x \rceil = x \Leftrightarrow x$ is an integer $\Leftrightarrow \lceil x \rceil = x$.

(ii) $\lceil x \rceil - x = [x$ is not an integer]

That is, $\lceil x \rceil - \lfloor x \rfloor = \begin{cases} 1, & \text{if } x \text{ is not an integer} \\ 0, & \text{otherwise} \end{cases}$

(iii) $x - 1 < \lfloor x \rfloor$ and $x + 1 > \lceil x \rceil \Rightarrow x - 1 < \lfloor x \rfloor \le x \le \lceil x \rceil < x + 1$

(iv) $\lfloor -x \rfloor = -\lceil x \rceil$ and $\lceil -x \rceil = -\lfloor x \rfloor$

(v)   For any real number x, $\lfloor x \rfloor \le x$ and $\lceil x \rceil \ge x$.

6.2.4   Some Rules on Floor and Ceiling Functions

In all the following cases, x is real and n is an integer.

1. $\lfloor x \rfloor = n \Leftrightarrow n \le x < n + 1$

2. $\lfloor x \rfloor = n \Leftrightarrow x - 1 < n \le x$

3. $\lceil x \rceil = n \Leftrightarrow n - 1 < x \le n$

4. $\lceil x \rceil = n \Leftrightarrow x \le n < x + n$

Example 6.10

The above rules can be illustrated, by taking $x = 4.5$

1. $\lfloor 4.5 \rfloor = 4 \Leftrightarrow 4 \le 4.5 < 5$
2. $\lfloor 4.5 \rfloor = 4 \Leftrightarrow 3.5 < 4 \le 4.5$
3. $\lceil 4.5 \rceil = 5 \Leftrightarrow 4 < 4.5 \le 5$
4. $\lceil 4.5 \rceil = 5 \Leftrightarrow 4.5 \le 5 < 5.5$

Example 6.11

Let X be the set of all statements in logic and Y denotes the set {T, F} where T and F are truth values. The assignment of truth values to each statement in X can be considered as a function from X to Y.

Example 6.12

(i)   Compiler transforms a program written in a high level language into a machine language.

(ii) The output from a computer is a function of its input.

Definition

Let f: X → Y be a function and let A ⊆ X. f_A: A → Y is called the restriction of f to A if $f_A(x) = f(x)$ for any x ∈ A. If g is the restriction of f then f is called the extension of g.

Example 6.13

Let f: R → R be such that $f(x) = x^2$. Then f_N: N → R is such that $f(n) = n^2$ is the restriction of f to N.

Example 6.14

Let X = {a, b, c} and Y = {0, 1}. List all the functions from X to Y.

Solution: The set X × Y = {(a, 0), (a, 1), (b, 0), (b, 1), (c, 0), (c, 1)} contains 6 elements.

Hence there are 2^6 of subsets of X × Y. Out of these subsets only the following $2^3 = 8$ subsets are functions.

$$f_0 = \{(a, 0), (b, 0), (c, 0)\}$$
$$f_1 = \{(a, 0), (b, 0), (c, 1)\}$$
$$f_2 = \{(a, 0), (b, 1), (c, 0)\}$$
$$f_3 = \{(a, 0), (b, 1), (c, 1)\}$$

$$f_4 = \{(a, 1), (b, 0), (c, 0)\}$$
$$f_5 = \{(a, 1), (b, 0), (c, 1)\}$$
$$f_6 = \{(a, 1), (b, 1), (c, 0)\}$$
$$f_7 = \{(a, 1), (b, 1), (c, 1)\}$$

We can observe that, in general, if X has m elements and Y has n elements then there will be n^m function from X to Y.

Example 6.15

For each positive integer n, we define a function $f_n: Z^+ \to N$ such that $f_n(x) = r$, where $x = r$ (mod n), $0 < r < n$. That is, r is the remainder obtained when x is divided by n.

Example 6.16

(Factorial function): $f: N \to Z^+$ such that $f(n) = n!$ For $n > 0$ and $f(0)$ is defined by $f(0) = 1$.

Example 6.17

(Hashing function): To determine to which list a particular record should be assigned, we create a hashing function from the set of keys to the set of list numbers. A unique identifier for a record is called a key.

For example, suppose 10,000 customer account records are to be stored and processed. The computer is capable of searching 100 items at a particular time. We create 101 linked lists for storage. We define a hashing function from the set of 7-digit account numbers to the set $\{0, 1, 2, ..., 100\}$ as $h(n) = n$ (mod 101). Thus

$$h(2473871) = 2473871 \ (\text{mod } 101) = 78.$$

This means the record with account number 2473871 be assigned to list 78. Range of h is $\{0, 1, 2, ..., 100\}$.

6.3   Composition of Functions and Inverse Functions

Definition

Let $g: S \to T$ and $f : T \to U$. The composition of f and g is a function fog $: S \to U$ defined by (fog) (s) = f(g(s)) for all s in S.

That is, fog = $\{s, u\} | s \in S, u \in U$ and $\exists t \in T$ and $t + g(s)$ and $u = f(t)\}$.

Example 6.18

Let $X = \{1, 2, 3\}$, $Y = \{a, b\}$ and $Z = \{p, q, r\}$. Let f: $X \to Y$ defined by f = $\{(1, a), (2, b), (3, a)\}$ and g: $Y \to Z$ defined by g = $\{(a, r), (b, q)\}$. Then gof = $\{(1, q), (2, q), (3, r)\}$.

Example 6.19

Let f: R $\rightarrow$ R and g: R $\rightarrow$ R where R is the set of real numbers. If $f(x) = x^2 - 2$ and $g(x) = x + 4$. Find gof and fog

Solution: $(gof)(x) = g(f(x)) = g(x^2 - 2) = (x^2 - 2) + 4 = x^2 + 2$; and

$(fog)(x) = f(g(x)) = f(x + 4) = (x + 4)^2 - 2 = x^2 + 8x + 14$

Definition

Let f: X $\rightarrow$ Y, g: Y $\rightarrow$ Z and h: Z $\rightarrow$ W are functions. Then the compositions are fog: X $\rightarrow$ Z and hog: Y $\rightarrow$ W. We can also form the compositions ho(gof) and (hog)of which are functions from X $\rightarrow$ W.

6.3.1   Observation

Note that composition of functions is associative: ho(gof) = (hog)of.

Example 6.20

Let $f(x) = x + 3$, $g(x) = x - 4$ and $h(x) = 5x$ are functions from R $\rightarrow$ R where R is the set of real numbers. Find fo(goh) and (fog)oh.

Solution: Now $\qquad$ fo(goh) (x) = f(goh)(x)

$$= f[g(h(x))]$$
$$= f[g(5x)]$$
$$= f(5x - 4)$$
$$= 5x - 4 + 3$$
$$= 5x - 1$$

Also, $\qquad$ (fog)oh (x) = (fog)(h(x))

$$= (fog)(5x)$$
$$= f(g(5x))$$
$$= f(5x - 4)$$
$$= 5x - 4 + 3$$
$$= 5x - 1$$

Therefore fo(goh) = (fog)oh

Example 6.21

Show that if $g : S \to T$ and $f : T \to U$ are one-one functions, then fog is also one-one.

Solution: Suppose that $(fog)(s) = (fog)(t)$ for s, t $\in$ S. By the definition of composition of maps, we have $f(g(s)) = f(g(t))$. Since f is one-one, we get $g(s) = g(t)$. Since g is one-one, we get $s = t$. Therefore fog is one-one.

Example 6.22

If $g : S \to T$ and $f : T \to U$ are onto, then so is fog.

Solution: Let u $\in$ U. To show that fog is onto, we have to find an element s in S such that $(fog)(s) = u$. Since f is onto, there exists t in T such that $f(t) = u$. Now since g is onto there exists s in S such that $g(s) = t$. It is clear that $(fog)(s) = f(g(s)) = f(t) = u$. Hence fog is an onto function.

Theorem 2

If f: $X \to Y$ and g: $Y \to Z$ are bijections then gof: $X \to Z$ is also a bijection.

Proof: Combination of the above two problems: 6.3.7 & 6.3.8.

Definition

A function $f : S \to T$ is said to have an inverse if there exists a function g from T to S such that $(gof)(s) = s$ for all s in S and $(fog)(t) = t$ for all t in T. We call the function 'g' the inverse of f. A function $f : S \to S$ is said to be an identity function if $f(s) = s$ for all s in S. The identity function on S is denoted by either I or I_S. Inverse of a function f, if it exists, is denoted by f^{-1}. Two functions $f : A \to B$ and $g : C \to D$ are said to be equal if A = C, B = D and $f(a) = g(a)$ for all elements a in A = C. If two functions f and g are equal, then we write $f = g$.

Theorem 3

Let $f : X \to Y$ be a function and I_x is the identity function of X, then $foI_x = I_xof = f$.

Proof: Now $foI_x(x) = f(I_x(x)) = f(x)$. Similarly we get $I_xof(x) = f(x)$.

The identity function is one-one and onto.

A function g is inverse of $f \Leftrightarrow$ fog and gof are identity functions.

Example 6.23

Find out two functions f and g defined from R to R, where R is the set of all real numbers such that fog $\neq$ gof.

Solution: Define f(x) = 2x and g(x) = x + 5 for all x in R.

Then (fog)(1) = f(g(1)) = f(1 + 5) = f(6) = 12

(gof)(1) = g(f(1)) = g(2) = 2 + 5 = 7

This shows that the two functions are not equal at 1.

Example 6.24

Prove that a function f has an inverse ⇔ f is one-one and onto.

Solution: Suppose the inverse of f : S → T is g : T → S.

By definition, gof(s) = s for all s in S and fog (t) = t for all t in T.

To show f is one-one, suppose a, b ∈ S such that f(a) = f(b).

By applying the function g on both sides, we get gof(a) = gof(b).

Since gof is identity, we get a = gof(a) = gof(b) = b.

Hence f is one-one.

To show f is onto, let t be an element of T. Write x = g(t).

Then x is in S and f(x) = f(g(t)) = fog(t) = t. Hence f is onto.

Converse: Suppose f is one-one and onto.

Define g : T → S as g(t) = s where f(s) = t. To verify that g is a function, suppose g(t) = a and g(t) = b. Then f(a) = t, f(b) = t.

So f(a) = f(b) which implies a = b (since f is one-one). Therefore g is a function. For all s in S, we have that gof(s) = g(f(s)) = g(t) = s (where t = f(s)).

Also for all t in T, we have fog(t) = f(g(t)) = f(s) = t. Hence g is an inverse of f.

Note:

Inverse of a function is unique. Identity function on a set is unique. Identity element in a set with respect to a binary operation is unique.

6.3.2   Notation

If S is a non-empty set, then we write

$$A(S) = \{f \mid f : S \to S \text{ is a bijection}\}.$$

Theorem 4

For f ∈ A(S), there corresponds an element f^{-1} in A(S) such that $fof^{-1} = 1 = f^{-1}of$.

Proof: Define f^{-1}: S → S by $f^{-1}(y) = x$, if f(x) = y

Then f^{-1} is a function and $(f^{-1}of)(x) = f^{-1}(f(x)) = f^{-1}(y) = x = I(x)$

So $f^{-1}of = I$. Similarly $(fof^{-1})(y) = f(f^{-1}(y)) = f(x) = y = I(y)$

This implies $fof^{-1} = I$. Therefore $fof^{-1} = I = f^{-1} of$.

Theorem 5

Let $f: X \to Y$ and $g: Y \to X$. Then $g = f^{-1}$ if and only if $gof = I_x$ and $fog = I_y$.

Proof: Let $gof = I_x$ and $fog = I_y$.

Then $g(f(x)) = x$ and $f(g(y)) = y$ for all $x \in X$ and $y \in Y$.

This means range of $f = Y$ and the range of $g = X$.

Hence both f and g are onto.

Now to show f is one one.

Suppose $f(x_1) = f(x_2) \Rightarrow x_1 = g(f(x_1)) = g(f(x_2)) = x_2$. Therefore f is one-to-one. Similarly g is one-to-one.

Hence both f and g are one-to-one and onto functions and so are invertible. Now

$$f^{-1}(y) = f^{-1}(f(g(y))) = f^{-1} (of) (g(y)) = I_x(g(y)) = g(y).$$

Hence $f^{-1} = g$. Similarly we can prove that $g^{-1} = f$.

Theorem 6

Let $f: X \to Y$ and $g: Y \to Z$ are invertible functions. Then

(i) $(f^{-1})^{-1} = f$

(ii) $(gof)^{-1} = f^{-1}og^{-1}$

Proof:

(i) To show that f^{-1} is one-one and onto.

Now $f^{-1}(y_1) = f^{-1}(y_2) \Rightarrow x_1 = x_2$ where $f(x_1) = y_1$ and $f(x_2) = y_2$, since f is onto.

$\Rightarrow \quad f(x_1) = f(x_2)$ (since f is one-one)

$\Rightarrow \quad y_1 = y_2$

Therefore f^{-1} is one-one.

Also, take $x \in X$. Then there exist a unique $y \in Y$ such that $f(x) = y$. That is there exists $y \in Y$ such that $f^{-1}(y) = x$. Hence f^{-1} is onto. Since f^{-1} is the inverse relation of f and vice versa, we get that $(f^{-1})^{-1} = f$.

(ii) Since f, g are one-one and onto, we have that (gof) is one one and onto. Hence (gof) is invertible. Also f, g are invertible. Hence $(gof)^{-1}$, f^{-1}, g^{-1} and $f^{-1}og^{-1}$ exist. Now $(gof)^{-1}$ and $f^{-1}og^{-1}$ are functions from Z to X.

Now for any $x \in X$, let $y = f(x)$ and $z = g(y)$.

Then $(gof)(x) = z$ and $(gof)^{-1}(z) = x$ for all $x \in X$, $y \in Y$, $z \in Z$.

Also $x = f^{-1}(y)$ and $y = g^{-1}(z)$ so that

$$(f^{-1}og^{-1})(z) = f^{-1}(g^{-1}(z)) = f^{-1}(y) = x, \text{ for all } x \in X, y \in Y, z \in Z.$$

Hence $(gof)^{-1}(z) = f^{-1}og^{-1}(z)$ for all $z \in Z$. Thus $(gof)^{-1} = f^{-1}og^{-1}$.

Example 6.25

Let $X = \{1, 2, 3, 4\}$ and $Y = \{a, b, c, d\}$.

(i)   For the function $f = \{(1, a), (2, a), (3, b), (4, d)\}$, the inverse relation $f^{-1} = \{(a, 1), (a, 2), (b, 3), (d, 4)\}$ is not a function since c has no relative and a has two relatives. Hence f is not invertible.

(ii) For the function $g = \{(1, d), (2, c), (3, b), (4, a)\}$, the inverse relation $g^{-1} = \{(a, 4), (b, 3), (c, 2), (d, 1)\}$ is a function. Hence g is invertible.

Example 6.26

Let $f: R \to R$ and $g: R \to R$ be defined as $f(x) = 2x+1$ and $g(y) = \dfrac{y}{3}$.

Verify whether or not $(gof)^{-1} = f^{-1}og^{-1}$.

Solution: Now consider

$$(gof)(x) = g(f(x)) = g(2x+1) = \frac{2x+1}{3} \text{ and } (gof)^{-1}(z) = \frac{3z-1}{2}.$$

Now $f^{-1}(y) = \dfrac{y-1}{2}$ and $g^{-1}(z) = 3z$.

Then $(f^{-1}og^{-1})(z) = f^{-1}(g^{-1}(z)) = f^{-1}(3z) = \dfrac{3z-1}{2}$.

Example 6.27

Show that the mapping $f: R \to R$ defined by $f(x) = ax + b$ where $a, b, x \in r$, $a \neq 0$ is invertible. Define its inverse.

Solution: Take $x_1, x_2 \in R$. Now $f(x_1) = f(x_2) \Rightarrow ax_1 + b = ax_2 + b \Rightarrow ax_1 = ax_2 \Rightarrow x_1 = x_2$.

Therefore f is one one.

Take $y \in R$. Now $y = f(x) \Rightarrow y = ax + b \Rightarrow x = \dfrac{(y-b)}{a}$. Therefore for $y \in R$, there exists $\dfrac{(y-b)}{a} \in R$ such that $f(\dfrac{(y-b)}{a}) = a(\dfrac{(y-b)}{a}) + b = y - b + b = y$.

This shows that f^{-1} exists and it is defined by $f^{-1}(y) = \dfrac{(y-b)}{a}$.

Example 6.28

If S contains more than two elements, then there exists f, g $\in$ A(S) such that fog $\neq$ gof.

Solution: Since S contains more than two elements, we have $|S| > 2 \Rightarrow |S| \geq 3$.

Let a b, c $\in$ S be three distinct elements. Define f: S $\to$ S by f(a) = b, f(b) = c, f(c) = a and f(x) = x for all x $\in$ S | {a, b, c}. Define g: S $\to$ S by g(a) = b, g(b) = a, and g(x) = x for all   x $\in$ S | {a, b}. Then f, g are bijections and hence f, g $\in$ A(S). Now (gof)(a) = g(f(a)) = g(b)=a and (fog)(a) = f(g(a)) = f(b) = c.

Therefore (gof)(a) = a $\neq$ c = (fog)(a). This shows that gof $\neq$ fog.

Example 6.29

If S is a non-empty set with $|S| \leq 2$, then for any two elements f, g $\in$ A(S), we have fog = gof.

Solution: If $|S| = 1$, then S = {x}. Now there exists only one bijection f: S $\to$ S defined by f(x) = x. So in this case, the result is clear. Now suppose that the set S contains two elements x and y. Define f: S $\to$ S and g: S $\to$ S by f(x) = x, f(y) = y, g(x) = y, g(y) = x. Clearly f, g are bijections and A(S) = {f, g}.

Since f is identity mapping on S, we have fog = log = g = goI = gof.

Example 6.30

If $|S| = n$, then show that $|A(S)| = n!$

Solution: Suppose S = {x_i | 1 $\leq$ i $\leq$ n}. If f $\in$ A(S), then f is a bijection.

To define f: S $\to$ S we have to define f(x_i) as an element of S for each 1 $\leq$ i $\leq$ n.

To define f(x_1) there are n possible ways (because f(x_1) $\in$ {x_1, x_2, ..., x_n}).

Since f is one-one, we have that f(x_1) $\neq$ f(x_2).

So after defining f(x_1), to define f(x_2) there are (n–1) ways, because f(x_2) $\in$ {$x_1, x_2, ..., x_n$} | {f(x_1)}. Thus f(x_1) and f(x_2) both can be defined in n(n–1) ways. Now for f(x_3)there are (n – 2) ways and so on.

Hence $f(x_1)$, $f(x_2)$, ..., $f(x_n)$ can be defined in $n(n - 1)(n - 2) ... 2 \times 1 = n!$ ways.

Therefore n! number of bijections can be defined from S to S. This means $A(S) = n!$.

Functions can be classified mainly into two ways.

(a)   Algebraic function

(b)   Transcendental function

6.3.3   Definition

A function which consists of a finite number of terms involving powers and roots of the independent variable x and the four fundamental operations (addition, subtraction, multiplication and division) is called algebraic function.

Example 6.31

(i)   A function of the form $a_0x^n + a_1x^{n-1} + ... + a_n$ where n is a positive integer and a_0, a_1, ..., a_n are real constants and $a_0 \neq 0$, is called a polynomial of x in degree n. (for example, $f(x) = 5x^3 + 4x^2 + 7x + 9 = 0$ is a polynomial of degree 3).

(ii) *Rational function*: A function of the form $\dfrac{f(x)}{g(x)}$ where $f(x)$ and $g(x)$ are polynomials in x and $g(x) \neq 0$, is called a rational function (for example, $\dfrac{x^2 + x + 1}{x + 3}$).

(iii) *Irrational function*: The functions involving radicals are called irrational functions (for example, $f(x) = \sqrt[3]{x} + 5$ is an irrational function.

Definition

A function which is not algebraic is called Transcendental function.

Example 6.32

Trigonometric functions and Inverse Trigonometric functions: The functions like sin x, cos x, tan x, sec x, cosec x, cot x; and $\sin^{-1}x$, $\cos^{-1}x$, $\tan^{-1}x$, $\sec^{-1}x$, $\text{cosec}^{-1}x$, $\cot^{-1}x$ where the angle x is measured in radians.

Example 6.33

Exponential and Logarithmic functions: A function $f(x) = a^x$ $(a > 0)$ satisfying the law $a^1 = a$ and $a^x.a^y = a^{x+y}$ is called the exponential function. The inverse of the exponential function is called the logarithmic function. If $y = a^x$ then $x = \log_a y$ is a logarithmic function.

Example 6.34

Find the inverse of the following functions:

(i) $f(x) = \dfrac{10}{\sqrt[5]{7-3x}}$ [JNTUH Nov 2008, Set No.3]

(ii) $f(x) = 4e^{(6x+2)}$ [JNTUH June 2010, Set No.1]

Solution:

(i)   Given that

$$f(x) = \frac{10}{\sqrt[5]{7-3x}}$$

For any x_1 and $x_2 \in R$,

$$f(x_1) = \frac{10}{\sqrt[5]{7-3x_1}}$$

$$f(x_2) = \frac{10}{\sqrt[5]{7-3x_2}}$$

$$f(x_1) = f(x_2) \Rightarrow \frac{10}{\sqrt[5]{7-3x_1}} = \frac{10}{\sqrt[5]{7-3x_2}}$$

$$\Rightarrow \frac{1}{\sqrt[5]{7-3x_1}} = \frac{1}{\sqrt[5]{7-3x_2}}$$

$$\Rightarrow \frac{1}{7-3x_1} = \frac{1}{7-3x_2}$$

$$\Rightarrow 7 - 3x_2 = 7 - 3x_1$$

$$\Rightarrow x_1 = x_2$$

Therefore f is one-one.

Take any $y \in R^+$, and put

$$y = \frac{10}{\sqrt[5]{7-3x}} \Rightarrow f(x) = y$$

$$\Rightarrow \log_e y = \log\left[\frac{10}{\sqrt[5]{7-3x}}\right]$$

$$= \log 10 \;-\; \log(7-3x)^{1/5}$$

$$= 1 - \frac{1}{5}\, \log\,(7-3x)$$

$$\Rightarrow \frac{1}{5} \log_e (7-3x) = 1 - \log_e y$$

$$\Rightarrow \log_e (7-3x) = 5 (1 - \log_e y)$$

$$\Rightarrow 7-3x = e^{5(1-\log y)}$$

$$= e^{5-5 \log y}$$

$$= e^5 . e^{-\log y^5}$$

$$= e^5 e^{\log \frac{1}{y^5}}$$

$$= \frac{1}{y^5} e^5$$

$$\Rightarrow 3x = 7 - \frac{e^5}{y^5}$$

$$x = \frac{1}{3}\left[7 - \frac{e^5}{y^5}\right]$$

Accordingly f is invertible,

Since $\quad x = f^{-1}(y)$

$$\Rightarrow f^{-1}(y) = \frac{1}{3}\left[7 - \frac{e^5}{y^5}\right]$$

(ii) Given function is

$$f(x) = 4 e^{(6x+2)}$$

For any $x_1, x_2 \in R,$

$$f(x_1) = f(x_2)$$

$$\Rightarrow 4 e^{(6x_1+2)} = 4 e^{(6x_2+2)}$$

$$\Rightarrow e^{6x_1+2} = e^{6x_2+2}$$

$$\Rightarrow e^{6x_1}.e^2 = e^{6x_2}.e^2$$

$$\Rightarrow e^{6x_1} = e^{6x_2}$$

$$\Rightarrow 6x_1 = 6x_2$$

$$\Rightarrow x_1 = x_2$$

Therefore f is one-one.

Take for an $y \in R$ suppose that

$$f(x) = y \Rightarrow x = f^{-1}(y)$$

So
$$y = 4\,e^{(6x+2)}$$

$$\Rightarrow e^{6x+2} = y/4$$

$$\Rightarrow (6x + 2)\log e = \log(y/4)$$

$$\Rightarrow 6x + 2 = \log(y/4)$$

$$6x = \log(y/4) - 2$$

$$x = \frac{1}{6}\big[(\log(y/4) - 2\big]$$

Therefore $f^{-1}(y) = \dfrac{1}{6}\left[\log\dfrac{y}{4} - 2\right]$

6.4   Permutation Functions

In this section we consider a permutation function of a set. A special type of permutation called cyclic permutation was discussed. We observe that a permutation can be expressed as a product (usual composition of mappings) of disjoint cycles (or transpositions).

Definition

Let $A = \{x_1, x_2\ x_3, ..., x_n\}$ be a set with n elements. A bijection (one to one and onto map) from A to A is called a permutation of A. Function values of a permutation p on A namely $p(x_1), p(x_2), ..., p(x_n)$ are given in the following form

$$p = \begin{pmatrix} x_1 & x_2 & ... & x_n \\ p(x_1) & p(x_2) & ... & p(x_n) \end{pmatrix}$$

Note:

A permutation is just a rearrangement of elements of A.

Example 6.35

Consider the set $A = \{1, 2, 3\}$.

(i)   There are $3! = 6$ permutations. These are

$$I_A = \begin{pmatrix} 1 & 2 & 3 \\ 1 & 2 & 3 \end{pmatrix},\ p_1 = \begin{pmatrix} 1 & 2 & 3 \\ 1 & 3 & 2 \end{pmatrix},\ p_2 = \begin{pmatrix} 1 & 2 & 3 \\ 2 & 1 & 3 \end{pmatrix},$$

$$p_3 = \begin{pmatrix} 1 & 2 & 3 \\ 2 & 3 & 1 \end{pmatrix}, \ p_4 = \begin{pmatrix} 1 & 2 & 3 \\ 3 & 1 & 2 \end{pmatrix}, \ p_5 = \begin{pmatrix} 1 & 2 & 3 \\ 3 & 2 & 1 \end{pmatrix}$$

(ii)   In the above example, the inverse of p_4 is p_3 and the inverse of p_2 is p_2.

Definition

If the set S contains n elements, then the group

$$A(S) = \{f: S \to S \mid f \text{ is a bijection}\}$$

has n! elements. Since S has n elements we denote A(S) by S_n and this $A(S) = S_n$ is called the symmetric group of degree n. If $\phi \in A(S) = S_n$, then ϕ is a one to one mapping of S onto itself.

Example 6.36

If $S = \{x_1, x_2, x_3, x_4\}$ and $\phi \in A(S)$ by $\phi(x_1) = x_2$, $\phi(x_2) = x_4$, $\phi(x_3) = x_1$, $\phi(x_4) = x_3$ is

denoted by $\phi = \begin{pmatrix} x_1 & x_2 & x_3 & x_4 \\ x_2 & x_4 & x_1 & x_3 \end{pmatrix}$ or $\begin{pmatrix} 1 & 2 & 3 & 4 \\ 2 & 4 & 1 & 3 \end{pmatrix}$. If $\theta = \begin{pmatrix} 1 & 2 & 3 & 4 \\ 3 & 1 & 2 & 4 \end{pmatrix}$ and

$\psi = \begin{pmatrix} 1 & 2 & 3 & 4 \\ 1 & 3 & 2 & 4 \end{pmatrix}$ then $\psi\theta = \begin{pmatrix} 1 & 2 & 3 & 4 \\ 2 & 1 & 3 & 4 \end{pmatrix}$ (verify). Here we use $\psi\theta(x) = \psi(\theta(x))$ (the

usual composition of mapping) for all $x \in S$.

Example 6.37

Permutation multiplication is not usually commutative. Let $\sigma = \begin{pmatrix} 1 & 2 & 3 & 4 \\ 4 & 1 & 2 & 3 \end{pmatrix}$ and

$\tau = \begin{pmatrix} 1 & 2 & 3 & 4 \\ 2 & 1 & 4 & 3 \end{pmatrix}$. Then $\sigma\tau = \begin{pmatrix} 1 & 2 & 3 & 4 \\ 1 & 4 & 3 & 2 \end{pmatrix}$ but $\tau\sigma = \begin{pmatrix} 1 & 2 & 3 & 4 \\ 3 & 2 & 1 & 4 \end{pmatrix}$. Therefore $\sigma\tau = \tau\sigma$.

Definition

Λ permutation $\sigma \in S_n$ is a cycle of length k if there exists elements $a_1, a_2, ..., a_k \in S$ such that $\sigma(a_1) = a_2$, $\sigma(a_2) = a_3$, ..., $\sigma(a_k) = a_1$ and $\sigma(x) = x$ for all other elements $x \in S$. We will write $(a_1, a_2, ..., a_k)$ to denote the cycle σ. Cycles are the building blocks of the permutations.

Example 6.38

Let P be a cyclic permutation of length 4 defined as

$$P = \begin{bmatrix} 1 & 2 & 3 & 4 & 5 \\ 3 & 2 & 4 & 5 & 1 \end{bmatrix} = (1 \ 3 \ 4 \ 5)$$

P can also be written as (3 4 5 1) or (4 5 1 3) or (5 1 3 4)

Example 6.39

Let $A = \{1, 2, 3, 4, 5, 6\}$. Compute $(2\ 1\ 3\ 5)\ O\ (1\ 6\ 2)$

Solution: We have

$$(2\ 1\ 3\ 5)\ O\ (1\ 6\ 2) =$$

$$\begin{bmatrix} 1 & 2 & 3 & 4 & 5 & 6 \\ 3 & 1 & 5 & 4 & 2 & 6 \end{bmatrix} O \begin{bmatrix} 1 & 2 & 3 & 4 & 5 & 6 \\ 6 & 1 & 3 & 4 & 5 & 2 \end{bmatrix} = \begin{bmatrix} 1 & 2 & 3 & 4 & 5 & 6 \\ 6 & 3 & 5 & 4 & 2 & 1 \end{bmatrix}$$

The composition is not cyclic.

Now $(1\ 6\ 2)\ O\ (2\ 1\ 3\ 5) = \begin{bmatrix} 1 & 2 & 3 & 4 & 5 & 6 \\ 3 & 6 & 5 & 4 & 1 & 2 \end{bmatrix}$. This is also not a cyclic.

Further $(2\ 1\ 3\ 5)\ O\ (1\ 6\ 2) \neq (1\ 6\ 2)\ O\ (2\ 1\ 3\ 5)$.

Example 6.40

The permutation $\sigma = \begin{pmatrix} 1 & 2 & 3 & 4 & 5 & 6 & 7 \\ 6 & 3 & 5 & 1 & 4 & 2 & 7 \end{pmatrix} = (1\ 6\ 2\ 3\ 5\ 4)$ is a cycle of length 6, whereas

$\tau = \begin{pmatrix} 1 & 2 & 3 & 4 & 5 & 6 \\ 1 & 4 & 2 & 3 & 5 & 6 \end{pmatrix} = (2\ 4\ 3)$ is a cycle of length 3. Also, not, every permutation is a

cycle. Consider the permutation $\begin{pmatrix} 1 & 2 & 3 & 4 & 5 & 6 \\ 2 & 4 & 1 & 3 & 6 & 5 \end{pmatrix} = (1\ 2\ 4\ 3)\ (5\ 6)$.

Example 6.41

Compute the product of cycles $\sigma = (1\ 3\ 5\ 2)$, $\tau = (2\ 5\ 6)$.

Solution: $\sigma\tau = (1\ 3\ 5\ 6)$

Note:

Two cycles $(a_1, a_2, ..., a_k)$ and $(b_1, b_2, ..., b_k)$ are said to be disjoint if $a_i \neq b_j$ for all i and j.

For instance, the cycles $(1\ 3\ 5)$ and $(2\ 7)$ are disjoint; however, the cycles $(1\ 3\ 5)$ and $(3\ 4\ 7)$ are not. Calculating their products, we get that

$$(1\ 3\ 5)\ (2\ 7) = (1\ 3\ 5)\ (2\ 7)$$

$$(1\ 3\ 5)\ (3\ 4\ 7) = (1\ 3\ 4\ 7\ 5)$$

Theorem 7

A permutation of a finite set that is not the identity or a cycle can be written as a product (composition) of disjoint cycles of length greater than or equal to 2.

Proof: Let θ (non identity with length ≥ 2) be a permutation.

Then its cycles are of the form $(s, s\theta, ..., s\theta^{i-1})$. Write

$$\psi = \text{the product of distinct cycles of } \theta.$$

Since each cycle forms an equivalence class, if we take two cycles, they are either equal or disjoint. Therefore any two distinct cycles are disjoint.

Suppose $\psi = c_1, c_2, ..., c_n$ where $c_1, c_2, ..., c_n$ are disjoint cycles. Let $s' \in S$. Now since $c_1, c_2, ..., c_n$ is a collection of disjoint cycles, we have that s' occurs in c_k for some $1 \leq k \leq n$. (clearly s' is not any other cycle, since any distinct cycles are disjoint). Also $s' c_k = s' \theta$.

If $i \neq k$, $s' c_i = s'$ (since s' is not in c_i). Hence $s'(\psi) = s'(c_1, c_2, ..., c_{k-1}, c_k, c_{k+1}, ..., c_n) = s'(c_1, c_2, ..., c_{k-1})(c_{k+1}, ..., c_n) = s' c_k(c_{k+1}, ..., c_n) = s' \theta$ (since $s' \sim s'\theta$, we have that $s'\theta$ is also not in c_i for $i \geq k + 1$). Therefore $s' \psi = s' \theta$ for all $s' \in S$. Hence $\psi = \theta$.

Example 6.42

Write $p = \begin{bmatrix} 1 & 2 & 3 & 4 & 5 & 6 & 7 & 8 \\ 4 & 3 & 2 & 5 & 1 & 8 & 7 & 6 \end{bmatrix}$ as a product of disjoint cycles.

Solution: Take $A = \{1, 2, 3, 4, 5, 6, 7, 8\}$. Start with element 1.

Now $p(1) = 4$, $p(4) = 5$, $p(5) = 1$ we get a cycle $(1\ 4\ 5)$.

Next we choose x such that $x \in A$ and x is not appeared in the cycle.

Choose 2. Now $p(2) = 3$,

$$p(3) = 2.$$

Thus we get a cycle $(2\ 3)$.

Next choose 6, we get the cycle $(6\ 8)$ and $p(7) = 7$.

Thus $p = (6\ 8) \circ (2\ 3) \circ (1\ 4\ 5)$.

Note:

The product is unique except for the order of the cycles.

The simplest permutation is a cycle of length 2. Such cycles are called transpositions.

Theorem 8

Every cycle can be written as a product of transpositions.

Proof: Take a cycle $(a_1, a_2, ..., a_n)$.

Now $(a_1\ a_2\ ...\ a_n) = (a_1\ a_n)\ O\ (a_1\ a_{n-1})\ ...\ O\ (a_1\ a_3)\ O\ (a_1\ a_2)$.

Therefore any cycle can be written as the product of transpositions.

Example 6.43

$(1\ 4\ 2\ 3\ 5) = (1\ 5)\ O\ (1\ 3)\ O\ (1\ 2)\ O\ (1\ 4)$.

Theorem **9**

Every permutations of a finite set with at least two elements can be expressed as a product of transpositions.

Definition

(i)   A permutation is said to be an odd permutation if it is the product of an odd number of transpositions (or 2–cycles).

(ii) A permutation is said to be an even permutation, if it is the product of an even number of transpositions (or 2–cycles).

Example 6.44

(i)   Consider the permutation (1 6) (2 5 3) = (1 6) (2 3) (2 5) = (1 6) (4 5) (2 3) (4 5) (2 5). There is no unique way to represent permutation as the product of transpositions.

   For instance, we can write the identity permutation as (1 2) (2 1), as (1 3) (2 4) (1 3) (2 4) and in many other ways.

(ii) No permutation can be written as the product of both an even number of transpositions and an odd number of transpositions.

   For instance, we could represent the permutations (1 6) by (2 3) (1 6) (2 3) or by (3 5) (1 6) (1 3) (1 6) (1 3) (3 5) (5 6) but (1 6) will always be the product of an odd number of transpositions.

Note:

(i)   The product of two even permutations is an even permutation.

(ii) The product of an even permutation and an odd one is odd (like wise for the product of an odd and even permutation).

(iii) The product of two odd permutations is an even permutation.

Theorem **10**

Let $A = \{a_1, a_2, ..., a_n\}$ be a finite set with n elements and $n > 2$. Then there are $\dfrac{n!}{2}$ odd permutations.

Proof: Let A_n be the set of all even permutations and B_n be the set of all odd permutations. Define f: $A_n \rightarrow B_n$ by

   $f(p) = q_0 \, o \, p$ for $p \in A_n$ and q_0 be a particular transposition.

   f is one-to-one:

For $p_1, p_2 \in A_n$,

$$\Rightarrow q_0 \text{ o } p_1 = q_0 \text{ o } p_2$$

$$\Rightarrow q_0 \text{ o } (q_0 \text{ o } p_1) = q_0 \text{ o } (q_0 \text{ o } p_2)$$

$$f(p_1) = f(p_2) \Rightarrow \Rightarrow (q_0 \text{ o } q_0) \text{ o } p_1 = (q_0 \text{ o } q_0) \text{ o } p_2$$

$$\Rightarrow I_A \text{ o } p_1 = I_A \text{ o } p_2 \text{ since } q_0 \text{ o } q_0 = I_A$$

$$\Rightarrow p_1 = p_2$$

Therefore f is one-to-one.

f is onto: Let $q \in B_n$. Then $q_0 \text{ o } q \in A_n$ and $f(q_0 \text{ o } q)$

$$= q_0 \text{ o } (q_0 \text{ o } q)$$

$$= (q_0 \text{ o } q_0) \text{ o } q$$

$$= I_A \text{ o } q$$

$$= q$$

Therefore f is onto.

Thus f is an one-to-one and onto function from a finite set A_n to a finite set B_n. Hence A_n and B_n have same number of elements.

We have $A_n \cap B_n = \phi$ and $A_n \cup B_n = n!$.

Thus $n! = |A_n \cup B_n| = |A_n| + |B_n| - |A_n \cap B_n| = 2|A_n| = 2|B_n|$.

Therefore $|A_n| = \dfrac{n!}{2} = |B_n|$

6.5   Recursive Functions

In this section, we discussed the concept of recursive function, which is a very elegant and powerful tool that can often be used to describe rather complex process in a very understandable way. We also give the computation of the greatest common divisor using the recursive function.

Recursive function is an important facility in many programming languages.

Recursion is the technique of defining a function, a set or an algorithm in terms of itself. That is, the definition will be in terms of previous values.

Definition

A function f: $N \to N$, where N is the set of non-negative integers is defined recursively if the value of f at 0 is given and for each positive integer n, the value of f at n is defined in terms of the values of f at k, where $0 \le k < n$.

This function f is also called as recursive function.

Note:

Observe that f defined (above) may not be a function. Hence, when a function is defined recursively it is necessary to verify that the function is well defined.

Example 6.45

The sequence 1, 4, 16, 64, ..., can be defined explicitly by the formula

$f(n) = 4^n$ for all integers $n \geq 0$.

The same function can also be defined recursively as follows:

$f(0) = 1, f(n + 1) = 4f(n)$, for $n > 0$

To prove that the function is well defined we have to prove existence and uniqueness of such function. In this case, existence is clear as $f(n) = 4^n$.

6.5.1   Theorem: (Recursion Theorem)

Let F be a given function from a set S into S. Let s_0 be fixed element of S. Then there exists a unique function f: $N \to N$ where N is the set of non-negative integers satisfying.

(i) $f(0) = s_0$

(ii) $f(n + 1) = F(f(n))$ for all integers $n \in N$.

(Here the condition (i) is called initial condition and (ii) is called the recurrence relation).

Example 6.46

Define n! recursively and compute 5! recursively.

Solution: We have f: $N \to N$. Then

(i) $f(0) = 1$

(ii) $f(n + 1) = (n + 1) f(n)$ for all $n \geq 0$.

Clearly $f(n) = n!$.

Now we compute 5! recursively as follows:

$$5! = 5.\ 4!$$
$$= 5.\ 4.\ 3!$$
$$= 5.\ 4.\ 3.\ 2!$$
$$= 5.\ 4.\ 3.\ 2.\ 1!$$
$$= 5.\ 4.\ 3.\ 2.\ 1.\ 0!$$
$$= 5.\ 4.\ 3.\ 2.\ 1.\ 1$$
$$= 120$$

Note:

Any sequence in arithmetic progression or geometric progression can be defined recursively.
Consider the sequence a, a + d, a + 2d, ... Then

$$A(0) = a, A(n + 1) = A(n) + d.$$

Consider another sequence a, ar, ar^2, ... Then

$$G(0) = a, G(n + 1) = r\,G(n).$$

Definition

The Fibonacci sequence can be defined recursively as

(i) $F_0 = 1 = F_1$

(ii) $F_{n+1} = F_n + F_{n-1}$ for n > 1

Then

$$F_2 = F_1 + F_0 = 2$$
$$F_3 = F_2 + F_1 = 3$$
$$F_4 = F_3 + F_2 = 5$$

Hence, there are two initial conditions.

Example 6.47

Define

$$f(x) = \begin{cases} \dfrac{x}{2} & \text{when x is even} \\[2mm] \dfrac{x-1}{2} & \text{when x is odd} \end{cases}$$

Solution: Define f: N → N such that f(0) = 0 and f(x + 1) = x – f(x).

Then f(6) = 5 – f(5) = 5 –[4 – f(4)]

$$= 5 - 4 + [3 - (3)]$$
$$= 5 - 4 + 3 - 2 + [1 - f(1)]$$
$$= 5 - 4 + 3 - 2 + 1 - [0 - f(0)]$$
$$= 3$$

and   f(5) = 4 – f(4)

$$= 4 - [3 - f(3)]$$
$$= 4 - 3 + 2 - [1 - f(1)]$$
$$= 4 - 3 + 2 - 1 + [0 - f(0)]$$
$$= 2$$

Example 6.48

Using recursion theorem, verify that the object defined by the recursive definition is a function. That is.,

   (i) $g(0) = 1$

   (ii) $g(n + 1) = 3[g(n)]^2 + 7$ for all $n > 0$

Solution: We obtain

   (i) $s_0 = 1$

   (ii) $f(k) = 3k^2 + 7$, where f: $N \to N$

        Then $g(0) = s_0$. And $g(n + 1) = f(g(n))$. Thus g is a well-defined function.

Definition

If m and n are two non-negative integers then the (greatest common divisor) g.c.d. (m, n) is defined as the largest positive integer d such that d divides both m and n. Euclidean algorithm computes the greatest common divisor (g.c.d) of two non-negative integers.

We can find g.c.d. (m, n) recursively as follows:

$$\text{g.c.d. } (m,n) = \begin{cases} \text{g.c.d. } (n, m) & \text{if } n > m \\ m & \text{if } n = 0 \\ \text{g.c.d. } (n, \text{mod } (m,n)) & \text{Otherwise} \end{cases}$$

where mod (m, n) is the remainder obtained when m is divided by n.

6.5.2 Observation

Consider the above definition.

   (i)   The first part interchanges the order of m and n if $n > m$.

   (ii)   Second part is the initial condition.

   (iii)   Third part is the recursive part mod (m, n) will become 0 in a finite number of steps.

Example 6.49

Calculate the g.c.d. (20, 6).

Solution: g.c.d. (20, 6) = g.c.d. (6, mod (20, 6)) (since $20 = 6.3 + 2$)

                     = g.c.d. (6, 2)

                     = (2, mod (6, 2))

                     = g.c.d. (2, 0)

                     = 2

Example 6.50

Calculate the g.c.d. (81, 36).

Solution: g.c.d. (81, 36) = g.c.d. (36, 9)

= g.c.d. (9, 0)

= 9

Example 6.51

Calculate the g.c.d. (22, 8)

Solution: g.c.d. (22, 8) = g.c.d. (8, mod (22, 8))

= g.c.d. (8, 6)

= g.c.d. (6, mod (8, 6))

= g.c.d. (6, 2)

= g.c.d. (2, 0)

= 2

Note:

The recursive definition can be extended to functions of more than one variable.

Consider the following example.

Example 6.52

Define $f(x, y) = x + y$ respectively.

Solution: Here, we keep x fixed and use recursion on y. We define

(i) $f(x, 0) = x$

(ii) $f(x, y + 1) = f(x, y) + 1$

Take x = 2, y = 3. Now $f(2, 3) = f(2, 2) + 1$

$= f(2, 1) + 1 + 1$

$= f(2, 0) + 1 + 1 + 1$

$= 2 + 1 + 1 + 1$

$= 5$

Example 6.53

Define $g(x, 0) = 0$, $g(x, y + 1) = g(x, y) + x$.

Take x = 3, y = 4.

then $g(3, 4) = g(3, 3) + 3$

$$= g(3, 2) + 3 + 3$$
$$= g(3, 1) + 3 + 3 + 3$$
$$= g(3, 0) + 3 + 3 + 3 + 3 = 12 \text{ (since } g(3, 0) = 0).$$

Exercises

Definitions and Examples of Functions

(i)   Show that the function f = {(1, 2), (2, 3), (3, 1)} defined on a set A = {1, 2, 3} is bijective function.

(ii)   Prove that of A = {1, 2, 3, 4}, B = {w, x, y, z} and f = {(1, w), (2, x), (3, y), (4, z)} is a bijective function.

(iii)   Let A and B be finite sets and f be a function from A to B. Then show that if f is one to one $\Rightarrow |A| \le |B|$.

Types of Functions

(i)   Let f and g are two functions from R to R, defined by f(x) = ax + b and $g(x) = 1 - x + x^2$. If (gof) (x) = $9x^2 - 9x + 3$. Determine a and b.

Ans: a = 3, b = – 1 and a = – 3, b = 2

(ii)   Set f: A → B and g: B → C be any two functions. Then show that if f and g are one to one so is gof.

(iii)   Set f, g, h are functions from Z to Z defined by

$$f(x) = x - 1, g(x) = 3x \text{ and}$$

$$h(x) = \begin{cases} 0 & \text{if x even} \\ 1 & \text{if x odd} \end{cases}$$

then find (fo(goh)) (x).

Ans:
$$h(x) = \begin{cases} -1 & \text{if x is even} \\ 2 & \text{if x is odd} \end{cases}$$

Composition of Functions and Inverse Functions

(i)   Set A = {1, 2, 3, 4} and B = {a, b, c, d}. Check whether the function f: A → B defined as

$$f = \{(1, a), (2, a), (3, c), (4, d),\} \text{ is invertable or not?}$$

Ans: No

(ii)   Find the inverse of the function f: A → B if

A = B = {1, 2, 3, 4, 5}, f = {(1, 3), (2, 2), (3, 4), (4, 5), (5, 1)}

Ans: f^{-1} = {(3, 1), (2, 2), (4, 3), (5, 4), (1, 5)}

(iii)   If a function f: A → B is invertable then show that it has a unique inverse.

Permutation Functions

(i)   Write all the permutations of the set

s = {1, 2, 3}.

(ii)   Show that every permutation can be uniquely expressed as a product of disjoint cycles.

(iii)   If s = {1, 2, 3, 4, 5, 6}. Compute (5, 6, 3) (4 1 3 5)

Ans: $\begin{pmatrix} 1 & 2 & 3 & 4 & 5 & 6 \\ 3 & 2 & 4 & 1 & 6 & 5 \end{pmatrix}$

(iv)   Show that the permutation

$\begin{pmatrix} 1 & 2 & 3 & 4 & 5 & 6 & 7 \\ 3 & 4 & 1 & 5 & 6 & 7 & 2 \end{pmatrix}$ is odd.

Recursive Functions

(i)   Obtain a recursive definition for the function f(n) = a_n for an = 5n

Ans: a_0 = 0 and $a_n = a_{n-1} + 5$ for n ≥ 1

(ii)   A function f(n) = a_n is defined recursively by a_0 = 4 and $a_n = a_{n-1} + n$ for n ≥ 1. Find f(n) in explicit form.

Ans: f(n) = a_n for n ≥ 0.

(iii)   For the function f(n) = a_n defined recursively by a_0 = 2, a_1 = 5, $a_n = 5a_{n-1} - 6a_{n-2}$ for n ≥ 2. Then prove that

$a_n = 2^n + 3^n$ for n ≥ 0.

CHAPTER - 7

Graph Theory - I

7.1   Basic Concepts of Graphs

Graph theory was born in 1736 with Euler's paper in which he solved the Kongsberg Bridges problem. In 1847, Kirchhoff (1824-87) developed the theory of trees to applications in electrical networks. Cayley discovered trees while he was trying to enumerate the isomers of (C_n H_{2n+2}). The last three decades have witnessed more interest in Graph Theory, particularly among applied mathematicians and engineers. Graph Theory has a surprising number of applications in many developing areas. The Graph Theory is also intimately related to many branches of mathematics including Group Theory, Matrix Theory, Automata and Combinatorics. One of the features of Graph Theory is that it depends very little on the other branches of mathematics. Graph Theory serves as a mathematical model for any system involving a binary relation. One of the attractive features of Graph Theory is its inherent pictorial character. The development of high-speed computers is also one of the reasons for the recent growth of interest in Graph Theory.

LEARNING OBJECTIVES

♦ *to know the fundamental concepts of Graph theory and related examples*

♦ *to understand the substructure namely subgraph*

♦ *to represent the graphs in different Matrix forms*

♦ *to identify the Isomorphic Graphs*

Definition

(i)   A linear graph (or simply a graph). G = (V, E) consists of a nonempty set of objects, V = {v_1, v_2, ...} called vertices and another set, E = {e_1, e_2, ...} of elements called edges such that each edge 'e_k' is identified with an unordered pair {v_i, v_j} of vertices. The vertices v_i, v_j associated with edge e_k are called the end vertices of e_k.

(ii)   An edge associated with a vertex pair {v_i, v_i} is called a loop (or) selfloop.

(iii) If there are more than one edge associated with a given pair of vertices, then these edges are called parallel edges (or) multiple edges.

Example 7.1

Consider the graph given here.

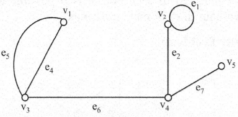

Fig. 7.1.1

This is a graph with five vertices and six edges. Here $G = (V, E)$ where

$$V = \{v_1, v_2, v_3, v_4, v_5\} \text{ and } E = \{e_1, e_2, e_4, e_5, e_6, e_7\}.$$

The identification of edges with the unordered pairs of vertices is given by

$$e_1 \leftrightarrow \{v_2, v_2\}, e_2 \leftrightarrow \{v_2, v_4\}, e_4 \leftrightarrow \{v_1, v_3\}, e_5 \leftrightarrow \{v_1, v_3\}, e_6 \leftrightarrow \{v_3, v_4\}.$$

Here 'e_1' is a loop and e_4, e_5 are parallel edges.

Definition

A graph that has neither self-loops nor parallel edges called a simple graph. Graph containing either parallel edges or loops is also referred as general graph. A graph 'G' with a finite number of vertices and a finite number of edges is called a finite graph. A graph 'G' that is not a finite graph is said to be an infinite graph.

Example 7.2

Consider the following three graphs

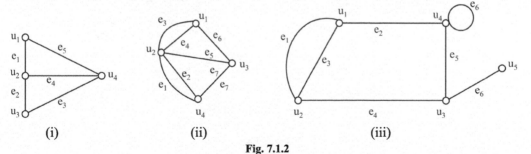

(i)   (ii)   (iii)

Fig. 7.1.2

It can be observed that the number of vertices, and the number of edges are finite. Hence these three graphs are finite graphs.

7.1.1   Example (Utilities Problem)

Suppose we have three houses and three utility outlets (electricity, gas and water) situated so that each utility outlet is connected to each house. Is it possible to connect each utility to each of the three houses without lines or main crossings?

Graphical Model of Utility Problem:

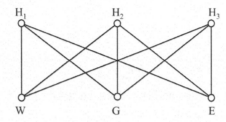

Fig. 7.1.3

We represent the above situation by a graph whose vertices corresponds to the houses and the utilities, and where an edge joins two vertices if and only if one vertex denotes a house and the other a utility.

Later we prove that the graph obtained here is non planar.

Example 7.3

Consider the two graphs given here. It can be understood that the number of vertices of these two graphs is not finite. So we conclude that these two figures represent infinite graphs.

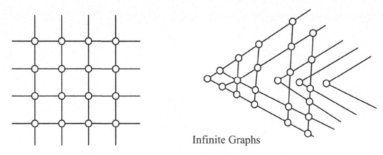

Infinite Graphs

Fig. 7.1.4

Definition

If a vertex v is an end vertex of some edge e, then v and e are said to be incident with (or on, or to) each other.

Example 7.4

Consider the graph given in the following. Here the edges e_2, e_6, e_7 are incident with the vertex u_4.

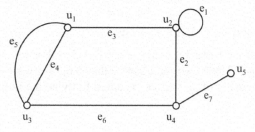

Fig. 7.1.5

Definition

    (i)   Two non-parallel edges are said to be adjacent if they are incident on a common vertex.

    (ii) Two vertices are said to be adjacent if they are the end vertices of the same edge.

Example 7.5

Consider the graph given in the example 7.4. Here the vertices u_4, u_5 are adjacent. The vertices u_1 and u_4 are not adjacent. The edges e_2 and e_3 are adjacent.

Definition

The number of edges incident on a vertex v is called the degree (or valency) of v. The degree of a vertex v is denoted by d(v). It is to be noted that a selfloop contributes two to the degree of the vertex.

Example 7.6

Consider the graph given in following diagram

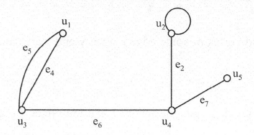

Fig. 7.1.6

Here $d(u_1) = 2$, $d(u_3) = d(u_4) = 3$; $d(u_2) = 3$; $d(u_5) = 1$

So, $d(u_1) + d(u_2) + d(u_3) + d(u_4) + d(u_5) = 2 + 3 + 3 + 3 + 1 = 12 = 2(6) = 2e$, where e denotes the number of edges. Hence we can observe that $d(u_1) + d(u_2) + d(u_3) + d(u_4) + d(u_5) = 2e$ (that is, the sum of the degrees of all vertices is equal to twice the number of edges).

Theorem 1

The sum of the degrees of the vertices of a graph G is twice the number of edges. That is, $\sum_{v_i \in V} d(v_i) = 2e$. (Here e is the number of edges).

Proof: (The proof is by induction on the number of edges 'e').

Case-I: Suppose $e = 1$. Suppose f is the edge in G with $f = uv$.

Then $d(v) = 1$, $\qquad d(u) = 1$

Therefore $\displaystyle\sum_{x \in V} d(x) = \sum_{x \in V|\{u,v\}} d(x) + d(u) + d(v) = 0 + 1 + 1 = 2 = 2 \times 1$

$$= 2 \times \text{(number of edges)}.$$

Hence the given statement is true for $n = 1$.

Now we can assume that the result is true for $e = k - 1$.

Take a graph G with k edges. Now consider an edge 'f' in G whose end points are u and v. Remove f from G. Then we get a new graph $G^* = G - \{f\}$.

Suppose $d^*(v)$ denotes the degree of vertices v in G^*. Now for any $x \notin \{u, v\}$, we have $d(x) = d^*(x)$, and $d^*(v) = d(v) - 1$, $d^*(u) = d(u) - 1$.

Now G^* has $k - 1$ edges. So by induction hypothesis $\displaystyle\sum_{v_i \in V} d*(v_i) = 2(k-1)$.

Now $\displaystyle 2(k-1) = \sum_{v_i \in V} d*(v_i) = \sum_{v_i \notin \{u,v\}} d*(v_i) + d*(u) + d*(v)$

$$= \sum_{v_i \notin \{u,v\}} d(v_i) + (d(u) - 1) + (d(v) - 1)$$

$$= \sum_{v_i \notin \{u,v\}} d(v_i) + d(u) + d(v) - 2 = \sum_{v_i \in V} d*(v_i) - 2$$

$$\Rightarrow 2(k-1) + 2 = \sum_{v_i \in V} d(v_i) \Rightarrow 2k = \sum_{v_i \in V} d(v_i)$$

Hence by induction we get that "the sum of the degrees of the vertices of the graph G is twice the numbers of edges".

Theorem 2

The number of vertices of odd degrees is always even.

Proof: We know that the sum of degrees of all the 'n' vertices (say, u_i, $1 \le i \le n$) of a graph G is twice the number of edges (e) of G.

So we have $\sum_{i=1}^{n} d(v_i) = 2e$ .....(7.1)

If we consider the vertices of odd degree and even degree separately, then

$$\sum_{i=1}^{n} d(v_i) = \sum_{v_j \text{ is even}} d(v_j) + \sum_{v_k \text{ is odd}} d(v_k)$$ .....(7.2)

Since the L.H.S of (7.2) is even (from (7.1)) and the first expression on the R.H.S is even, we have that the second expression on RHS is always even.

Therefore $\sum_{v_k \text{ is odd}} d(v_k)$ is an even number .....(7.3)

In (7.3), each $d(v_k)$ is odd. The number of terms in the sum must be even to make the sum an even number. Hence the number of vertices of odd degree is even.

Example 7.7

Show that the number of people who dance (at a dance where the dancing is done in couples) an odd number of times is even.

Solution: Suppose the people are vertices. If two people dance together, then we can consider it as an edge. Then the number of times a person v danced is $\delta(v)$. By Theorem 2, the number of vertices of odd degree is even. Therefore the number of people who dance odd number of times is even.

Definition

A vertex having no incident edge is called an isolated vertex. In other words, a vertex v is said to be an isolated vertex if the degree of v is equal to zero.

Example 7.8

Consider the graph given in following. The vertices v_4 and v_7 are isolated vertices.

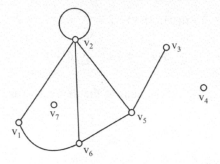

Fig. 7.1.7

Definition

A vertex of degree one is called a pendent vertex or an end vertex. Two adjacent edges are said to be in series if their common vertex is of degree two.

Example 7.9

In the Example 7.7 the vertex 'v_3' is of degree 1, and so it is a pendent vertex. The two edges incident on u_1 are in series.

Definition

The minimum of all the degrees of the vertices of a graph G is denoted by $\delta(G)$, and the maximum of all the degrees of the vertices of G is denoted by $\Delta(G)$. If $\delta(G) = \Delta(G) = k$, that is, if each vertex of G has degree k, then G is said to be k-regular or regular of degree k. 3-regular graphs are called cubic graphs.

Examples 7.10

(i)   Consider the graph G given in Figure below. It is easy to observe that the degree of every vertex is equal to 3. Hence the graph G is a regular graph of degree 3.

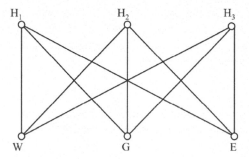

Fig. 7.1.8

(ii)   The following graph is a regular graph of degree-4.

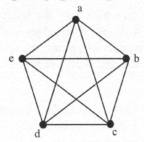

Fig. 7.1.9

Definition

If v_1, v_2, ..., v_n are the vertices of G, then the sequence $(d_1, d_2, ..., d_n)$ where d_i = degree (v_i), is the degree sequence of G. Usually, we order the vertices so that the degree sequence is monotone increasing, that is, so that $\delta(G) = d_1 \leq d_2 \leq ... \leq d_n = \Delta(G)$.

Example 7.11

The vertex c of the graph G in Example 7.10 is degree 5 while the degree of c in G^1 is 3. The degree sequence of G is (2, 2, 3, 5) while the degree sequence of G^1 is (2, 2, 3, 3).

Definition

A graph G = (V, E) is said to be a null graph if E = ϕ.

Example 7.12

The graph G given in the following; contains no edges and hence G is a null graph.

V_5 o   o V_6

V_1 o V_2 o   o V_3 o V_4

Definition

In a non-directed graph G a sequence P of zero more edges of the form $\{v_0, v_1\}$, $\{v_1, v_2\}$, ..., $\{v_{n-1}, v_n\}$, (in this repetition of vertex is allowed) is called a path from v_0 to v_n. The vertex v_0 is called the initial vertex and v_n is the terminal vertex, and they both are called endpoints of path P.

We denote this path P as a $v_0 - v_n$ path. If $v_0 = v_n$ then it is called a closed path, and

if $v_0 \neq v_n$ then it is called an open path.

Definition

A path P may have no edges at all, in which case, the length of P is zero, P is called a trivial path, and $V(P) = \{v_0\}$. A path P is simple if all edges and vertices on the path are distinct except possibly the endpoints. Two paths in a graph are said to be edge-disjoint if they share no common edges; they are vertex-disjoint if they share no common vertices.

Note:

An open simple path of length n has $n + 1$ distinct vertices and n distinct edges, while a closed simple path of length n has n distinct vertices and n distinct edges. The trivial path is taken to be a simple closed path of length zero.

Definition

A path of length ≥ 1 with no repeated edges and whose endpoints are equal is called a circuit. A circuit may have repeated vertices other than the endpoints; a cycle is a circuit with no other repeated vertices except its endpoints.

7.1.2   Observation

A simple path is certainly a path and the converse statement need not be true.

Theorem 3

In a graph G, every u-v path contains a simple u-v path.

Proof: If a path is a closed path, then it indeed contains the trivial path.

　　　　Assume that P is an open u-v path.

Use induction on the length of path: If P has length one, then P is itself a simple path.

Induction hypo: Suppose that all opens u-v paths of length k, where $1 \leq k \leq n$, contains a simple u-v path.

　　Now suppose that P is the open u-v path $\{v_0, v_1\}, ..., \{v_n, v_{n+1}\}$ where $u = v_0$ and $v = v_{n+1}$. It may be that P has repeated vertices, but if not, then P is a simple u-v path.

　　If there are repeated vertices in P, let i and j be distinct positive integers where $i < j$ and $v_i = v_j$. If the closed path v_i-v_j is removed from P, an open path P' is obtained having length $\leq n$ since at least the edge $\{v_i, v_{i+1}\}$ was deleted from P. Therefore by the inductive hypothesis, P^1 contains a simple u-v path. Therefore P contains simple u-v path.

7.1.3   Example: Konigsberg Bridges (or Seven Bridges)

This is one of the best known examples of graph theory. This problem was solved by Leonhard Euler (1707-1783) in 1736 by using the concept of Graph Theory. He is the originator of the Graph Theory.

Example 7.13

There were two islands 'C' and 'D' connected to each other and to the banks 'A' and 'B' with seven bridges as shown in the following diagram.

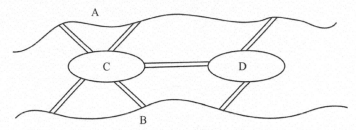

Fig. 7.1.10

The problem was to start at any of the four land areas of the city A, B, C, D walk over each of the seven bridges once and only once, and returns to the starting point.

Euler represented this situation by means of a graph (it is given in the following figure):

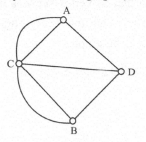

Fig. 7.1.11

Euler proved that a solution for this problem does not exist.

7.1.4   Properties of Konigsberg Bridges Graph

  (i)   The graph contains multiple edges. Hence the graph is not a simple graph.

  (ii)   All the vertices are of odd degree.

  (iii) It is a finite graph.

  (iv) There is no isolated vertex.

  (v)   There is no pendent vertex.

  (vi) It is not a null graph

Definition

A simple graph of order ≥ 2 in which there is an edge between every pair of vertices is called a complete graph or a full graph.

A complete graph with $n \geq 2$ vertices is denoted by K_n.

Example 7.14

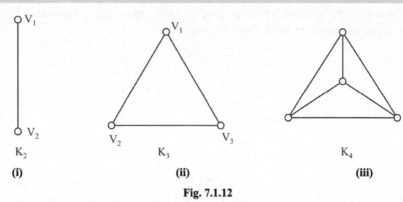

Fig. 7.1.12

Definition

Suppose a simple graph G is such that its vertex set V is the union of two mutually disjoint non-empty sets V_1 and V_2 which are such that every edge in G joins a vertex in V_1 and a vertex in V_2. Then G is called a bipartite graph.

If E is the edge set of this graph, the graph is denoted by $G = (V_1, V_2, E)$. The sets V_1 and V_2 are called bipartites of V.

Example 7.15

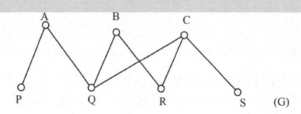

Fig. 7.1.13

Here $V_1 = \{A, B, C\}$, $V_2 = \{P, Q, R, S\}$

Definition

A bipartite graph $G = (V_1, V_2, E)$ is called a complete bipartite graph if there is an edge between every vertex in V_1 and every vertex in V_2.

Example 7.16

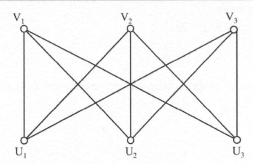

Fig. 7.1.14

Give an example of a bipartite graph

[JNTUH, Nov 2010, Set No. 1]
[JNTUH, Nov 2008, Set No. 4]

Solution: A graph G is said to be a bipartite graph, if its vertex set V can be partitioned into disjoint sub sets V_1 and V_2 such that every edge in G has one end vertex in V_1 and another vertex is in V_2.

Example 7.17

Consider the graph G = (V, E) where V = $\{v_1, v_2, v_3, v_4, v_5, v_6\}$

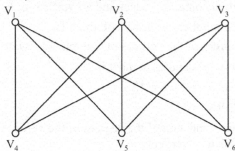

Fig. 7.1.15

graph G is a bipartite with $V_1 = \{ v_1, v_2, v_3\}$ and $V_2 = \{ v_4, v_5, v_6\}$ as partition of V such that each edge of G has one end vertex in V_1 and another in V_2. The graph given here is related to utilities example 7.1.1.

How many edges does a graph have if it has vertices of degree 4, 3, 3, 2, 2? Draw such a graph.

[JNTUH, June 2010, Set–2]

Solution: Let G be a graph with vertices degree 4, 3, 3, 2, 2

Since sum of the degrees of the vertices is twice to its number of edges.

That is $d(\vartheta) = 2e$, where 'e' denotes number of edges.

By the hypothesis, we have

$$4 + 3 + 3 + 2 + 2 = 2e$$

$$\Rightarrow \quad 14 = 2e$$

$$\Rightarrow \quad e = 7$$

So the number of edges of graph G with its vertices degree 4, 3, 3, 2, 2 is 7.

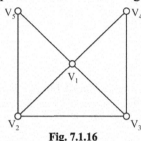

Fig. 7.1.16

In this graph $d(v_1) = 4$, $d(v_2) = 3$, $d(v_3) = 3$, $d(v_4) = 2$, $d(v_5) = 2$

Example 7.18

How many vertices will the graph contains 6 edges and all vertices of degree 3.

[JNTUH, June 2010, Set No. 2]

Solution: Given that graph G contains 6 edges and all vertices of G are of degree 3.

Now we have to find the number of vertices of graph G.

Since the sum of the degrees of the vertices of a graph G is twice to its number of edges.

That is $\Sigma d(\vartheta) = 2e$, where 'e' denotes number of

$$\Rightarrow \Sigma d(\vartheta) = 2 \times 6 = 12$$

$$\Rightarrow \Sigma d(\vartheta) = 12$$

Since each vertex degree 3 and sum of the degrees of the vertices is 12 gives that $(3 + 3 + 3 + 3 = 12)$ graph G contains four vertices.

Example 7.19

If G is non-directed graph with 12 edges. Suppose that G has 6 vertices of degree 3 and the rest have degree less than 3. Determine the minimum number of vertices.

[JNTUH, Nov 2008, Set No. 2]

Solution: Let G be a non-directed graph with 12 edges.

Suppose that G has 6 vertices of degree 3.

Since the sum of the degrees of the vertices is twice to its number of edges.

That is $d(\vartheta) = 2e$

By the hypothesis $d(\vartheta) = 2 \times 12 = 24$

[Moreover there are 6 vertices of degree 3]

$\Rightarrow$ $\Sigma d(u) + \Sigma d(w) = 24$ where $\Sigma d(u)$ denotes the sum of the degrees of vertices of degree 3.

$\Rightarrow$ $18 + \Sigma d(w) = 24$

$\Rightarrow$ $\Sigma d(w) = 24 - 18 = 6$

Now the sum of the degrees of the vertices of degree less than 3 is equals to 6.

If degree of each of these vertices is equals to 2 then the sum is equals to 6.

Hence the minimum number of vertices is $6 + 3 = 9$.

Example 7.20

How many vertices will be in a graph containing 16 edges and all vertices of degree 2.

[JNTUH, June 2008, Set No. 4]

Solution: Let G be a graph with 16 edges and each vertex of degree 2.

We have to find number of vertices of graph G. Since the sum of the degrees of the vertices of a graph G is twice to its number of edges.

That is $\qquad \displaystyle\sum_{v \in |v|} d(v) = d(\vartheta)$

$\Rightarrow \qquad \displaystyle\sum_{v \in |v|} d(v) = d(\vartheta) = 2 \times 16 = 32 \Rightarrow \sum_{v \in |v|} 2 \Rightarrow 2|v| = 32 \Rightarrow |v| = 16$

So the graph G contains 16 vertices with each vertex of degree 2.

7.2   Subgraphs

Definition

A graph H is said to be a sub-graph of a graph G if all the vertices and all the edges of H are in G, and each edge of H has the same end vertices in H as in G.

Example 7.21

The graphs H and K are subgraphs of graph G.

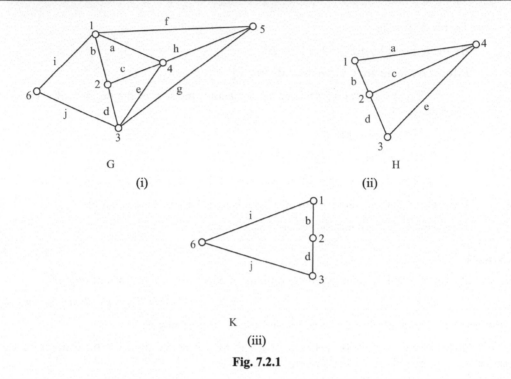

Fig. 7.2.1

7.2.1   Observations

Here we can observe the following facts:

   (i)   Every graph is a sub graph of itself;

  (ii)   A sub graph of a sub graph of G is a sub graph of G;

 (iii)   A single vertex in a graph G is a sub graph of G; and

 (iv)   A single edge in G together with its end vertices is a subgroup of G.

Definition

Two sub graphs G_1 and G_2 of a graph G are said to be edge-disjoint if G_1 and G_2 do not have any edges in common. The subgraphs that do not have vertices in common are said to be vertex disjoint.

Example 7.22

Observe the two graphs given in Figures A and B. These two graphs are subgraphs of the graph given in the Figure C. There are no common edges in these two subgraphs. Hence these two subgraphs are edge disjoint subgraphs.

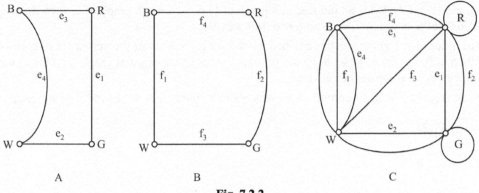

A   B   C

Fig. 7.2.2

Definition

If H is a sub-graph of G, then the complement of H in G, denoted by $\overline{H}(G)$, is the sub graph G-E(H); that is, the edges of H are deleted from those of G.

Example 7.23

Consider the graph G, subgraph H of G, and the complement of H in G.

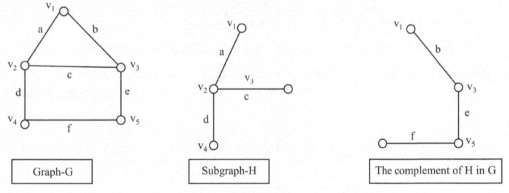

| Graph-G | Subgraph-H | The complement of H in G |

Fig. 7.2.3

Let $G = (V_1, E)$ be an undirected graph, with $G_1 = (V_1, E_1)$ a sub graph of G. Under what conditions is G, not an induced sub graph of G?

[JNTUH, June 2008, Set No. 3]

Solution: A graph $G_1 = (V_1, E_1)$ is a sub graph of $G = (V, E)$ provided that $V_1 \subseteq V$ and $E_1 \subseteq E$ and for each $e \in E$, both end points of e are in V_1.

Graph $G_1 = (V_1, E_1)$ is an induced sub graph of G if G_1 is a sub graph of G such that E_1 contains all the edges of G with both end points in V_1.

An induced sub graph can be obtained by choosing a sub set of the vertex set of graph G and then defining the edge set to be the set of all edges of the original graph with both ends are in the choosen sub set of vertices.

If an edge in E with end points in V_1 was not in E_1 then G_1 is not an induced sub graph.

Example 7.24

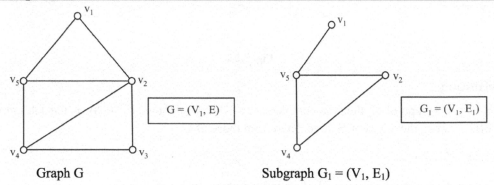

Graph G   Subgraph $G_1 = (V_1, E_1)$

Fig. 7.2.4

where $V_1 = \{v_1, v_2, v_4, v_5\}$ & $E_1 = \{\overline{v_1v_5}, \overline{v_4v_5}, \overline{v_2v_5}, \overline{v_2v_4}\}$.

Since $\overline{v_1v_2}$ is in E but not in E_1. Hence G_1 is not an induced subgraph of G.

For the graph shown in the below, find a subgraph that is not an induced graph.

[JNTUH, June 2008, Set No. 3]

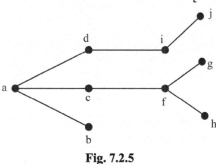

Fig. 7.2.5

Solution: We have to find a sub graph of the given graph which is not an induced sub graph.

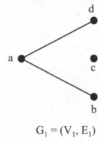

$$G_1 = (V_1, E_1)$$

Fig. 7.2.6

where $V_1 = \{a, b, c, d\}$ & $E_1 = \{\overline{ab}, \overline{ad}\}$.

Here the edge $\overline{ac}$ is not in G_1. Hence G_1 is not an induced sub graph.

7.3   Matrix Representation of Graphs

A diagrammatic representation of a graph has limited usefulness. Further more such a representation is only possible when the number of nodes and edges in reasonably small.

In this section, we shall present an alternative method of representing graphs using matrices. Such a method of representation has several advantages.

Given a simple diagraph $G = (V, E)$. It is necessary to assume such kind of ordering of the nodes of the graph in the sense that a particular node is called as first node, another second node, and so on. Our matrix representation of G depends upon the ordering of the nodes.

Definition

Let $G = (V, E)$ be a simple digraph in which $V = \{V_1, V_2, ..., V_n\}$ and the nodes are assumed to be ordered from V_1 to V_n. An $n \times n$ matrix A whose elements a_{ij} are given by

$$a_{ij} = \begin{cases} 1 & \text{if}(v_i, v_j) \in E \\ 0 & \text{otherwise} \end{cases}$$

is called the adjacency matrix of the graph G.

Example 7.25

Find the incidence matrix representation of the following graphs.

(i)

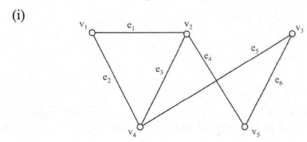

Fig. 7.3.1

Ans:

$$
\begin{array}{c}
\begin{array}{cccccc} e_1 & e_2 & e_3 & e_4 & e_5 & e_6 \end{array} \\
\begin{array}{c} v_1 \\ v_2 \\ v_3 \\ v_4 \\ v_5 \end{array}
\begin{bmatrix}
1 & 1 & 0 & 0 & 0 & 0 \\
1 & 0 & 1 & 1 & 0 & 0 \\
0 & 0 & 0 & 1 & 1 & 1 \\
0 & 1 & 1 & 0 & 1 & 0 \\
0 & 0 & 0 & 1 & 1 & 0
\end{bmatrix}
\end{array}
$$

(ii)

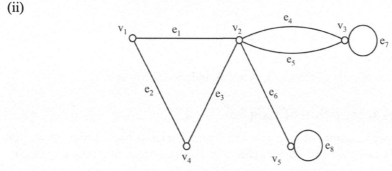

Fig. 7.3.2

Ans:

$$
\begin{array}{c}
\begin{array}{cccccccc} e_1 & e_2 & e_3 & e_4 & e_5 & e_6 & e_7 & e_8 \end{array} \\
\begin{array}{c} v_1 \\ v_2 \\ v_3 \\ v_4 \\ v_5 \end{array}
\begin{bmatrix}
1 & 1 & 0 & 0 & 0 & 0 & 0 & 0 \\
1 & 0 & 1 & 1 & 1 & 1 & 0 & 0 \\
0 & 0 & 0 & 1 & 1 & 0 & 1 & 0 \\
0 & 1 & 1 & 0 & 0 & 0 & 0 & 0 \\
0 & 0 & 0 & 0 & 0 & 1 & 0 & 1
\end{bmatrix}
\end{array}
$$

Theorem 4

Let A be the adjacency matrix of a digraph G. The element in the ith row and j^{th} column of A^n where n is a non negative integer, is equal to the number of paths of length 'n' from the ith node to the jth node.

Definition

Let G = (V, E) be a simple digraph in which $|v| = n$ and the nodes of G are assumed to be ordered. An $n \times n$ matrix p whose elements are given by

$$
P = \begin{cases} 1 & \text{if there exist a path from } v_i \text{ to } v_j \\ 0 & \text{otherwise} \end{cases}
$$

is called the path matrix of the graph G.

7.3.1   Incidence Matrices

Let G be an undirected graph with vertices $v_1, v_2, \ldots, v_n$ and $e_1, e_2, \ldots, e_n$ are the edges. Then the incidence matrix with respect to this ordering of V and E is the $n \times m$ matrix $M = [b_{ij}]$

where

$$[b_{ij}] = \begin{cases} 1 & \text{when edge } e_j \text{ is incident with } v_i \\ 0 & \text{otherwise} \end{cases}$$

Example 7.26

Find the incidence matrix of the following graph:

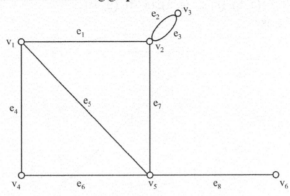

Fig. 7.3.3

Solution: The incidence matrix representation of the given graph is:

$$\begin{array}{c} \\ v_1 \\ v_2 \\ v_3 \\ v_4 \\ v_5 \\ v_6 \end{array} \begin{array}{cccccccc} e_1 & e_2 & e_3 & e_4 & e_5 & e_6 & e_7 & e_8 \\ \begin{bmatrix} 1 & 0 & 0 & 1 & 1 & 0 & 0 & 0 \\ 1 & 1 & 1 & 0 & 0 & 0 & 1 & 0 \\ 0 & 1 & 1 & 0 & 0 & 0 & 0 & 0 \\ 0 & 0 & 0 & 1 & 0 & 1 & 0 & 0 \\ 0 & 0 & 0 & 0 & 1 & 1 & 1 & 1 \\ 0 & 0 & 0 & 0 & 0 & 0 & 0 & 1 \end{bmatrix} \end{array}$$

Example 7.27

Let $A = \{a, b, c, d\}$ and $R = \{(a, b), (b, c), (c, a), (a, d), (d, a)\}$. Find digraphs for R^K for $0 \leq k \leq 6$.

[JNTUH, June 2010, Set No. 2]

Solution: Given set $A = \{a, b, c, d\}$ and relation in A is

$R = \{(a, b), (b, c), (c, a), (a, d), (d, a)\}$.

$RoR = \{(a, d), (b, a), (c, b), (c, d), (a, a)\} = R^2$

$RoRoR = \{(a, a), (b, b), (b, d), (c, c), (c, a), (a, b), (a, d)\} = R^3$

$RoRoRoR = \{(a, b), (a, d), (b, c), (b, a), (c, a), (c, b), (c, d). (a, c), (a,a)\} = R^4$

$RoRoRoRoR = \{(a, c), (a, a), (b, a), (b, b), (b, d), (c, b), (c, d), (c, c), (c, a), (a, b), (a, d)\}$
$\qquad\qquad = R^5$

RoRoRoRoRoR = {(a, a), (a, b), (a, d), (b, b), (b, d), (b, c), (b, a), (c, c), (c,a), (c, b), (c,d), (a, c)}

$$= R^6$$

Now the digraphs of the relation R^K, $0 \le k \le 6$ are as follows:

The digraph of R^k when k = 1

Then relation R = {(a, b), (b, c), (c, a), (a, d), (d, a)}.

The digraph of R is

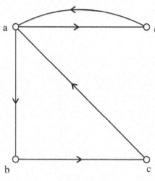

Fig. 7.3.4

The digraph of R^k, when k = 2

The relation $R^2 = \{(a, d), (b, a), (c, d), (a, a)\}$

The digraph of R^2 is

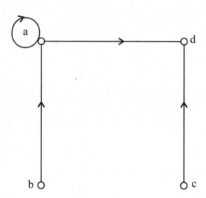

Fig. 7.3.5

The digraph of R^3 (k = 3)

The relation $R^3 = \{(a, a), (b, b), (b, d), (c, c), (c, a), (a, b), (a, d)\}$.

The digraph of R^3 is

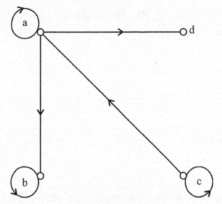

Fig. 7.3.6

The digraph of R^4 (k = 4)

The relation R^4 = {(a, b), (a, d), (b, c), (b, a), (c, a), (c, b), (c, d), (a, c), (a, a)}

The digraph of R^4 is

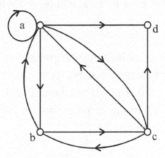

Fig. 7.3.7

The digraph of R^5 (k = 5)

The relation R^5 = {(a, c), (a, a), (a, b), (a, d), (b, a), (b, b), (b, d), (c, b), (c, d), (c, c), (c, a)}

The digraph is

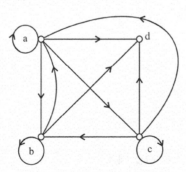

Fig. 7.3.8

The digraph of $R^6 = (k = 6)$

The relation $R^6 = \{(a, a), (a, b), (a, d), (b, b), (b, d), (b, c),$ $(b, a), (c, c), (c, a), (c, b), (c, d), (a, c)\}$.

The digraph of R^6 is

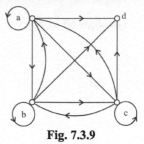

Fig. 7.3.9

7.4   Isomorphic Graphs

Definition

Two graphs G and G^1 are said to be isomorphic to each other if there is an one-to-one correspondence between their vertices and an one-to-one correspondence between their edges such that the incident relationship must be preserved.

That is, two graphs $G = (V, E)$ & $G^1 = (V^1.E^1)$ are said to be isomorphic if there exist one-one and onto functions f: $V \rightarrow V^1$ and g: $E \rightarrow E^1$ such that $g(v_iv_j) = f(v_i)f(v_j)$ for any edge v_iv_j in G.

Example 7.28

Consider the two graphs given in following diagrams A and B. Observe that these are isomorphic. The correspondence between these two graphs is as follows. $f(a_i) = v_i$ for $1 \leq i \leq 5$ and $g(i) = e_i$ for $1 \leq i \leq 6$. Except the labelling of their vertices and edges of the isomorphic graphs, they are same, perhaps may be drawn differently.

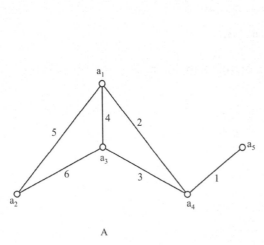

A

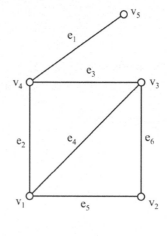

B

Fig. 7.4.1

Example 7.29

Observe that the three graphs given in figures A, B and C are isomorphic.

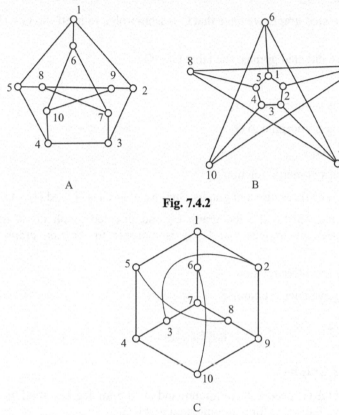

Fig. 7.4.2

Fig. 7.4.3

Note:

If there is an isomorphism between two graphs G and H, then G and H must have:

(i)   The same number of vertices,

(ii) The same number of edges,

(iii) An equal number of vertices of a given degree, and

(iv) The incident relationship must be preserved.

Example 7.30

Prove that isomorphism is an equivalence relation on diagraphs.

[JNTUH, June 2010, Set No. 4]

Solution: Let G_1, G_2 be any two diagraphs

Define $G_1 \sim G_2$ if and only if G_1 is isomorphic to G_2. We have to show '$\sim$' is an equivalence relation.

Since for any directed graph, we have that G is isomorphic to itself. So $G \sim G$. Hence $\sim$ is reflexive relation.

Let G_1, G_2 be two directed graphs, such that $G_1 \sim G_2$.

Now $G_1 \sim G_2$

$\Rightarrow$ G_1 is isomorphic to G_2

$\Rightarrow$ G_2 is isomorphic to G_1

$\Rightarrow$ $G_2 \sim G_1$

Therefore '$\sim$' is a symmetric relation.

Let G_1, G_2 and G_3 be three directed graphs. Suppose that $G_1 \sim G_2$ and $G_2 \sim G_3$.

Then G_1 is isomorphic to directed graph G_2 and directed graph G_2 is isomorphic to directed graph G_3. This implies that G_1 is isomorphic to directed graph G_3. That is $G_1 \sim G_3$.

Therefore, $\sim$ is a transitive relation.

Hence $\sim$ is an equivalence relation.

Exercises

Basic Concepts of Graphs

1.   Let $\delta(G)$ and $\Delta(G)$ denote the minimum and maximum degrees of all vertices of G respectively. Show that for a non-directed graph G,

$$\delta(G) \leq 2, \frac{|E|}{|V|} \leq \Delta(G)$$

2.   How many vertices will the following graphs have if they contain:

(i)   16 edges and all vertices of degree 2

(ii)   21 edges, 3 vertices of degree 4, and the other vertices of degree 3.

Ans: (i) 16 (ii) 13

3.   For any simple graph G, prove that the number of edges of G is less than or equal to $\frac{1}{2n(n-1)}$, where n is the number of vertices of G.

Sub Graphs

1. For the following graph, find two edge–disjoint subgraph and two vertex disjoint subgraphs.

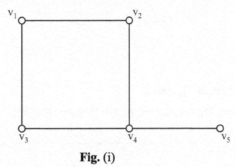

Fig. (i)

Ans:

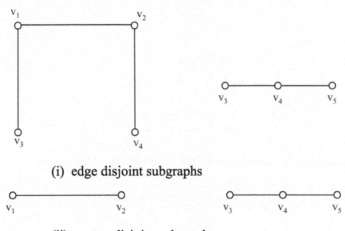

(i) edge disjoint subgraphs

(ii) vertex disjoint subgraphs

2. Can a finite graph be isomorphic to one of its subgraphs other than itself?

Ans: No

3. Let G be the graph shown below. Verify whether $G_1 = (V_1, E_1)$ is a subgraph of G in the following case.

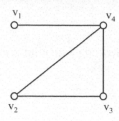

$$V_1 = \{v_1, v_2, v_4\}$$

$$E_1 = \{(v_1, v_4), (v_1, v_3)\}$$

Matrix Representation of Graphs

1. Write the matrix representation of a following graph:

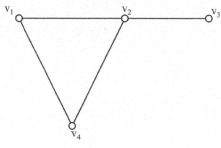

Ans:

$$
\begin{array}{c}
\quad\ \ v_1\ v_2\ v_3\ v_4 \\
\begin{array}{c}
v_1 \\ v_2 \\ v_3 \\ v_4
\end{array}
\begin{bmatrix}
0 & 3 & 2 & 0 \\
3 & 0 & 1 & 1 \\
2 & 1 & 0 & 2 \\
0 & 1 & 2 & 1
\end{bmatrix}
\end{array}
$$

2. Write the matrix representation of the following directed graph:

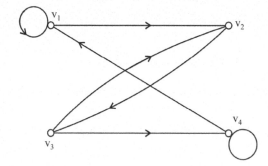

$$\begin{array}{c} \quad v_1\ v_2\ v_3\ v_4 \\ \begin{array}{c} v_1 \\ v_2 \\ v_3 \\ v_4 \end{array} \begin{bmatrix} 1 & 1 & 0 & 0 \\ 0 & 0 & 1 & 0 \\ 0 & 1 & 0 & 1 \\ 1 & 0 & 0 & 1 \end{bmatrix} \end{array}$$

Ans:

3.   Draw a graph with the following adjacency matrix:

(i) $\begin{bmatrix} 0 & 0 & 1 & 1 \\ 0 & 0 & 1 & 0 \\ 1 & 1 & 0 & 1 \\ 1 & 0 & 1 & 0 \end{bmatrix}$ (ii) $\begin{bmatrix} 0 & 1 & 0 \\ 1 & 0 & 1 \\ 0 & 1 & 0 \end{bmatrix}$

Ans: (i)   (ii)

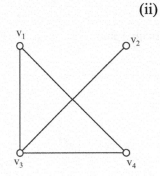   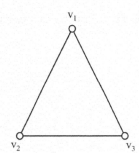

4.   Find the adjacency matrix of the following graph.

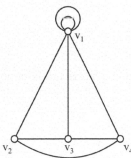

$$\begin{array}{c} \quad v_1\ v_2\ v_3\ v_4 \\ \begin{array}{c} v_1 \\ v_2 \\ v_3 \\ v_4 \end{array} \begin{bmatrix} 2 & 1 & 1 & 1 \\ 1 & 0 & 1 & 1 \\ 1 & 1 & 0 & 1 \\ 1 & 1 & 1 & 0 \end{bmatrix} \end{array}$$

Ans:

Isomorphic Graphs

1. Show that the following two graphs are Isomorphic

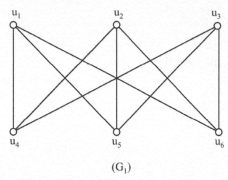 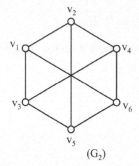

(G_1) (G_2)

2. Verify that the following two graphs are Isomorphic

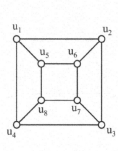 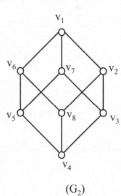

(G_1) (G_2)

Ans: Two graphs are Isomorphic.

3. Show that the following graphs are not Isomorphic

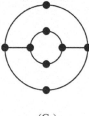 

(G_1) (G_2)

CHAPTER - 8

Graph Theory - II

8.1   Paths and Circuits

(i)   A directed path in a diagraph $A = (V, E)$ is a sequence of zero or more edges $e_1, e_2, \ldots, e_n$ in E such that for each $2 \leq i \leq n$, e_i is to the vertex, that is of the form (v_{i-1}, v_i) for each $1 \leq i \leq n$. Such a path is said to be from v_0 to v_n, and its length is 'n'.

In this case, v_0 and v_n are called the end points of the path.

(ii) A non directed path in G is a sequence of zero or more edges $e_1, e_2, \ldots, e_n$ in E for which there is a sequence of vertices $v_0, v_1, \ldots, v_n$ such that $e_i = (v_{i-1}, v_i)$ (or) $e_i = (v_i, v_{i-1})$ for each $1 \leq i \leq n$. A path is simple if all edges and vertices on the path are distinct, except that v_0 and v_n may be equal.

(iii) A path of length $n \geq 1$ with n_0 repeated edges and whose end points are equal is a circuit.

A simple circuit is called a cycle.

(iv) A path of length zero is known as a non-trivial path.

(v)   A path $e_1, e_2, \ldots, e_n$ is said to traverse a vertex 'v' if one (or more) of the e_i's is to v (or from 'v'), and 'v' is not serving as one of the end points of the path or more precisely, if $e_i = (v, w)$, then $2 \leq i \leq n$ or if $e_i = (w, v)$, then $1 \leq i \leq n-1$.

LEARNING OBJECTIVES

♦ to know different types of Paths

♦ to identify different types of graphs (Eularian Graph, Hamiltonian Graph, etc.)

Example 8.1

The following diagram includes some of each kind of path and circuits.

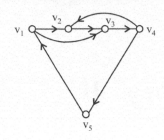

Fig. 8.1.1

Solution:

(i)   There are two simple directed paths from v_1 to v_4.

They are:

     a) $(v_1, v_2), (v_2, v_3), (v_3, v_4)$

     b) $(v_1, v_3), (v_3, v_4)$

(ii)   There are several more simple non directed paths from v_1 to v_4 including (v_1, v_2), (v_2, v_4).

(iii) There are number of nontrivial directed cycles, including $(v_1, v_2), (v_2, v_3), (v_3, v_4)$, $(v_4, v_5), (v_5, v_1)$ and a larger number of non-directed cycles, including all the directed cycles as well as cycles such as $(v_1, v_2), (v_2, v_3), (v_3, v_1)$.

Theorem 1

If $G = (V, E)$ is a diagraph, then for $n \geq 1$, $(u, v) \in E^n$ if and only if there is a directed path of length n from u to v in G.

     (E^n = E, E, E, ... E, the composition of n copies of E).

Proof: Let $G = (V, E)$ be a diagraph.

Now we shall prove the theorem by mathematical induction on 'n'.

Case I: For n = 1

Then $E^n = E$

By the definition of path that $(u, v) \in E$ if there is a path of length 1, from u to v. Since a path of length 1 is an edge of G.

Case II: For n > 1.

     Assume that the theorem is true for n–1. Now we shall prove the theorem into two parts.

Part I: Let $(v_0, v_1), (v_1, v_2), \ldots , (v_{n-1}, v_n)$ be a directed path from v_0 to v_n.

Then $(v_0, v_{n-1}) \in E^{n-1}$, (by induction).

By the definition $E^n = E^{n-1}$, E and $(v_{n-1}, v_n) \in E$ so that $(v_0, v_n) \in E^n$.

Part II: Suppose $(v_0, v_n) \in E^n$.

Since $E^n = E^{n-1}.E$, there exists some v_{n-1} such that $(v_0, v_{n-1}) \in E^{n-1}$ and $(v_{n-1}, v_n) \in E$. By the inductive hypothesis, there is a directed path $(v_0, v_1), (v_1, v_2), ..., (v_{n-2}, v_{n-1})$ of length n–1, from v_0 to v_{n-1}.

Adding (v_{n-1}, v_n) to this path gives the path of length n.

The proof is complete.

8.1.1   Corollary

If $G = (V, E)$ is a digraph then for any two vertices u and v in V, $(u, v) \in E$ if and only if there is a non-trivial directed path from u to v in G.

Proof: This is a consequence of the previous theorem.

If there is a directed path from u to v in G of some length $n \geq 1$, then $(u, v) \in E^n$, so that $(u, v) \in E^+$.

Conversely, if $(u, v) \in E^+$, then $(u, v) \in E^n$ for some $n \geq 1$, and so the above theorem shows that there is a directed path of length 'n' from u to v.

Definition

(i) A pair of vertices in a directed graph are said to be weakly connected if there is a non-directed path between them.

(ii) A pair of vertices in a directed graph are unilaterally connected if there is a directed path between them.

(iii) A pair of vertices in a directed graph are said to be strongly connected if there is a directed path from u to v and from v to u.

(iv) A graph is said to be connected if every pair of vertices in the graph is connected.

(v) A sub graph G^1 of a graph G is a connected component if it is a maximal connected sub graph. (That is, there is no connected subgraph of G that properly contains G^1).

(vi) A graph which is not connected is known as a disconnected graph.

Example 8.2

Give an example of a connected graph G where removing any edge G results a disconnected graph.   [JNTUH, Nov 2008, Set No. 4]

Solution: Let G be a tree.

Then G is a connected graph without circuits.

Now by removing any edge from G results in a disconnected subgraph.

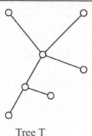

Hence every tree T is a connected graph and by removing any edge in T results a disconnected graph.

Tree T

The graph T given here is a tree. It is clear that if we remove any edge then we get a disconnected graph.

Fig. 8.1.2

8.2   Eularian Graphs

Definition

(i)   An Euler path in a multigraph is a path that includes each edge of the multigraph exactly once and intersects each vertex of the multigraph at least once.

(ii) A multigraph is said to be traversable if it has an Euler path.

(iii) An Euler circuit is an Euler path whose end points are identical. [That is, if an Euler path is a sequence of edges e_1, e_2, ..., e_k corresponding to the sequence of pairs of vertices (v_1, v_2), (v_2, v_3), ..., (v_{k-1}, v_k), then the edges are all distinct and $v_1 = v_k$].

(iv) A multigraph is said to be an Eularian multigraph if it has an Euler circuit.

Theorem 2

A non-directed multigraph has an Euler path if and only if it is connected and has 'O' or exactly 2 vertices of add degree. In the latter case, the two vertices of odd degree are the end points of every Euler path in the multigraph.

Proof: (only if part):

Let G be a multigraph G have an Euler path.

It is clear that G must be connected. Moreover, every time the Euler path meets a vertex it traverses two edges which are incident on the vertex and which have not been traced before.

Except for the two end points of the path, the degree of all other vertices must be even.

If the end points are distinct, then their degrees are odd.

If the two end points co-incide, their degrees are even and then the path becomes an Euler circuit.

(If part)

Let us construct an Euler path by starting at one of the vertices of odd degree and traversing each edge of G exactly once.

If there are no vertices of odd degree we will start at an arbitrary vertex.

For every vertex of even degree, the path will enter the vertex and leave the vertex by tracing an edge that was not traced before.

Thus the construction will terminate at a vertex with an odd degree, or return to the vertex where it started.

This tracing will produce an Euler path if all edges in G are traced exactly once this way.

If not all edges in G are traced, we will remove those edges that have been traced and obtain the sub graph say G^1 induced by the remaining edges.

The degrees of all vertices in this subgraph must be even and at least one vertex must intersect with the path, since G is connected.

Starting from one of these vertices, now we can construct a new path, which in this case will be a cycle, since all degrees of vertices are now even.

This path will be joined into the previous path.

This will be repeated until a path that traverses all edges in G is obtained.

From the above theorem, we get the following corollaries:

8.2.1 Corollary

(i) A non-directed multigraph has an Euler circuit if and only if it is connected and all of its vertices are of even degree.

(ii) A directed multigraph G has an Euler path if and only if it is unilaterally connected and the in-degree of each vertex is equal to its out-degree, with the possible exception of two vertices, for which it may be that the in-degree of one larger than its out-degree and the in-degree of the other is one less than its out degree.

(iii) A directed multigraph G has an Euler circuit if and only if G is unilaterally connected and the in-degree of every vertex in G is equal to its out-degree.

8.3 Hamiltonian Graphs

Definition

A graph G is said to be Hamiltonian if there exists a cycle containing every vertex of G.

Such a cycle is called as a Hamiltonian cycle.

Note:

A Hamiltonian graph is a graph containing a Hamiltonian cycle.

Definition

A Hamiltonian path is a simple path that contains all vertices of G but where the end points may be distinct.

Note:

A Hamiltonian cycle always provides a Hamiltonian path by deletion of any edge.

Example 8.3

Show that the following graph is a Hamiltonian graph:

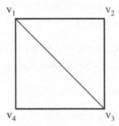

Fig. 8.3.1

Solution: Clearly the cycle shown in thick lines is a Hamiltonian cycle, which does not contain the edge $v_1 v_3$. So the graph is Hamiltonian graph.

8.3.1   Some Basic Rules for Constructing Hamiltonian Paths and Cycles

Rule 1: If G has 'n' vertices, then a Hamiltonian path must contain exactly n–1 edges, and a Hamiltonian cycle must contain exactly 'n' edges.

Rule 2: If a vertex 'v' in G has degree k, then a Hamiltonian path must contain at least one edge incident on 'v' and at most two edges incident on 'v'. A Hamiltonian cycle will, of course, contain exactly two edges incident on 'v'. In particular both edges incident on a vertex of degree two will be contained in every Hamiltonian cycle. sum: there cannot be three or more edges incident with one vertex in a Hamiltonian cycle.

Rule 3: No cycle that does not contain all the vertices of G can be formed when building a Hamiltonian path or cycle.

Rule 4: Once the Hamiltonian cycle we are building has passed through a vertex 'v', then all other unused edges incident on 'v', can be deleted because only two edges incident on 'v' can be included in a Hamiltonian cycle.

Example 8.4

The path through the vertices of G shown in the diagram given here in the order of appearance in the English alphabet forms a Hamiltonian path. However G has no

Hamiltonian cycle since if so, any Hamiltonian cycle must contain the edges $\{v_1, v_2\}$, $\{v_1, v_5\}$, $\{v_3, v_4\}$, $\{v_4, v_5\}$, $\{v_6, v_7\}$ and $\{v_5, v_7\}$. But there would be three edges of the cycle incident on the vertex v_5.

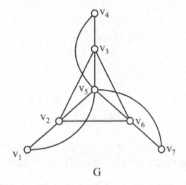

Fig. 8.3.2

Theorem (Grinberg) 3

Let G be a simple plane graph with 'n' vertices. Suppose that C is a Hamiltonian cycle in G, then with respect to the cycle C,

$$\sum_{i=3}^{n}(i-2)\,(r_i-r_i') = 0$$

where

r_i = the number of regions of G in the interior of C whose boundary contains exactly 'i' edges.

r_i' = the number of regions of degree 'i' in the exterior of C.

Theorem (Dirac's Theorem) 4

A simple graph with 'n' vertices ($n \geq 3$) in which each vertex has degree at least $\dfrac{n}{2}$ has a Hamiltonian cycle.

8.3.2   Corollary

If G is a complete simple graph on n–vertices ($n \geq 3$), then G has a Hamiltonian cycle.

8.3.3   Example (City-Route Puzzle)

Hamilton made a regular dodecahedron of wood (please see the graph given below) whose 20 corners were marked with the names of cities, and the routes are the edges of the graph.

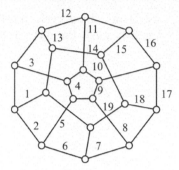

Fig. 8.3.3

Now the problem is to start from a city and find a route along the edges of the dodecahedron, that passes through every city exactly once and return to the city of origin.

The graph of dodecahedron is given above. This can be represented by the graph given below:

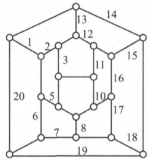

Fig. 8.3.4

Here the closed walk (of edges)

1 2 3 4 5 6 7 8 9 (10) (11) (12) (13) (14) (15) (16) (17) (18) (19) (20) is a Hamiltonian circuit.

Theorem 5

Let G be a complete graph with 'n' vertices, where 'n' is an odd number greater than or equal to 3. Then there are $\dfrac{n-1}{2}$ edge-disjoint Hamiltonian circuits.

8.3.4   Example (The seating arrangement problem)

Consider the seating arrangement problem. First we construct the graph of this problem.

(i) Represent a member 'x' by a vertex, and the possibility of sitting next to another member 'y' by an edge between x and y.

(ii) Every member is allowed to sit next to any other member. So G is a complete graph of nine vertices. (Here 'nine' is the number of people to be seated around the table). Clearly every seating arrangement around the table is a Hamiltonian circuit.

(iii) On the first day of their meeting, they can sit in any order, and it will be a Hamiltonian circuit (H_1, say).

On the second day, they are to sit such that every member must have different neighbors. So we have to find another Hamiltonian circuit (H_2, say) in G, with an entirely different set of edges from those in H_1 (that is, H_1 and H_2 are edge-disjoint Hamiltonian circuits).

(iv) By above theorem, we know that the number of edge-disjoint Hamiltonian circuits in G is $\dfrac{n-1}{2} = \dfrac{9-1}{2} = 4$.

Therefore, we can conclude that there exist 'four' such arrangements among 'nine' people.

The figures A, B, C and D, were obtained by following the procedure mentioned in the proof of the above theorem.

Observe that figure E is same as fig A.

Therefore all the four distinct Hamiltonian circuits (that is, different seating arrangements) were shown in the following figures A, B, C and D.

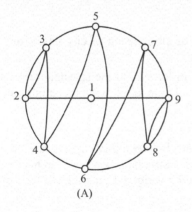

(A)

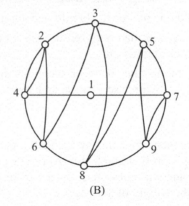

(B)

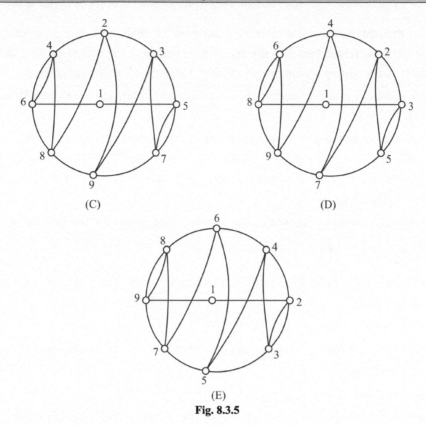

(C)

(D)

(E)

Fig. 8.3.5

8.3.5 Travelling – Salesman Problem

This problem is related to Hamiltonian circuits.

Problem: A salesman required to visit a number of cities (each of city has a road to every other city) during his trip.

Given the distance between the cities. In what order should the salesman travels so as to visit every city precisely once and return to his home city, with the minimum mileage travelled?

Solution:

(i) Represent the cities by varieties, and the roads between them by edges. Then we get a graph. In this graph, for every edge 'e' there corresponds a real number 'w' (e) (the distance in miles, say). Such a graph is called a weighted graph. Here 'w' (e) is called as the weight of the edge 'e'.

(ii) If each of the cities has a road to every other city, we have a complete weighted graph. This graph has numerous Hamiltonian circuits, and we are to select the Hamiltonian circuit that has the smallest sum of distances (or weights).

(iii) The number of different Hamiltonian circuits (may not be edge–disjoint) in a complete graph of 'n' vertices is equal to $\dfrac{(n-1)!}{2}$.

[Reason: Start from a vertex.

To go from the first vertex to second vertex, we can choose any one of the (n – 1) edges. To go from the second vertex to third vertex, we can choose any one of the (n – 2) edges. To go from the third vertex to fourth vertex, we can choose any one of the (n – 3) edges, and so on.

Since these selections are independent, and each Hamiltonian circuit has been counted twice, we have that the number of Hamiltonian circuits is $\dfrac{(n-1)(n-2)...2.1}{2} = \dfrac{(n-1)!}{2}$]

(iv) First we list all the $\dfrac{(n-1)!}{2}$ Hamiltonian circuits that are possible in the given graph.

Next calculate the distance travelled on each of these Hamiltonian circuits. Then select the Hamiltonian circuit with the least distance.

This provides a solution for the travelling salesman problem.

Example 8.5

Write the difference between Hamiltonian graphs and Euler graphs.

[JNTUH, June 2010, Set No. 4]

Solution:

   (i) Euler graph contains Euler line.

  (ii) Hamiltonian graph contains Hamiltonian path.

 (iii) Euler line should contain all the edges once and only once. There is no restriction on the edges of a Hamiltonian path.

 (iv) Hamiltonian path must contain all the vertices once and only once. There is no restriction on the vertices of an Euler line.

  (v) In Euler graph every vertex must be of every degree. This condition may not be satisfied by Hamiltonian graphs. For example the graph given in Figure is a Hamiltonian graph but not an Euler graph.

 (vi) Every complete graph is Hamiltonian, but it may not be an Euler graph. For example, K5 is a Hamiltonian graph, but it is not an Euler graph.

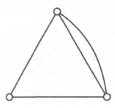

Fig. 8.3.6

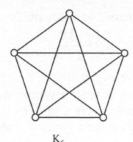

K_5

Fig. 8.3.7

Example 8.6

Write the rules for constructing Hamiltonian paths and cycles.

[JNTUH, Nov 2008, Set No. 2]

Solution:

(i) A graph with vertex degree one cannot have a Hamiltonian circuit, since in a Hamiltonian circuit each vertex is incident with two edges in the circuit.

(ii) If G is a simple graph with n vertices and $n \geq 3$ such that the degree of every vertex in G is at least $n/2$, then G has a Hamiltonian circuit.

(iii) If G is a simple graph with n vertices $n \geq 3$. Such that deg (u) + deg (v) $\geq$ n for every pair of non-adjacent vertices u and v in G, then G has a Hamiltonian circuit.

8.4 Multiple Graphs

Definition

A multigraph is a graph G = (V, E) in which E may contain multiple edges. That is, E may contain more than one loop.

Note:

Graphs (or multigraphs) are represented as follows: Each vertex 'v' in V is represented by a dot (or) a small circle and each edge 'e', e = (v_1, v_2) is represented by a curve connecting its end points v_1 and v_2.

Definition

(i) A multigraph G = (V, E) is finite if both V and E are finite.

If V is finite then obviously E is also finite. So it is sufficient to say that 'V' is finite.

(ii) A multigraph is said to be traversable if it can be drawn without any break in the curve and without repeating any edges.

Such type of path must be a trail as no edge is used twice. It is called a traversable trail.

A traversable multigraph must be finite and connected.

Example 8.7

The multigraph is traversable; its traversable trail is shown below:

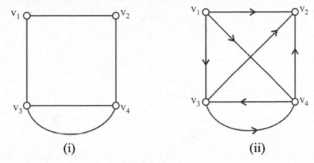

(i)   (ii)

Fig. 8.4.1

Definition

A directed multigraph $G = (V, E)$ consists of a set 'V' of vertices, a set E of edges, and a function g from E to $\{(u, v) : u, v \in V\}$. The edges e_1 and e_2 are multiple edges if

$$g(e_1) = g(e_2)$$

Example 8.8

The following graph is a directed multigraph.

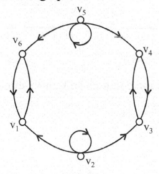

Fig. 8.4.2

Exercises

Paths and Circuits

(i) Determine the total number of different paths of length 2 in the following graph:

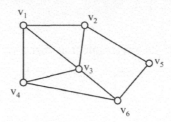

Ans: 19

(ii) Show that the following graph is not a circuit.

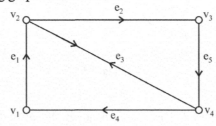

(iii) Show that the following graph is a circuit.

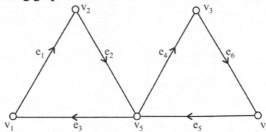

Eularian Graphs

(i) Show that the following graph contains an Euler's circuit.

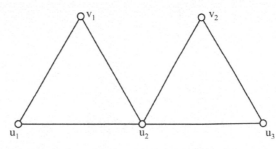

(ii) Is there a graph with even number of vertices and odd number of edges that contains an Euler circuit?

Ans: Yes

(iii) Show that a connected graph with exactly two vertices of odd degree has an Euler trail.

Hamiltonian Graphs

(i) Prove that, if G is a bipartite graph with an odd number of vertices, then G is non-Hamiltonian.

(ii) Show that the following graph is Hamiltonian.

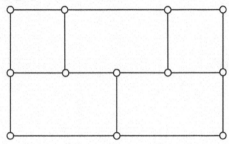

(iii) Identify five different Hamiltonian cycles in the following graph.

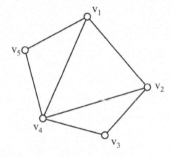

Multiple Graphs

(i) Give an example for a multiple graph.

Ans:

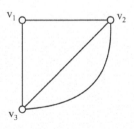

(ii) Which of the following graph is a multiple graph?

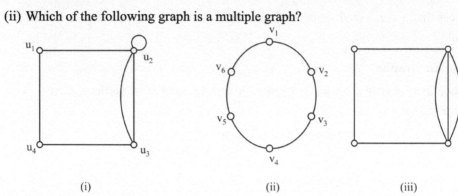

(i)   (ii)   (iii)

Ans: (iii) is multiple graph.

CHAPTER - 9

Graph Theory- III

9.1 Planar Graphs

In this lesson, we consider the embedding (drawing without crossings) of graphs on surfaces, especially on the plane. We study the planar graphs, which has great significance from a theoretical point of view, that is., whether it is possible to draw a graph G in a plane without its edges crossing over. Planarity and other related concepts are useful in many practical situations. In designing printed circuits, it is desirable to have as few lines cross as possible.

We also observed the non-planar graphs, called Kuratoski's graphs. We provided a useful characterization of Kuratoski's graphs. That is., a graph is non-planar if it contains a sub graph which is isomorphic to either of the graphs K_5 or K_3. We gave an illustration to certain results in this lesson for better understanding.

Definition

(i) A graph G is said to be a planar graph if there exists some geometric representation of G which can be drawn on a plane such that no two of its edges intersect.

(ii) A graph that cannot be drawn on a plane without a cross over between its edges is called a non-planar graph.

(iii) A drawing of a geometric representation of a graph on any surface such that no edges intersect is called an embedding.

We can define the planar graph as follows:

A graph G is said to be planar if it can be drawn on a plane without any crossovers; otherwise G is said to be non-planar.

Example (Utilities Problem) 9.1

Suppose we have three houses and three utility outlets (electricity, gas and water) situated so that each utility outlet is connected to each house. Is it possible to connect each utility to each of the three houses without lines or mains crossing?

Graphical model of utility problem: We represent the above situation by a graph whose vertices correspond to the houses and the utilities, and where an edge joins two vertices if and only if one vertex denotes a house and the other a utility. Later we prove that the graph obtained is non planar. (Refer the graph given below).

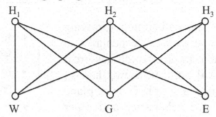

Fig. 9.1.1

9.1.1 Observation

1. If we have drawn a cycle in the plane, then any edge not on the cycle must be either inside the cycle, outside the cycle, or the edge must cross over one of the edges of the cycle.

2. The roles of being inside or outside the cycle are symmetric that is, the graph can be redrawn so that edges and vertices formerly outside the cycle are now inside the cycle and vice versa.

   The following fig. indicates various possible configurations for the edges {a, e} and {c, g} relative to the cycle: a-b-c-d-e-f-g-h-a.

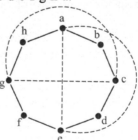

Fig. 9.1.2

Note:

A plane graph G divides the plane into number of regions [also called windows, faces or meshes] as shown in following graph:

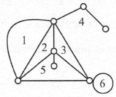

Fig. 9.1.3

A region characterized by the set of edges or the set of vertices forming its boundary. In the graph, the numbers 1, 2, 3, 4, 5, 6 stand for the regions.

Definition

The portion of the plane lying outside a graph embedded in plane is called an infinite [or unbounded or outer or exterior] region for that particular plane representation.

Example 9.2

The graph (b) is the planar representation of the graph (a).

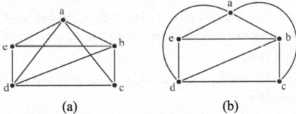

(a)   (b)

Fig. 9.1.4

Definition

Given a plane graph G, we can define another multi-graph G* as follows: Corresponding to each region r of G there is a vertex r* of G, and corresponding to each edge e of G there is an edge e* of G*; two vertices r* and s* are joined by the edge e in G if and only if corresponding regions r and s are separated by the edge e in G. In particular, a loop is added at a vertex r* of G* for each cut-edge of G that belongs to the boundary of the region r. The multi-graph G* is called the dual of G.

Example 9.3

Consider the graph given in Fig.9.1.5(a). If we draw the edge f outside the quadrilateral while the other edges are unchanged, then we get the Fig. 9.1.5(b).

The graph given in Fig. 9.1.5(b) can be embedded in a plane.

So the graph given in Fig. 9.1.5(b) is planar. Therefore the graph given in Fig. 9.1.5(a) is a planar graph.

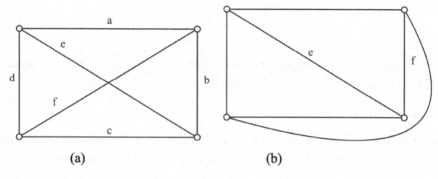

(a)   (b)

Fig. 9.1.5

Example (Tracing of the dual) 9.4

Place r* in the corresponding region r of G. (think of r* as the capital of the region r). Then draw each e* in such a way that it crosses the corresponding edge e of G exactly once (and crosses no other edge of 0).

The following figure illustrates, where the dual edges are indicated by dashed lines and the dual vertices by asterisks.

Observe that if e is a loop of G, then e* is a cut edge of G* and conversely.

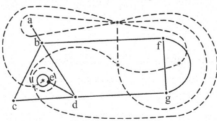

Fig. 9.1.6

Note: Observe the graph K_5 (Kuratowski's first graph) given here. This is a complete graph on 5 vertices.

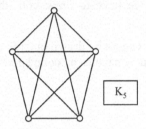

Fig. 9.1.7

Theorem **1**

The complete graph of 5 vertices (denoted by K_5) is a non-planar graph.

Proof: Suppose the five vertices of the computer graph are v_1, v_2, v_3, v_4, v_5.

Since the graph is complete, we get a circuit going from

v_1 to v_2 to v_3 to v_4 to v_5 to v_1

This is, we have a pentagon (given in Graph-a).

Now this pentagon must divide the plane of the paper into two regions, one inside and the other outside.

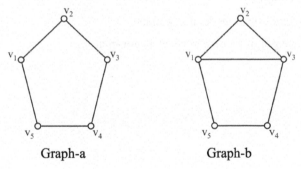

Graph-a   Graph-b

Fig. 9.1.8

We have to connect v_1 and v_3 by an edge.

This edge may be drawn inside (or) outside the pentagon (without intersecting the 5 edges of Graph-a).

Let us select a line from v_1 to v_3 inside the pentagon. (if we choose outside, then we end up with a similar argument) (Observe the Graph-b).

Now we have to draw an edge from v_2 to v_4, and another one from v_2 to v_5.

Since neither of these edges can be drawn inside the pentagon without crossing over the edges that have already drawn, we have to draw both these edges outside the pentagon (observe the Graph-c).

The edge connecting v_3 and v_5 cannot be drawn outside the pentagon without crossing the edge between v_2 and v_4. So v_3 and v_5 have to be connected with an edge inside the pentagon (observe the Graph-d).

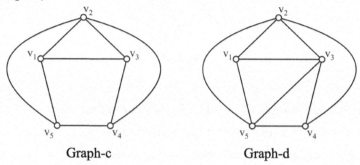

Graph-c   Graph-d

Fig. 9.1.9

Now we have to draw an edge between v_1 and v_4.

It is clear that this edge cannot be drawn either inside (or) outside the pentagon without a cross-over (Observe Graph-e).

Thus this graph cannot be embedded in a plane.

Hence the complete graph K_5 on 5 vertices is non-planar.

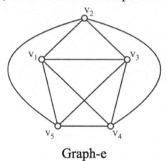

Graph-e

Fig. 9.1.10

Theorem 2

$K_{3,3}$ is a non-planar graph

Proof: Observe the graph $K_{3,3}$

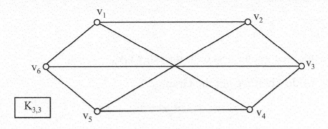

Fig. 9.1.11

The vertex set $V = \{v_i \mid 1 \le i \le 6\}$, and there are edges

$$\overline{v_1 v_2}, \ \overline{v_2 v_3}, \ \overline{v_3 v_4}, \ \overline{v_4 v_5}, \ \overline{v_5 v_6}, \ \overline{v_6 v_1}$$

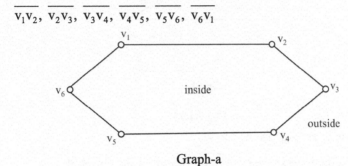

Graph-a

Fig. 9.1.12

So plane of the paper is divided into two regions, one inside and the other outside.

Since v_1 is connected to v_4, we can add the edge $\overline{v_1 v_4}$ in either inside or outside (without intersecting the edges already drawn). Let us draw $\overline{v_1 v_4}$ inside.

(If we choose outside, then we end up with the same argument. Now we have the Graph-b).

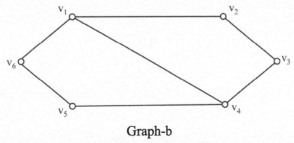

Graph-b

Fig. 9.1.13

Next we have to draw an edge $\overline{v_2 v_5}$ and also another edge $\overline{v_3 v_6}$.

First we draw $\overline{v_2 v_5}$.

If we draw it inside, we get a cross over the edge $\overline{v_1 v_4}$.

So we draw it outside, then we get the Graph-c.

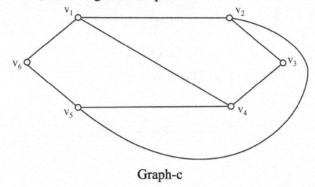

Graph-c

Fig. 9.1.14

Still we have to draw an edge from v_3 to v_6.

If $\overline{v_3 v_6}$ drawn inside, it cross the edge $\overline{v_1 v_4}$ (observer the Graph-d).

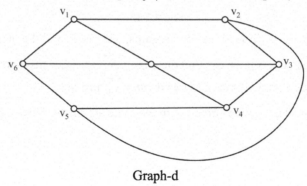

Graph-d

Fig. 9.1.15

So we cannot draw it inside. We select the case of drawing $\overline{v_3 v_6}$ cross the edge $\overline{v_2 v_5}$ (observe the Graph-e).

Thus $\overline{v_3 v_6}$ cannot be drawn either inside or outside without a cross over.

Hence the given graph is not a planar graph.

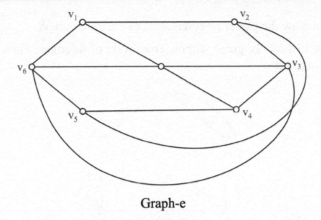

Graph-e

Fig. 9.1.16

Example 9.5

Draw a planar representation of the following graph.

[JNTUH, Nov 2010, Set No. 2]

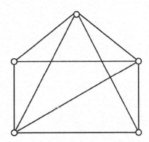

Fig. 9.1.17

Solution: Consider the given graph G and let us label the vertices as follows:

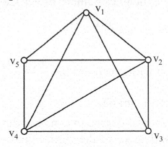

Fig. 9.1.18

Now we have to draw the planar representation of the given graph.

That is, we have to draw the graph without crossovers of its edges. Then the given graph will get the following form.

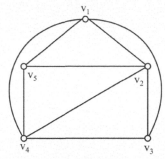

Fig. 9.1.19

This is a planar representation of the given graph G.

Example 9.6

Determine the number of edges in

   (i) K_n

   (ii) $K_{m,n}$

   (iii) C_n

   (iv) P_n

[JNTUH, June 2010, Set No. 3]

Solution:

   (i) Since K_n is a complete graph with 'n' vertices, and each vertex has degree 'n – 1'. Since the sum of the degrees of the vertices is twice to its number of edges.

   (that is, $\Sigma d(v) = 2e$ where 'e' denotes number of vertices).

   This gives us $n(n-1) = 2e \Rightarrow e = \dfrac{n(n-1)}{2}$

   Therefore a complete graph with n vertices has $\dfrac{n(n-1)}{2}$ edges.

   (ii) $K_{m,n}$ is a complete and bipartite graph.

   In $K_{m,n}$ the number of vertices of degree m is n, and the number of vertices of degree n is m. Since the sum of the degrees of the vertices is twice to its number of edges in a graph.

We have m.n + n.m = 2e where 'e' denotes number of edges in graph G. So 2m.n = 2e and hence e = m.n

Therefore, the graph $K_{m,n}$ contains 'mn' edges.

(iii) C_n denotes a cycle (circuit) with 'n' vertices. Since in any cycle each vertex is of degree is 2, and so in C_n, each vertex degree is 2. Now the sum of the degrees of the vertices is twice to its number of edges.

$\Rightarrow \Sigma d(v) = 2e$ where 'e' denotes number of edges in G.

$\Rightarrow 2n = 2e$

$\Rightarrow e = n$

In C_n, there are n edges.

(iv) P_n denotes a path with 'n' vertices. Since any path with 'n' vertices contains n − 1 edges, we have that P_n contains 'n − 1' edges.

9.2   Euler's Formula

If G is a connected planar graph, then any drawing of G in the plane as a plane graph will always form |R| = |E| − |V| + 2 regions, including the exterior region, where |R|, |E|, and |V| denote the number of regions, the number of edges, and the number of vertices of G.

This remarkable formula is known as Euler's formula, discovered by Euler in 1752.

Theorem (Euler's Formula) 3

If G is a connected plane graph, then |V| − |E| + |R| = 2.

Proof:

Case I: We shall prove this theorem first by observing the result for a tree. By convention, a tree determines only one region. We know already that the number of edges of a tree is one less than the number of vertices.

Thus, for a tree the formula |V| − |E| + |R| = 2 holds.

Moreover, we note that a connected plane graph G with only one region must be a tree since otherwise there would be a circuit implies an internal region and an external region.

Case II: We shall prove the general result by induction on the number 'n' of regions determined by G.

By case I the result is true for n = 1. Assume that the result is true for n ≥ 1. Then suppose that G is connected plane graph that determines n + 1 regions.

Delete an edge common to the boundary of two separate regions.

The resulting graph G' (say) has the same number of vertices, one fewer edge, but also one fewer region since two previous regions have been consolidated by the removal of the edge.

So if $|E'|$, $|V'|$, $|R'|$ are the number of edges, vertices and regions respectively for G', $|E| = |E| - 1$, $|R'| = |R| - 1$, $|V'| = |V|$. But $|V| - |E| + |R| = |V'| - |E'| + |R'|$.

By the induction hypothesis, $|V'| - |E'| + |R'| = 2$.

Therefore $|V| - |E| + |R| = 2$.

The proof is complete.

Definition

A connected plane graph is said to be polyhydral if degree (n) $\geq$ 3 for each region n $\in$ R (G), in addition degree (v) $\geq$ for each vertex v $\in$ V (G).

9.2.1 Corollary

In a connected plane graph G, with $|E| > 1$ then

  (i) $|E| \leq 3 |V| - 6$

  (ii) There is a vertex 'v' of G such that degree (v) $\leq$ 5.

Proof: Let G be a connected plane graph. By Euler's formula, we have

$|R| + |V| = |E| + 2$, and since G is simple, $3 |R| \leq 2 |E|$ (or) $|R| \leq \frac{2}{3} |E|$.

Hence $\frac{2}{3} |E| + |V| \geq |R| + |V| = |E| + 2$

Thus, $|V| - 2 \geq \frac{1}{3} |E|$ (or) $3 |V| - 6 \geq |E|$.

$\Rightarrow |E| \leq 3 |V| - 6$

Theorem 4

A complete graph K_n is planar if and only if $n \leq 4$.

Proof: It is easy to see that K_n is planar for $n = 1, 2, 3, 4$.

Now it is enough to show that K_n is non-planar if $n \geq 5$, and for this it suffices to show that K_5 is non-planar. For proof see theorem.

Theorem 5

A complete bipartite graph $K_{m,n}$ is planar if and only if $m \leq 2$ or $n \leq 2$.

Proof: It is easy to see that $K_{m,n}$ is planar if $m \leq 2$ or $n \leq 2$.

Now let $m \geq 3$, and $n \geq 3$.

To prove that $K_{m,n}$ is non-planar, it is enough to prove that $K_{3,3}$ is non-planar. For proof see theorem.

Example 9.7

Prove that there does not exist a polyhedral graph with exactly seven edges.

Solution: Suppose G be a polyhedral graph with 7 edges.

i.e., $|E| = 7$

then $3 |R| \leq 2 |E| - 14$

(Since each region R has degree ≥ 3, so that $3 |V| \leq 2 |E| = 14$. Thus, $|R| \leq 4$ and $|V| \leq 4$.

By Euler's formula,

$|R| + |V| = |E| + 2 = 9$ and then $8 \geq |R| + |V| = |E| + 2 = 9$, a contradiction.

So a polyhedral graph with 7 edges does not exist.

9.3   Graph Colouring, Covering and Chromatic Numbers

Definition

(i)   A coloring of a simple graph is the assignment of the given colors to each vertex of the graph so that no two adjacent vertices are assigned the same color.

(ii)   An n-coloring of a graph G is a coloring of G using n colors. If G has an n-coloring, then G is said to be n-colorable.

Example 9.8

The following graph G shows 4-coloring as well as 3-coloring.

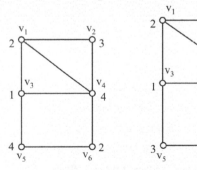

Fig. 9.3.1

Definition

The chromatic number of a graph is the least number of colors needed for a coloring of the graph G. This is denoted by $\chi(G)$

If $\chi(G) = k$, we say that G is k-chromatic.

Example 9.9

What is the chromatic number of the following graph?

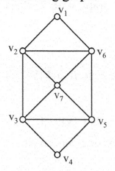

Fig. 9.3.2

Solution: Since the vertices v_1, v_2, and v_6 must be assigned different colors, the chromatic number of the given graph must be at least 3. Now we have to check whether 3 colors will be sufficient to color the graph.

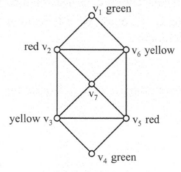

Fig. 9.3.3

Suppose we assign green to v_1, red to v_2 and yellow to v_6.

Then v_7 can be colored green as it is adjacent to v_2 and v_6.

Further v_3 can be colored yellow, as it is adjacent to v_2 and v_7.

Similarly v_5 can be colored red and v_4 can be colored green.

This completes the coloring of the given graph with exactly 3 colors.

9.3.1 The Scheduling Problem

Suppose that the state legislature has a list of 21 standing committees. Each committee is supposed to meet one hour each week. What is wanted is a weekly schedule of committee meeting times that uses as few different hours in the week as possible so as to maximize the time available for other legislative activities.

The one constraint is that no legislator should be scheduled to be in two different committee meetings at the same time.

The question is: What is the minimum number of hours needed for such a schedule?

Theorem 6

The minimum number of hours for the schedule of committee meeting in our scheduling problem is $\chi(G_o)$.

[Here G_o is committee graph that has a vertex for each committee and has an edge between vertices corresponding to committees with a common member].

Definition

A graph G is a k-critical if $\chi(G) = K$ and $\chi(G–V) < \chi(G)$ each vertex V of G.

Note: A K-chromatic graph has a K-critical subgraph.

Theorem 7

Let G be a K-critical graph, then

  (i) G is connected

 (ii) The degree of each vertex of G is atleast $K – 1$, that is, $\delta(G) \geq K – 1$.

(iii) G cannot be expressed in the form $G_1 \cup G_2$, where G_1 and G_2 are graphs which intersect in a complete graph. In particular G contains no cut vertices.

9.3.2 Rules for Graph Coloring

1. $\chi(G) \leq |V|$, where $|V|$ is the number of vertices in G.
2. A triangle always requires three colors, so $\chi(K_3) = 3$. In general, if K_n is the complete graph on n-vertices, then $\chi(K_n) = n$.
3. If any subgraph of G requires K colors, then $\chi(G) \geq K$.
4. If degree (v) = d, then atmost d colors are required to color the vertices that are adjacent to v.
5. $\chi(G) =$ maximum $\{ \chi(C)|C$ is a connected component of G$\}$.

6. Every K-chromatic graph has atleast K-vertices v such that the degree (v) ≥ K − 1.

7. Let G be a graph, then $\chi(G) \leq 1 + \Delta(G)$, where $\Delta(G)$ is the largest degree of any vertex of G.

8. When we are building a k-coloring of a graph G, we may delete all vertices of degree less than K (along with their incident edges). In general when attempting to build a K-coloring of a graph, it is desirable to start by K-coloring a complete subgraph of K-vertices and then successively finding vertices adjacent to K − 1 different colors, thereby forcing the color choice of such vertices.

9. The following are equivalent:

   (i) A graph G is 2-colorable.

   (ii) G is bipartite.

   (iii) Every cycle of G has even length.

10. If $\delta(G)$ is the minimum degree of any vertex of G, then $\chi(G) \geq \dfrac{|V|}{|V| - \delta(G)}$ where $|V|$ is the number of vertices in G.

Example 9.10

Discuss Graph coloring problem with required example.

[JNTUH, Nov 2008, Set No. 4]

Solution: Let G = (V, E) be a graph.

If we assign the colors to each vertex of the graph such that no two of adjacent vertices have same color. Such a process is called as coloring of the graph G.

A vertex coloring by assigning K-colors to color the vertices is called K-vertex coloring.

Moreover every graph with n vertices is n-colorable.

Example 9.11

Observe the graph given in Fig. 9.3.4. This graph is a 3-colorable graph.

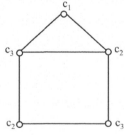

Fig. 9.3.4

The graph given in Fig. 9.3.5 is a 3-colorable graph.

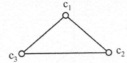

Fig. 9.3.5

The graph given in Fig. 9.3.6 is a two colorable graph.

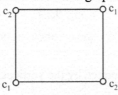

Fig. 9.3.6

Example 9.12

State and prove five color theorem

[JNTUH, Nov 2010, Set No. 4]

Statement: Every planar graph is 5-colorable.

Proof: We prove this theorem by mathematical induction on number of vertices n of the planar graph G. If G be a graph with 1, 2, 3, 4 or 5 vertices, then we can color the vertices with five colors. Therefore, the theorem is true for all planar graphs with $n \le 5$.

Suppose $n > 5$. Suppose the theorem is true for all planar graphs with less than n vertices. That is, all the planar graphs with less than 'n' vertices are five colorable.

Since G is a planar graph with $n > 5$, G has vertex such that $d(v) \le 5$.

Take $G' = G - v$, then $d(v) = 5$.

Now by mathematical induction, we have G' is five colorable. Suppose c_1, c_2, c_3, c_4 and c_5 are the five colors assumed to give for the vertices in G'.

If the vertices adjacent to the vertex v use less than five colors, then we use one of the remaining colors to vertex v. In this case, we got the coloring for the planar graph G.

Suppose $d(v) = 5$

That is, vertex v is adjacent to five vertices. Suppose that these five vertices are assigned five different colors.

Assume that the vertices moving in counter clockwise direction are v_1, v_2, v_3, v_4 and v_5.

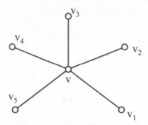

Fig. 9.3.7

Suppose that c_1, c_2, c_3, c_4 and c_5 are the colors assigned to the vertices v_1, v_2, v_3, v_4 and v_5 respectively.

Let H_1 be the subgraph of G, generated by the vertices colored with c_1 and c_3.

Clearly H_1 contains vertices v_1 and v_3.

If vertices v_1 and v_3 are in different components of H, then interchange the colors c_1 and c_3 in the component containing v_1 without disturbing the coloring of G-v.

Then assume the color c_3 to the vertices v_1 and v_3. Now our graph is a proper coloring with five colors as v colored with color c_1.

If v_1 and v_3 belonging to the same component of H_1, then there exists a v_1, v_3 path (since every component is connected) in G all of whose vertices are colored with c_1 or c_3.

Now the paths v_1 v_3 and v_1 v v_3 produces a cycle in G_1 which encloses the vertex v_2 or both the vertices v_4 and v_5.

In any case, there exists no path between v_2 and v_4 all of whose vertices are colored with c_2 or c_4.

Hence, if we let H_2 denote the subgraph G-v induced by the vertices colored c_2 or c_4, then v_2 and v_4 belong to different components of H_2.

Thus we interchange colors of the vertices in the component of H_2 containing v_2.

Now we arrived a five coloring of G-v in which no vertex adjacent to v is colored c_2.

Thus we may assign the color c_2 to vertex v and obtain a five-coloring of G.

Exercises

Planar Graphs

1. Show that K_4 is a planar graph.

2. Show that Q_3 is a planar graph.

3. A planar simple graph G has 30 vertices, each of degree 3. Determine the number of regions into which this planar graph can be splitted.

Ans: R = E − V + 2 = 45 − 30 + 2 = 17

4. Show that the following graph G is non-planar.

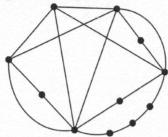

Euler's Formula

1. A connected planar simple graph G with E-edges and V vertices contain no simple circuit of length 4 or less, then show that

$$E \le \frac{5}{3}V - \frac{10}{3} \text{ if } V \ge 4$$

2. Verify Euler's formula for the following graph.

Ans: Satisfied Euler's formula

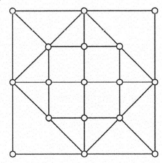

3. Let G be a 4-regular connected planar graph having 16 edges. Find the number of regions of G.

Ans: 10.

Graph Colouring, Covering and Chromatic Numbers

1. Prove that the vertices of every connected simple planar graph can be properly colored with five colors.

2. Find the chromatic number of each of the following graphs:

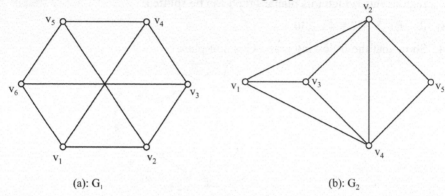

(a): G_1 (b): G_2

Ans: (a) 2; (b) 4

3. If $\Delta(G)$ is the maximum of the degree of the vertices of a graph G, then prove that $\chi(G) \leq 1 + \Delta G$, where $\chi(G)$ is chromatic number of G.

4. Prove that a graph G is 2-chromatic if and only if it is a non-null bipartite graph.

CHAPTER - 10

Graph Theory - IV

10.1   Fundamental Concepts of Trees

Trees are extensively used as models in areas like computer science, chemistry, geology, electrical networks and botany etc. We shall now describe such model based on trees. In saturated hydrocarbons, the molecules, where atoms are represented by vertices and bonds between them by edges. In graph models of saturated hydrocarbons, each carbon atom represented by a vertex of degree 4, and hydrogen atom is represented by a vertex of degree 1. So there are $3n + 2$ vertices in a graph representing a compound of the form $C_n H_{2n+2}$. Trees are also useful in design of wide range of algorithms.

The concept of a 'tree' plays a vital role in the theory of graphs. First we introduce the definition of 'tree', study the some of its properties and its applications. We also provide equivalent conditions for a tree.

LEARNING OBJECTIVES

♦ *to know Fundamental Concepts of Trees*

♦ *to understand the concepts: Directed Tree, Binary Tree and Decision Tree*

♦ *to find the Spanning Trees of a given graph by using different algorithms*

Definition

A connected graph without circuits is called a tree.

Example 10.1

Trees with one, two, three and four vertices are given in the Fig. 10.1.1.

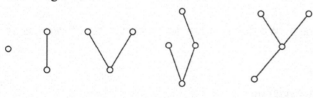

Fig. 10.1.1

241

Example 10.2

Consider the two trees $G_1 = (V, E_1)$ and $G_2 = (V, E_2)$ where

$V = \{a, b, c, d, e, f, g, h, i, j\}$

$E_1 = \{(a, c), (b, c), (c, d), (c, e), (e, g), (f, g), (g, i), (h, i), (i, j)\}$

$E_2 = \{(c, a), (c, b), (c, d), (c, f), (f, e), (f, i), (g, d), (h, e), (j, g)\}$

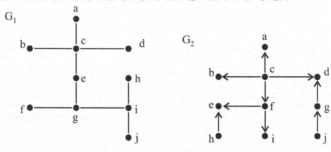

Fig. 10.1.2

Neither of these two trees is a directed tree.

If vertex c is designated as the root of each tree, vertex j is a level 4 in G_1 and at level 3 in G_2.

Note 1:

Directed trees are conventionally drawn with the root at the top and all edges going from the top of the page toward the bottom so that the direction of edges is sometimes not explicitly shown.

Note 2:

(i) Since a tree is a graph, we have that a tree contains at least one vertex.

(ii) A tree without any edge is referred to as a null tree.

(iii) Since we are considering only finite graphs, we have that the trees considered are also finite.

(iv) A tree is always a simple graph.

(v) A vertex of degree of 1 is called a pendent vertex.

Note 3:

Let $G = (V, E)$ be a disconnected graph. We define a relation $\sim$ on the set of vertices as follows:

$v \sim u \Leftrightarrow$ there is a walk from v to u.

Then this relation $\sim$ is an equivalence relation.

Let $\{V_i\}_{i \in \Delta}$ be the collection of all equivalence classes. Now $V = \bigcup\limits_{i \in \Delta} V_i$.

Write $E_i = \{e \in E \mid \text{an end point of e is in } V_i\}$ for each i.

Then (V_i, E_i) is a connected subgraph of G for every $i \in \Delta$.

This connected subgraph (V_i, E_i) of G is called a connected component (or component) of G for every $i \in \Delta$.

The collection $\{(V_i, E_i)\}_{i \in \Delta}$ of subgraphs of G is the collection of all connected components of G.

10.1.1   Note (Formation of Components)

If G is a connected graph, then G contains only one connected component and it is equal to G.

Now suppose that G is a disconnected graph. Consider a vertex v in G. If each vertex of G is joined by some path to v, then the graph is connected, a contradiction. So there exists at least one vertex which is not joined by any path to v.

The vertex v and all the vertices of G that are joined by some paths to v together with all the edges incident on them form a component (G_1, say).

To find another component, take a vertex u (from G) which is not G_1. The vertex u and all the vertices of G that are joined by some paths to u together with all the edges incident on them form a component (G_2, say).

Continue this procedure to find the components. Since the graph is a finite graph, the procedure will stop at a finite stage. In this way, we can find all the connected components of G. It is clear that, a component itself is a graph.

Definition

Let G be a connected graph. A cut-set is a subset C of the set of all edges of G whose removal from the graph G leaves the graph G disconnected; and removal of any proper subset of C does not disconnect the graph G.

(Equivalently, cut-set can also be defined as a minimal set C of edges in a connected graph G whose removal reduces the rank of the graph by one).

Example 10.3

Observe graph in Fig. 10.1.3(a). If we remove $\{a, c, d, f\}$ from the graph, then we get the subgraphs given in Fig. 10.1.3(b).

So in the Graph-A, the subset $\{a, c, d, f\}$ of edges, is a cut-set.

Also there are many other cut-sets such as $\{a, b, g\}$, $\{a, b, e, f\}$, $\{d, h, f\}$.

Also edge set $\{k\}$ is also a cut-set.

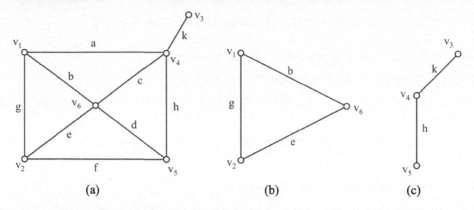

Fig. 10.1.3

Theorem 1

T is a tree $\Leftrightarrow$ there is one and only one path between every pair of vertices.

Proof: Suppose T is a tree. Then T is a connected graph and contains no circuits. Since T is connected, there exists at least one path between every pair of vertices in T. Suppose that between two vertices a and b of T, there are two distinct paths. Now, the union of these two paths will contain a circuit in T, a contradiction (since T contains no circuits).

This shows that there exists one and only one path between a given pair of vertices in T.

Converse: Let T be a graph.

Assume that there is one and only one path between every pair of vertices in G. This shows that T is connected. If possible suppose that T contains a circuit. Then there is at least one pair of vertices a, b such that there are two distinct paths between a and b. But this is a contradiction to our assumption.

So T contains no circuits. Thus T is a tree.

Theorem 2

A tree G with 'n' vertices has (n – 1) edges.

Proof: We prove this theorem by induction on the number vertices n.

If n = 1, then G contains only one vertex and no edge.

So the number of edges in G is n – 1 = 1 – 1 = 0

Suppose the induction hypothesis that the statement is true for all trees with less than 'n' vertices. Now let us consider a tree with 'n' vertices.

Let 'e_k' be any edge in T whose end vertices are v_i and v_j. Since T is a tree, by theorem 10.1.10, there is no other path between v_i and v_j. So by removing e_k from T, we get a disconnected graph.

Furthermore, T-e_k consists of exactly two components (say T_1 and T_2). Since T is a tree, there were no circuits in T and so there were no circuits in T_1 and T_2.

Therefore T_1 and T_2 are also trees.

It is clear that $|V(T_1)| + |V(T_2)| = |V(T)|$ where V(T) denotes the set of vertices in T.

Also $|V(T_1)|$ and $|V(T_2)|$ arc lcss than n.

Therefore by the induction hypothesis, we have

$$|E(T_1)| = |V(T_1)| - 1 \text{ and } |E(T_2)| = |V(T_2)| - 1$$

Now $|E(T)| - 1 = |E(T_1)| + |E(T_2)| = |V(T_1)| - 1 + |V(T_2)| - 1$

$\Rightarrow$ $|E(T)| = |V(T_1)| + |V(T_2)| - 1$

$$= |V(T)| - 1 = n - 1$$

This completes the proof.

10.1.3   Corollary

If T is a tree (with two or more vertices), then there exists at least two pendant (a vertex of degree 1) vertices.

Proof: Let T be a tree with $|V| \geq 2$.

Let $v_0 e_1 v_1 e_2 v_2 e_3 \ldots v_{n-1} e_n v_n$ be a longest path in T (since T is finite graph, it is possible to find a longest path).

Now we wish to show that $d(v_0) = 1 = d(v_n)$.

If $d(v_0) > 1$, then there exists at least one edge e with end point v_0 such that $e \neq e_1$.

If $e \in \{e_1, e_2, \ldots e_n\}$, then $e = e_i$ for some $i \neq 1$.

So either $v_{i-1} = v_0$ or $v_i = v_0$ $\Rightarrow$ v_0 repeated in the path, a contradiction.

Hence $e \notin \{e_1, e_2, e_3, \ldots e_n\}$.

Now e, $e_1, e_2, e_3, \ldots e_n$ is a path of length $n + 1$, a contradiction.

Hence $d(v_0) = 1$.

In a similar way, we can show that $d(v_n) = 1$.

Hence v_0, v_n are two pendant vertices.

Alternate Proof:

Let n = the number of vertices in G. Then G has $n - 1$ edges. Now

$$\sum_{i=1}^{n} \deg(v_i) = 2|E| = 2(n-1) = (2n-2)$$

Now if there is only one vertex, say v_1 of degree 1, then

$$\deg(v_i) \geq 2 \quad \text{for } i = 2, 3, \ldots, n \quad \text{and}$$

$$\sum_{i=1}^{n} \deg(v_i) = 1 + \sum_{i=2}^{n} \deg(v_i) \geq 1 + 2n - 2 = 2n - 1$$

But $2n - 2 \geq 2n - 1$ or $-2 \geq -1$, a contradiction.

Therefore there are at least two vertices of degree 1.

Theorem 3

If 2 non-adjacent vertices of a tree T are connected by adding an edge, then the resulting graph will contain a cycle.

Proof: If T has n vertices, then T has $n - 1$ edges and then if an additional edge is added to the edges of T the resulting graph G has n vertices and n edges. Hence G cannot be a tree by theorem 2.

But, the addition of an edge has not affected the connectivity.

Hence G must have a cycle.

Theorem 4

Any connected graph with 'n' vertices and $n - 1$ edges is a tree.

Proof: Let 'G' be a connected graph with n vertices and $n - 1$ edges. It is enough to show that G contains no circuits.

If possible, suppose that G contains a circuit.

Let 'e' be an edge in that circuit. Since 'e' in a circuit, we have that $G - e$ is still connected.

Now $G - e$ is connected with 'n' vertices, and so it should contain at least $n - 1$ edges, a contradiction (to the fact that $G - e$ contain only $(n - 2)$ edges).

So G contains no circuits.

Therefore, G is a tree.

Definition

A connected graph is said to be minimally connected if the removal of any one edge from the graph provides a disconnected graph.

Example 10.4

   (i)   Graph given in Fig. 10.1.4(a) is not minimally connected.

  (ii)   Graph given in Fig. 10.1.4(b) is minimally connected.

 (iii)   Any circuit is not minimally connected.

 (iv)   Every tree is minimally connected.

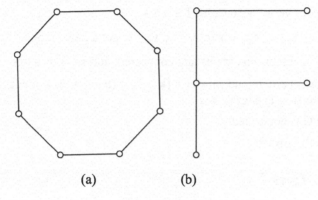

(a)   (b)

Fig. 10.1.4

Theorem 5

A graph G is a tree ⇔ it is minimally connected.

Proof: Assume that G is a tree.

Now we have to show that G is minimally connected.

In a contrary way, suppose that G is not minimally connected.

Then there exists an edge 'e' such that G – e is connected.

That is, e is in some circuit, which implies G is not a tree, a contradiction.

Hence G minimally connected.

Converse: Suppose that G is minimally connected.

Now it is enough to show that G contains no circuits.

In a contrary way, suppose G contains a circuit.

Then by removing one of the edges in the circuit, we get a connected graph, a contradiction (to the fact that the graph is minimally connected).

This shows that G contains no circuits. Thus G is a tree.

Note: To interconnect 'n' given distinct points, the minimum number of line segments needed is n – 1.

Theorem 6

If a graph G contains n vertices, n – 1 edges and no circuits, then G is a connected graph.

Proof: Let G be a graph with 'n' vertices, n – 1 edges and contains no circuits.

In a contrary way, suppose that G is disconnected.

G consists of two or more circuitless components (say, $g_1, g_2, ..., g_k$).

Now $k \geq 2$. Select a vertex v_i in g_i, for $1 \leq i \leq k$.

Add new edges $e_1, e_2, ..., e_{k-1}$ where $e_i = \overline{v_i \; v_{i+1}}$ to get a new graph G^*.

It is clear that G^* contains no circuits and connected, and so G^* is a tree.

Now G^* contains n vertices and $(n – 1) + (k – 1) = (n + k – 2) \geq n$ edges, a contradiction (since a tree contains $(n – 1)$ edges).

This shows that G is connected.

This completes the proof.

10.2 Directed Trees

The question now arises: Under what conditions does a diagraph have a directed spanning tree? The answer requires us to give a characterization of directed trees and of quasi-strongly connected graphs.

Definition

Two vertices u and v of a directed graph said to be quasi-strongly connected if there is a vertex w from there is a directed path to u and a directed path to v. The graph G is said to be quasi-strongly connected if each pair of vertices of G is quasi-strongly connected.

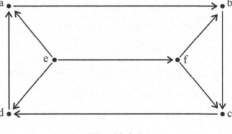

Fig. 10.2.1

Note:

It should be clear that if a directed graph G is quasi-strongly connected, then the underlying non-directed graph will be connected.

The digraph in the Fig. 10.2.1 is quasi-strongly connected.

Theorem 7

Let G be a digraph. Then the following are equivalent:

1. G is a quasi-strongly connected.

2. There is a vertex r in G such that there is a directed path from r to all the remaining vertices of G.

Proof: (2) $\Rightarrow$ (1) holds by definition.

(1) $\Rightarrow$ (2): Suppose that G is quasi-strongly connected and consider its vertices v_1, v_2, ..., v_n. There is a vertex w_2 from which there is a path to v_1 and a path to v_2.

Similarly, there is a vertex w_3 from which there is a path to w_2 and to v_3, and so on until finally we conclude that there is a vertex w_n from which there is a path to w_{n-1} and a path to v_n.

Clearly, there is a directed path from w_n to each vertex v_i of G since w_n is connected to v_1, v_2, ..., v_{n-1} through w_{n-1}. This proves (2).

Now we present a number of equivalent characterizations of a directed tree.

Theorem 8

Let G be a digraph with n >1 vertex. Then the following statements are equivalent:

1. G is a directed tree.
2. There is a vertex r in G such that there is a unique directed path from r to every other vertex of G.
3. G is quasi-strongly connected and G-e is not quasi-strongly connected for each edge e of G.
4. G is quasi-strongly connected and contains a vertex r such that the in-degree of r is zero (that is, $\deg^+(r) = 0$) and $\deg^+(v) = 1$ for each vertex $v \neq r$.
5. G has no circuits (that is, G has no directed circuits and no non-directed circuits) and has a vertex r such that $\deg^+(r) = 0$ and $\deg^+(v) = 1$ for each vertex $v \neq r$.
6. G is quasi-strongly connected without circuits.
7. There is a vertex r such that $\deg^+(r) = 0$ and $\deg^+(v) = 1$ for each vertex $v \neq r$, and the underlying non-directed graph of G is a tree.

Proof:

(1) $\Rightarrow$ (2): The root r in G satisfies the property.

(2) $\Rightarrow$ (3): By theorem 8, G is a quasi-strongly connected. Suppose quasi-strong connectivity is not destroyed when some edge (u, v) is removed. Then there is a vertex w such that there are two directed path one to u and one to v, neither of which uses the edge (u, v).

Thus, in G there are two distinct directed paths from w to v and hence two distinct directed paths from the root r of G to v. This contradicts (2).

(3) $\Rightarrow$ (4) is straightforward.

10.3   Binary Trees and Decision Trees

In this section, we study binary trees and applications of binary trees. Since the natural way their vertices correspond to an initial segment of the positive integers, complete binary trees can be represented very efficiently on computers. They are applied in a number of excellent algorithms, including "Heap Sort", priority queue implementation, and algorithms for the efficient ordering of data in hash tables. We also considered height balanced trees which are important in computer science.

Definition

A binary tree is a directed tree $T = (V, E)$, together with an edge-labelling $f: E \to \{0, 1\}$ such that every vertex has at most one edge incident from it labelled with 0 and at most one edge incident from it labelled with 1.

Each edge (u, v) labelled with 0 is called a left edge; in this case, u is called the parent of v and v is called the left child of u.

Each edge (u, v) labelled with 1 is called a right edge; in this case u is called the parent of u, but v is called the right child of u. The subtrees of which the left and right children of a vertex u are the roots are called the left and right subtrees of u, respectively. We represent a binary tree by a triple (V, E, f).

Note:

  (i)   Every vertex in a binary tree has a unique parent, a unique left child, and a unique right child, if it has any at all. That each vertex has a unique parent (if any) follows from the definition of tree, where it is required that there be a unique path from the root to each vertex.

  (ii)   Each vertex has a unique left child and a unique right child (if any) follows from the labelling of the edges of the tree with 0's and 1's. (At most one edge from the parent can have a 0 label and at most one edge can have a 1).

  (iii) Every vertex other than the root has a parent. Since every vertex u in a tree must have a path to it from the root and the last vertex before v on such a path must be the parent of v.

  (iv)   See the following figure for an illustration.

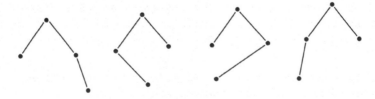

Fig. 10.3.1

Example 10.5

Details such as edge labels and the direction of edges are usually represented only implicitly in drawings of binary trees. The convention is that for each vertex v the root of u's left subtree lies below v and to its left on the page, whereas the root of u's right subtree lies below v and to its right on the page. Fig. 10.3.2 shows an example of a binary tree drawn with and without edge labels and directed edges.

There are a few special kinds of binary trees that are important in computer applications; one of these is the complete binary tree.

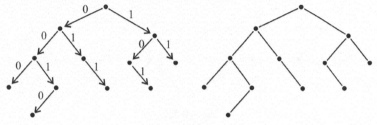

Fig. 10.3.2

Definition

Let T be binary tree. Every vertex v in T has a unique level-order index, defined as follows:

If v is the root, let index (v) = 1; if v is the left child of a vertex u, let index (v) = [index (u)]2; otherwise, if v is the right child of some vertex u, let index (u) = 1+[index (u)]2.

Note:

   (i) The index of a child is obtained by doubling the index of its parent and adding the label on the edge that goes from the parent to the child.

   (ii) The level of any vertex v can be computed as log2 ⌊(index (v))⌋.

Example 10.6

The binary tree shown in Fig 10.3.3, the level-order index written at the location of each vertex. Left children all have even indices, and right children have odd indices. The string corresponding to vertex number 18 is 0010. Putting a 1 in front of this, we get 10010, which is the base two representation of 18.

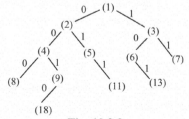

Fig. 10.3.3

Definition

A complete binary tree is a binary tree for which the level-order indices of the vertices form a complete interval 1, ..., n of the integers. That is, if such a tree has n vertices there is a vertex in the tree with index i for every i from 1 to n.

Example 10.7

Consider the two binary trees shown. Only the first one is complete. In particular, the second tree has ten vertices but has no vertex with index 10.

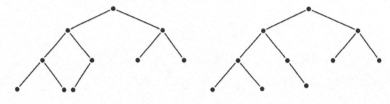

Fig. 10.3.4

10.3.1   Lemma

In a complete binary tree with n vertices the indices of the vertices in the kth level comprise the complete interval 2^k through $2^{k+1} - 1$, or from 2^k through n if n is less than $2^{k+1} - 1$.

Proof: The proof is by induction on k.

For k = 0 and n = 0 the lemma holds vacuously.

For k = 0 and n > 0 there is exactly one vertex with index $2^0 = 2^{0+1} - 1 = 1$, and that is the root, which is also the only vertex at level 0.

Induction Hypothesis: For larger values of k, we assume the lemma holds for k – 1, if n < 1 2k, the lemma holds vacuously. Otherwise we invoke the definition of level.

The vertices in level k are exactly those at distance k from the root.

The vertices in level k – 1 are exactly those at distance k – 1 from the root.

It follows that the vertices at level k are precisely the children of the vertices at level k – 1, which the inductive hypothesis asserts are those with indices 2^{k-1} through $2^k - 1$. By the definition of level-order index, the children have indices in the range 2^k through $2^{k+1} - 1$. This complete interval, or the initial segment of it up through n, must be in T, by the definition of complete binary tree.

Definition

A height balanced binary tree is a binary tree such that the heights of the left and right sub trees of every vertex differ by at most one.

Example 10.8

The tree shown in figures in Example 10.5 and example 10.7 is a height-balanced binary tree of height 4.

Deleting vertex (13) would result in a tree that would no longer be height balanced, since it would have a left subtree of height 3 and a right subtree of height 1.

Deleting vertex (8) would also result in a tree that would no longer be height balanced, since vertex (4) would have a left subtree of height – 1 and a right subtree of height 1.

Note:

Every complete binary tree is also a height-balanced binary tree.

Theorem 9

There are at least $\dfrac{1}{\sqrt{5}}\left[\dfrac{1+\sqrt{5}}{2}\right]^{h+3} - 2$ vertices in any height-balanced binary tree with height h.

Proof: This bound can be obtained from a recurrence.

Let V(h) = the least achievable number of vertices in a (nonempty) height balanced binary tree of height h.

Then clearly V(0) = 1 and V(1) = 2

For h > 1, we observe that there must be a root and two subtrees, possibly empty.

The height of one subtree must be h – 1.

The height of the other may be h – 1 or h – 2.

It is clear that it is not possible to construct a height-balanced subtree of height h – 1 with fewer vertices than are required to construct a height-balanced subtree of height h – 2.

A height-balanced tree of height h with the fewest possible vertices thus consists of a root, one height-balanced subtree of height h – 1 and one height-balanced subtree of height h – 2. The total number of vertices in such a tree is

$$V(h) = 1 + V(h-1) + V(h-2).$$

This recurrence is similar to the recurrence for the Fibonacci numbers.

Apply similar techniques, we obtain the solution.

$$V(h) = \left(\frac{5+2\sqrt{5}}{5}\right)\phi^h + \left(\frac{5-2\sqrt{5}}{5}\right)(1-\phi)^h - 1$$

where $\phi = \dfrac{1+\sqrt{5}}{2}$ and $1-\phi = \dfrac{1-\sqrt{5}}{2}$.

We can obtained a solution of this recurrence relation as

$$V(h) = \frac{1}{\sqrt{5}}(\phi)^{h+3} - \frac{1}{\sqrt{5}}(1-\phi)^{h+3} - 1$$

$$= F_{h+2} - 1, \text{ where } F_n \text{ is the } n^{th} \text{ Fibonacci number.}$$

This means that $\dfrac{5+2\sqrt{5}}{5} = \dfrac{1}{\sqrt{5}}\phi^3$ and $\dfrac{5-2\sqrt{5}}{5} = -\dfrac{1}{\sqrt{5}}(1-\phi)^3$

Now we compare this bound with the lower bound of 2^h vertices for complete binary trees of height h. Since ϕ is between 1.61803 and 1.61804, we have

$$-1 < (1-\phi)^{h+3} < 1$$

This implies $\dfrac{-1}{\sqrt{5}}(1-\phi)^{h+3} > -1$ so that $V(h) > \dfrac{1}{\sqrt{5}}(\phi)^{h+3} - 2$.

Thus, V(h) is greater than $(1.89)(1.61)^h - 2$; that is, we have an exponential lower bound on the number of vertices in a height balanced tree comparable to the bound for a complete binary tree.

Definition

A binary search tree is a binary tree with a vertex labeling $1: V \to A$, where A $\{a_1, a_2, ..., a_n\}$ is a totally ordered set with $a_1 < a_2 < ... < a_n$, and where the labeling 1 satisfies the properties.

(i) For each vertex is in the left subtree of a vertex v, $1(u) < 1(v)$.

(ii) For each vertex u in the right subtree of a vertex v, $1(u) > 1(v)$.

Example 10.9

Consider a sequence of numbers 17, 23, 4, 7, 9, 19, 45, 6, 2, 37, 99. Now let us build a binary search tree for the set A obtained by sorting the numbers in their proper order.

First label the root with 17; then since 23 is larger than 17, make a right child for the root and label it 23.

Next, since 4 is less than the root, label a left child of the root with the label 4.

Continue to the next number 7 in the list.

Since 7 is less than the root 17, 7 will appear as the label of some vertex in the left subtree of the root, yet 7 is greater than 4, so label with 7 the right child of the vertex 4.

Next, use the label 9. Again 9 must be the label of a vertex in the left subtree of the root since 9 < 17. Moreover, 9 is greater than 4, so 9 must be a label in the right subtree of 4. Likewise 9 > 7, so 9 is a label for the right child of 7.

Next, 19 is greater than 17, so 19 will label some vertex in the right subtree of 17; but 19 is less than 23, the right child of 17, so label the left child of 23 with the label 19. Next consider 45; 45 > 17 so 45 is a label for a vertex in the right subtree of 17; 45 > 23, so label the right child of 23 with 45.

Continuing as above, we see that 6 must be the label for the left child of the vertex 7, 2 is the left child of 4, 37 and 99 are respectively the left and right children of 45. Thus, we have the following binary tree:

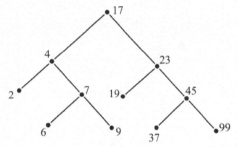

Fig. 10.3.5

10.4   Spanning Trees and Properties

In this section, we will study the tree as a subgraph of another graph. A given graph may have numerous subgraphs. If e is the number of edges in G, then there are 2^e distinct subgraphs are possible. Obviously some of these subgraphs will be trees. Out of these trees we particularly interested in certain type of trees, called spanning trees.

Definition

A tree T is said to be a spanning tree of a connected graph G if T is a subgraph of G and T contains all the vertices of G.

In other words: A subgraph H of a graph G is called a spanning tree of G if

    (i) H is a tree, and

    (ii) H contains all the vertices of G.

A spanning tree that is a directed tree is called a directed spanning tree of G.

Example 10.10

Consider the digraph G (V, E) where V = {a, b, c, d, e} and E = {(a, c), (b, a), (b, b), (b, c), (c, d), (c, e), (d, c), (d, d), (e, b)} (shown in the Fig. 10.4.1).

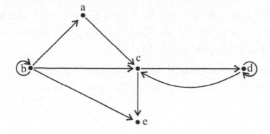

Fig. 10.4.1

A spanning tree of the above graph is T = (V, {(b, a), (a, c), (c, d), (c, e)}, (shown in Fig. 10.4.2).

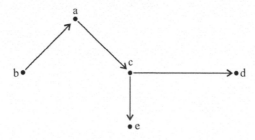

Fig. 10.4.2

Example 10.11

Consider the graph G given in Fig. 10.4.3(a). Graph T (given in Fig. 10.4.3(b)) is a spanning tree of G.

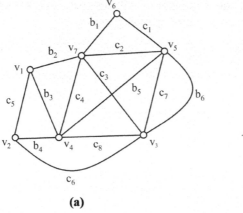

(a)

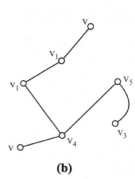

(b)

Fig. 10.4.3

Note:

(i) Since spanning trees are the largest (with the maximum number of edges) trees among all the trees in G, we have that a spanning tree is also called a maximal tree subgraph or maximal tree of G.

(ii) Spanning is defined only for a connected graph. (Because, a tree is always connected).

(iii) However, each component of a disconnected graph does have a spanning tree.

Thus a disconnected graph with k components contains a spanning forest (collection of trees is called a forest) consisting of k spanning trees.

Example 10.12

Consider the graph G in Fig. 10.4.4(a). Suppose that this graph represents a communication network in which the vertices correspond to stations and the edges corresponding to communication links. The largest number of edges that can be deleted while still allowing the stations to communicate with each other (Fig. 10.4.4(b)).

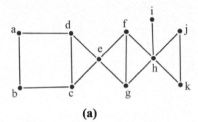

(a)   (b)

Fig. 10.4.4

Note:

Let G be a connected graph with n vertices and m edges, a spanning tree of G must have n − 1 edges. Hence, the number of edges that must be removed before a spanning tree is obtained must be m − (n − 1) = m − n + 1. This number is called the circuit rank of G.

Theorem 10

A non directed graph G is connected if and only if G contains a spanning tree. Indeed, if we successively delete edges of cycles until no further cycles remain, then the result is a spanning tree of G.

Proof: If G has a spanning tree T, then there is a path between any pair of vertices in G along the tree. Thus G is a tree.

Converse: Let G be a connected graph.

If G has no circuit, then G is a spanning tree.

If G has a circuit, then delete an edge from this circuit and till leaves the graph connected.

If there are more circuits, repeat the operation till an edge from the last circuit is deleted, leaving the graph connected, circuitless and contains all the vertices of G.

Thus the subgraph obtained is a spanning tree of G.

Hence every connected graph has at least one spanning tree.

Definition

A tree, in which one vertex (called the root) is distinguished from all the other vertices, is called a rooted tree. In a rooted tree, the root is generally marked in a small triangle.

Example 10.13

Distinct rooted trees with four vertices, were given in Fig. 10.4.5.

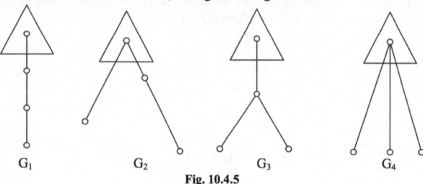

G_1 G_2 G_3 G_4

Fig. 10.4.5

Definition

Let T be a rooted tree with designated root v_0. Suppose that u and v are vertices in T and $v_0 - v_1 -, \ldots, - v_n$ is a simple path in T. Then

  (i) v_{n-1} is the parent of v_n

  (ii) $v_0, v_1, \ldots, v_{n-1}$ are the ancestors of v_n

  (iii) v_n is a child of v_{n-1}

  (iv)   If u is an ancestor of v, then u is a descendant of u

  (v)   If u has no children, then u is a leaf of T.

  (vi)   If v is not a leaf of T, then u is an internal vertex of T.

  (vii)   The subgraph of T consisting of u and all its descendants, with designated as a root, is the subtree of T rooted at v.

Example 10.14

Consider the tree of graph as the root of G_1 Fig. 10.1.2, a is designated as the root, then b, d, f, h, and j are leaves of T; i is the parent of h and j; f, h, i, and j are the descendants of g; a and c are the ancestors of e; and the children of c are b, d, and e. Moreover, the vertices a, c, e, g, and i are intern vertices to the tree rooted at a.

If, on the other hand, we let c be the root, then a becomes a leaf, and e the parent of a.

Properties of spanning tree:

   (i) Spanning tree is a spanning subgraph.

  (ii) Spanning tree contains no cycles.

 (iii) The number of edges in a spanning tree is $n - 1$ where n is the number of vertices.

 (iv) If T is a spanning subgraph of a graph G then if we add any edge to T then we get a circuit (or cycle).

  (v) Spanning tree T is a maximal tree subgraph (we can also note that T is a maximal tree of G).

 (vi) We can get a spanning tree T of a given graph G only if G is connected.

10.5   Algorithms for Spanning Trees

10.5.1   Breadth-First Search Algorithm (BFS): (for a Spanning Tree)

An algorithm based on the proof of the Theorem 10 could be designed to produce a spanning tree for a connected graph. Recall from the proof that the entire one need do is destroy cycles in the graph by removing and from a cycle until no cycles remain.

Algorithm:

Input: A connected graph G with vertices labelled $v_1, v_2, \ldots, v_n$

Output: A spanning tree T for G.

Method:

Step **1:** (Start) Let v_1 be the root of T. Form the set $V = \{v_1\}$.

Step **2:** (Add new edges). Consider the vertices of V in order consistent with the original labeling. Then for each vertex $x \in V$, add the edge $\{x, v_k\}$ to T where k is the minimum index such that adding the edge $\{x, v_k\}$ to T does not produce a cycle. If no edge can be added, then stop; T is a spanning tree for G. After all the vertices of V have been considered in order, go to Step 3.

Step **3:** (Update V). Replace V by all the children v in T of the vertices x of V where the edges $\{x, v\}$ were added in Step 2. Go back and repeat Step 2 for the new set V.

10.5.2   Depth-First Search Algorithm (DFS): (for a Spanning Tree)

Input: A connected graph G with vertices labeled $v_1, v_2, \ldots, v_n$

Output: A spanning tree T for G.

Method:

Step **1:** (Visit a vertex) Let v_1 be the root of T, and set $L = v_1$. (The name L stands for the vertex last visited.)

Step **2:** (Find an unexamined edge and an unvisited vertex adjacent to L.) For all vertices adjacent to L, choose the edge $\{L, v_k\}$, where k is the minimum index such that adding $\{L, v_k\}$ to T does not create a cycle.

If no such edge exists, go to Step 3; otherwise, add edge $\{L, v_k\}$ to T and set $L = v_k$; repeat Step 2 at the new value for L.

Step **3:** (Backtrack or terminate.) If x is the parent of L in T, set $L = x$ and apply Step 2 at the new value of L. If, on the other hand, L has no parent in T (so that $L = v_1$) then the depth-first search terminates and T is a spanning tree for G.

10.5.3   Minimal Spanning Trees

The applications of spanning trees are many and varied, and in order to gain some appreciation for this fact, we will describe what is sometimes called the connector problem. Suppose that we have a collection of n cities, and that we wish to construct a utility, communication, or transportation network connecting all of the cities. Assume that we know the cost of building the links between each pair of cities and that, in addition, we wish to construct the network as cheaply as possible.

The desired network can be represented by a graph by regarding each city as a vertex and by placing an edge between vertices if a link runs between the two corresponding cities. Moreover, given the cost of constructing a link between cities v_i and v_j, we can assign the weight c_{ij} to the edge $\{v_i, v_j\}$. The problem, then, is to design such a network so as to minimize the total cost of construction. If M is the graph of a network of minimal cost, it is essential that M be connected for all of the cities are to be connected by links. Moreover, it is also necessary that there be no circuits in the graph M, for otherwise we can remove an edge from a circuit and thereby reduce the total cost by the cost of construction of that edge. Hence, a graph of minimal cost must be a spanning tree of the graph of the n vertices.

10.5.4   Kruskal Algorithm: (Finding Shortest Spanning Tree)

Step **1:** List all the edges of G in order of non-decreasing weight. Now we select an edge e_1 of G such that $\omega(e_1)$ is as small as possible and 'e_1' is not a loop.

Step **2:** Select next smallest edge from the set of all remaining edges of G such that the selected edge do not form a circuit with the edges that have already been chosen.

***Step* 3:** We continue this process of taking smallest edges among those not already chosen, provided no circuit is formed with those, that have been chosen already.

[If edges e_1, e_2, ..., e_i have been chosen, then chose e_{i+1} from $E - \{e_1, e_2, ..., e_i\}$ in such way that graph with $\{e_1, e_2, ..., e_{i+1}\}$ is a cyclic and $\omega(e_1)$ is as small as possible].

***Step* 4:** If a graph G has 'n' vertices, then we will stop this process after choosing n − 1 edges. These edges form a subgraph T, which is not cyclic. (Thus T is a shortest spanning tree of G).

Example 10.15

Let us illustrate breadth-first search on the graph given below:

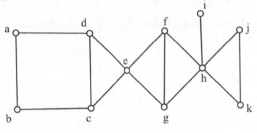

We select the ordering of the vertices abcdefghijk. Then we select a as the first vertex chosen in the spanning tree T and designate it as the root of T. Thus, at this stage, T consists of the single vertex a. Add to T all edges $\{a, x\}$, as x runs in order from b to k, that do not produce a cycle in T. Thus, we add $\{a, b\}$ and $\{a, d\}$. These edges are now called tree edges for the breadth-first search tree.

Now repeat the process for all vertices on level one from the root by examining each vertex in the designated order. Thus, since b and d are at level one, we first examine b.

For b, we include the edge $\{b, c\}$ as a tree edge. Then for d, we reject the edge $\{d, c\}$ since its inclusion would produce a cycle in T. But we include $\{d, e\}$.

Next, we consider the vertices at level two.

Reject the edge $\{c, e\}$; include $\{e, f\}$ and $\{e, g\}$.

Then repeat the procedure again for vertices on level three.

Reject $\{f, g\}$, but include $\{f, h\}$.

At g, reject $\{f, g\}$ and $\{g, h\}$.

On level four, include $\{h, i\}$, $\{h, j\}$, and $\{h, k\}$.

Next, we attempt to apply the procedure on level five at i, j, and k, but no edge can be added at these vertices so the procedure ends. Therefore the spanning tree T includes the vertices a, b, c, d, e, f, g, h, i, j, and k, and the edges $\{a, b\}$, $\{a, d\}$, $\{b, c\}$, $\{d, e\}$, $\{e, f\}$, $\{e, g\}$, $\{f, h\}$, $\{h, i\}$, $\{h, j\}$, and $\{h, k\}$.

10.5.5   Prim's Algorithm

***Step* 1:** Choose any vertex v_1 in G.

Step 2: Choose an edge $e_1 = v_1 v_2$ of G such that $v_1 \neq v_2$ and e_1 has the smallest weight among the edges of G incident with v_1 [For convenience, we can form a table indicating the weights].

Step 3: If the edges e_1, e_2, ..., e_k have been already choosen involving end vertices v_1, v_2, ..., v_{k+1}. Choose an edge e_{r+1} where $e_{r+1} = v_1 v_k$ with $v_1 \in \{v_1, v_2, ..., v_{k+1}\}$ and $v_k \notin \{v_1, v_2, ..., v_{k+1}\}$ such that e_{r+1} has the smallest weight among the edges of G with precisely one end in $\{v_1, v_2, ..., v_{k+1}\}$.

Note that after adding e_{r+1}, the graph should be a cycle.

Step 4: The process will stop after choosing the k – 1 edges, otherwise repeat the Step 3.

Example 10.16

Consider the connected weighed graph G.

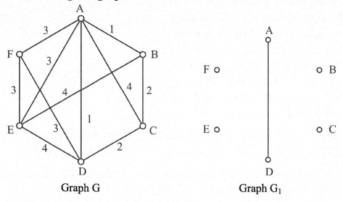

Graph G   Graph G_1

Step 1: Take the spanning null subgraph G_1 of G.

Let us choosen $V_1 = A$. The edge AD is incident on V_1 and has the smallest weight among the edges incident on $V_1 = A$.

     write $G_1 = G_0 + AD$

Step 2: $G_2 = G_4 + AB$

Step 3: In this step we select DC and write $G_3 = G_2 + DC$

Step 4: We select AF and write $G_4 = G_3 + AF$

Step 5: We select FE and write $G_5 = G_4 + FE$

Step 6: Since G_5 is a spanning tree, the process stops here.

     write $T = G_5$

Theorem 11

Let G be a connected graph where the edges of G are labeled by non-negative numbers. Let T be an economy tree of G obtained from Kruskal's Algorithm. Then T is a minimal spanning tree.

Example 10.17

Determine a railway network of minimal cost of the cities in Figure using Kruskal's algorithm.

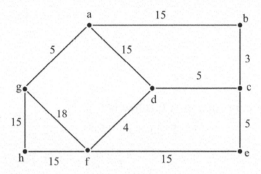

Solution: We collect the lengths of edges into a Table 10.1.

Step 1: Choose the edges {b, c}, {d, f}, {a, g}, {c, d}, {c, e}.

Step 2: Then we have options: we may choose only one of {a, b} and {a, d} for the selection of both creates a circuit. Suppose that we choose {a, b}.

Step 3: Next we may choose only one of {g, h} and {f, h}. Suppose we choose {f, h}.

Table 10.1

Edge	Cost
{b, c}	3
{d, f}	4
{a, g}	5
{c, d}	5
{c, e}	5
{a, b}	15
{a, d}	15
{f, h}	15
{g, h}	15
{e, f}	15
{f, g}	18

Step 4: Now we have a spanning tree as illustrated below.

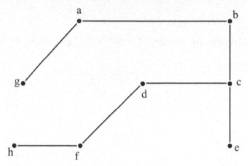

The minimal cost for construction of this tree is

$$3 + 4 + 5 + 5 + 5 + 15 + 15 = 52$$

Example 10.18

Apply DFS and find the spanning tree of the following graph.

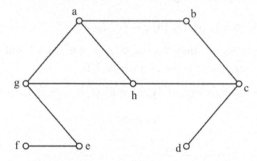

Solution: Let 'a' be the starting vertex for the required spanning tree T.

Since vertex 'a' is adjacent to vertices g, h, b.

We choose vertex 'b' and join 'a' and 'b' [because b is the first letter among b, g, h].

Then the graph obtained as follows:

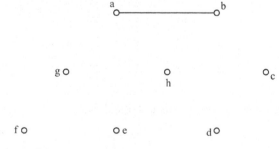

Now we start at vertex 'b'.

The vertex set adjacent to vertex 'b' which do not appear (consider) previous is {c}.

Now join the vertex 'b' and vertex 'c' by edge. We get the following graph.

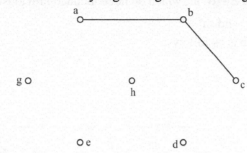

Now we arrived at vertex 'c'.

The vertex set which is adjacent to vertex 'c' and do not consider previously is {d, h}.

Now we join the vertex 'c' and 'd'. Then the graph obtained is

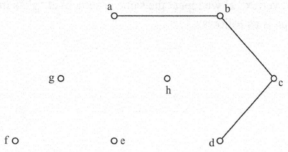

Now we arrived at vertex 'd'.

At vertex 'd' we don't find any vertex adjacent to it, which do not previously considered. So we take back track and we start from vertex 'c'.

Now at vertex 'c' the adjacent vertex set which do not contain previously considered is {h}.

Now join the vertices 'c' and 'h'.

Then the new graph is as follows:

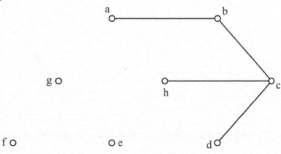

Now we are at vertex 'h'.

The vertex set adjacent to 'h', which do not contained previously visited vertices is {g}.

Join the vertices 'g' and 'h'.

Then the new graph is as follows:

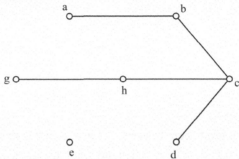

Now we arrived at vertex 'g' we repeat the same argument at 'g' as in above.

Then the new graph is as follows:

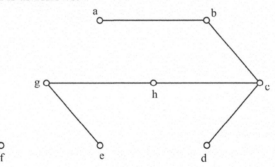

Thus we are at vertex 'e' and repeat the process. We obtain the following:

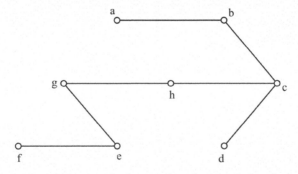

Hence we at vertex 'f' and we stop since we got the required spanning tree.

Example 10.19

What do you mean by spanning tree? Explain DFS method for finding a spanning tree for the graph.

[JNTUH, Nov 2010, Set No. 2]

Solution:

Part 1: A spanning tree is a tree that contains all vertices of the given graph G.

Part 2: Now we explain DFS method:

Let G be a connected simple graph. We have to find a spanning tree.

Let G be a connected graph with vertices $v_1, v_2, ..., v_n$.

Step 1: Let v_1 be the root and write $v = v_1$.

Step 2: Now choose a vertex from $v_2, v_3, ..., v_n$ which are adjacent to v_1 with least index among them.

Let it be v_2.

Now let $v = v_2$. Find all the vertices adjacent to v_2 which do not appeared previously. Let it be v_3 and join v_2 and v_3. Repeat the process till the adding do not form any cycle.

If we do not have no such edge exists (which do not form a cycle) we go to step 3.

Step 3: If u is the parent of v_i in T, let $v_i = v$ and repeat the process as in step 2.

If such a vertex not available, then we stop with a required spanning tree.

Example 10.20

Determine the spanning tree of the following using DFS.

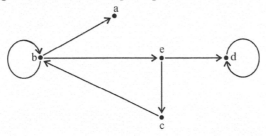

[JNTUH, Nov 2008, Set No. 1]

Solution: Given directed graph G is

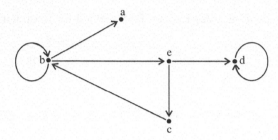

We have to find a spanning tree of the given graph G.

Let 'b' be the starting vertex of spanning tree T.

At vertex 'b' {a, e} is the set of all adjacent vertices of b. By vertex labeling we choose the edge 'ba'.

Now at vertex 'a' we cannot move further. So take track back and start from vertex 'b'. Now join the edge 'be'.

At vertex 'e', {c, d} is the vertex set adjacent to 'e'.

By following the labeling of vertices we join the edge 'ec'.

Now we are at vertex 'c'. But there is no further move from vertex 'c', so we take back track and start from vertex 'e'.

Now join 'ed'.

At present we are at vertex 'd'.

At vertex 'd' there is no further move and no back track from 'e'.

Hence we are now with the following spanning tree.

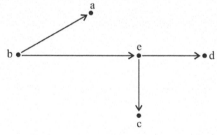

Example 10.21

Apply BSF on the following graph.

[JNTUH, Nov 2008, Set No. 3]

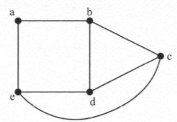

Solution: Given graph G is

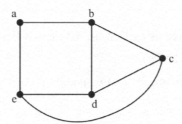

We have to find a spanning tree from given graph G with ordering a b c d e.

Start constructing a spanning tree from vertex 'a'.

Then {b, e} is the set of vertices adjacent to 'a'. Join the edges ab and ae.

Now at the vertex 'b' {c ,d} is the set of vertices adjacent to the vertex 'b'.

Join the edges 'bc' and 'bd'.

At vertex 'e' {c, d} is the vertex set adjacent to 'e'.

But if we add any edge 'ec' or 'ed' produces a cycle in G – a contradiction.

Therefore the required spanning tree is

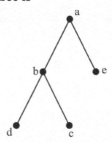

Example 10.22

We find the spanning tree of the given graph using Depth-First Search and find Breadth First Search.

Given graph G is

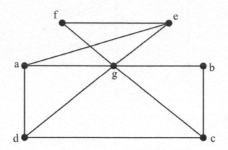

First we find the spanning tree using Depth-First Search (DFS).

Given vertex set v = {a b c d e f g}.

First take v = a as the initializing vertex for spanning tree T.

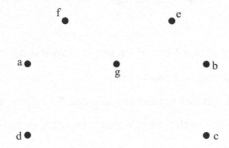

Clearly the vertex v = a is adjacent to {d, e, g}. We choose the vertex 'd' by step 2 of Algorithm.

Join vertices 'a' and 'd'. Then the graph is given below.

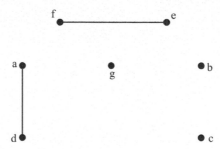

Now at vertex 'd', it is adjacent to {a, c, g}. Since a is previously appear, so we choose 'c' as next vertex by step 2 of the algorithm.

Join 'd' and 'c', then the graph is

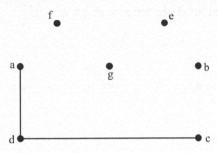

At vertex 'c' it is adjacent to {d, g, b}. By step 2 of the Algorithm choose the vertex 'b' adjacent to the vertex 'c'.

Join 'c' and 'b', then the graph is

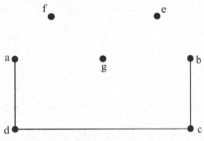

Now we arrive at vertex 'b'.

The vertices adjacent to 'b' are {c, g}.

By step 2 of the DFS Algorithm, we choose vertex 'g'.

Join 'b' and 'g'. Then the new graph is

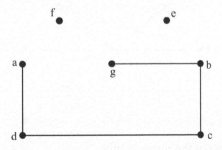

At the vertex 'g' the adjacent vertices are {a, b, c, d, f, e}. But a, b, c, d are already appeared.

By step 2 of the algorithm, we choose vertex 'e'.

Join vertex 'g' and vertex 'e'. Then the new graph is

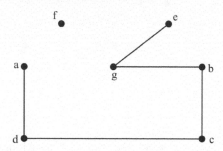

By the same argument in the above, we choose vertex 'f' and join vertex 'e' and vertex 'f'. We get a new graph

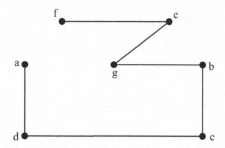

which is the required spanning tree obtained by DFS method.

Example 10.23

Find the spanning tree of the following graph using DFS Method.

[JNTUH, June 2010, Set No. 1]

Solution: Take the vertex set of the given graph

V = {a, b, c, d, e, f, g}

Let us start to construct the spanning tree 'T' with vertex 'a'. The set of vertices adjacent to 'a' is {d, e, g}.

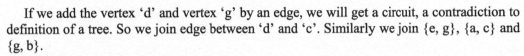

Join the vertices 'd', 'e', 'g' with vertex 'a' by edges.

These vertices are at level 1.

Now there are no vertices adjacent to vertex 'a'.

Now we start from vertex 'd'.

The vertex set adjacent to vertex 'd' is {c, g}.

If we add the vertex 'd' and vertex 'g' by an edge, we will get a circuit, a contradiction to definition of a tree. So we join edge between 'd' and 'c'. Similarly we join {e, g}, {a, c} and {g, b}.

Then the graph is given below:

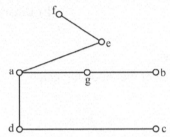

which is the required spanning tree with root 'a' by BFS method.

Exercises

Trees

(i)   Show that the complete graph K_n is not a tree when $n > 2$.

(ii) Show that, in a tree, if the degree of every non-pendant vertex is 3, the number of vertices in the tree is even.

(iii) Let F be a forest with K–components (trees). If n is the number of vertices and m is the number of edges in F, then prove that $n = m + k$.

Binary Trees

(i)   Find the number of vertices and the number of leaves in a complete binary tree having 10 internal vertices.

Ans: 11

(ii) A complete binary tree has 20 leaves. How many vertices does it have?

Ans: 39

(iii)   Which of the following is binary tree? A complete binary tree?

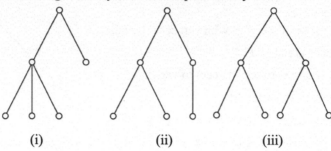

(i)   (ii)   (iii)

Ans:

(i) Not a binary tree.
(ii) Binary tree, not complete binary tree
(iii) Complete binary tree.

Spanning trees and Properties

(i)   Find all the spanning trees of the following graph

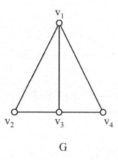

G

Ans:

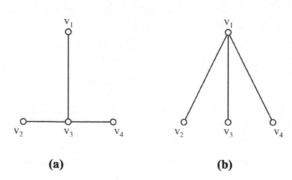

(a)   (b)

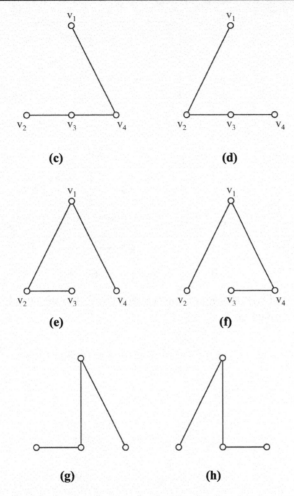

(c)

(d)

(e)

(f)

(g)

(h)

(ii)   Show that a Hamilton path is a spanning tree.

(iii) Prove that a graph G is a tree, if and only if G is loop-free and has exactly one spanning tree.

Algorithms for Spanning Trees

(i) Find the Depth–First Search Spanning tree for the graph shown below if the order of the vertices is given

   (i)   a, b, c, d, e, f, g, h

   (ii) h, g, f, e, d, c, b, a

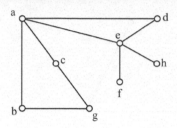

Ans:

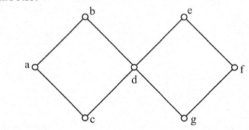

(i)   (ii)

(ii)   Find (i) DFS and (ii) BFS Spanning tree for the following graph, where in the order of vertices is alphabetic.

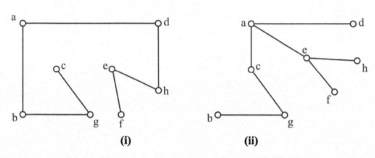

Ans:

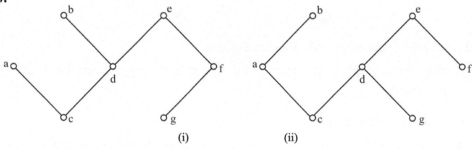

(i)   (ii)

Minimum Spanning Trees

(i)   Find the minimal spanning tree of the following graph:

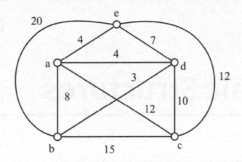

Ans:

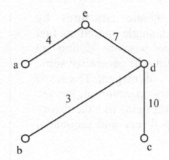

CHAPTER - 11

Algebraic Structures

Basic Algebraic Structures

We begin our study of algebraic structures by investigating sets associated with single operations that satisfy certain axioms. So we wish to define an operation on a set in a way that it will generalize some familiar structures such as the set of integers. The set of algebraic systems with one binary operation on the set, have useful applications in several fields like the theory of finite state machines, coding theory and sequential machines.

11.1 Some Algebraic Properties

Definitions

(i) Let $(P, \geq)$ be a Poset and $A \subseteq P$. An element $x \in P$ is called a lower bound for A if $a \geq x$, for all $a \in A$. A lower bound x of A is called a greatest lower bound (infimum) of A if $x \geq y$ for all lower bounds y of A.

(ii) An element $x \in P$ is called an upper bound for A if $x \geq a$, for all $a \in A$. An upper bound x is called a least upper bound (supremum) of A if $b \geq a$ for all upper bounds b of A.

Example 11.1

(i) Consider the Poset $(Z^+, \leq)$, where $\leq$ denotes divisibility. Let $A = \{1, 2, 3, 4, 6, 8, 12, 24\} = D_{24}$. Clearly A is a subset of Z^+. Now the set of upper bounds set of A = $\{24, 48, 72, \ldots\}$. Here 24 is the least upper bound and 1 is the glb.

(ii) For any subset A of R (the set of all real numbers), we have that inf A = min A and sup A = max A.

11.2 Lattice as an Algebraic System

In this section we discuss the algebraic structure 'lattice'. Properties of lattices were discussed. Lattices with universal lower and universal upper bounds considered. Diagramic representations of lattices are observed. Two equivalent forms of lattice are defined. Some characterizations of complemented, distributive lattices are obtained. These concepts play important role in logical circuits and Boolean algebras.

Definition

A poset $(L, \leq)$ is said to be a lattice (or lattice ordered) if supremum of x and y; and infimum of x and y exist for every pair $x, y \in L$.

Note:

(i) Every chain is lattice ordered.

(ii) Let $(L, \leq)$ be a lattice ordered set; and $x, y \in L$. Then we have the following:
$x \leq y \Leftrightarrow \sup(x, y) = y \Leftrightarrow \inf(x, y) = x$.

Definition

A lattice $(L, \wedge, \vee)$ is a set L with two binary operations $\wedge$ (called as meet or product) and $\vee$ (called as join or sum) which satisfy the following laws, for all $x, y, z \in L$:

1. $x \wedge y = y \wedge x$, and $x \vee y = y \vee x$ (Commutative laws).
2. $x \wedge (y \wedge z) = (x \wedge y) \wedge z$, and $x \vee (y \vee z) = (x \vee y) \vee z$ (Associative laws).
3. $x \wedge (x \vee y) = x$; and $x \vee (x \wedge y) = x$ (Absorption laws).

Example 11.2

Let Z^+ be the set of positive integers. Define a relation 'D' on Z^+ by $aDb \Leftrightarrow a$ divides b for any $a, b, \in Z^+$. Then (Z^+, D) is a lattice, in which, $a \wedge b = \gcd\{a, b\}$ and $a \vee b = \text{lcm}\{a, b\}$.

Definition

Let $(L, \geq)$ be a lattice. If every non-empty subset of L has greatest lower bound and least upper bound, then L is said to be a complete lattice.

Example 11.3

(i) Let P be the set of all integers with usual ordering. Clearly it is a lattice. The set of all even integers is a subset of P and it has no upper bound or lower bound. Hence P is not a complete lattice.

(ii) If $P = \{i \mid 1 \leq i \leq n\}$ and $\geq$ is the usual ordering of integers, then P is a complete lattice.

Definition

A subset S of a lattice L is called a sublattice of L if S is a lattice with respect to the restriction of $\wedge$ and $\vee$ from L to S. So a subset S of L is a sublattice of the lattice L $\Leftrightarrow$ S is "closed" with respect to $\wedge$ and $\vee$ (that is, $s_1, s_2 \in S \Rightarrow s_1 \wedge s_2 \in S$ and $s_1 \vee s_2 \in S$).

Example 11.4

The lattice $(D_n, \leq)$ is a sublattice of $(Z^+, \leq)$ where $\leq$ is the divisibility relation.

Definition

Let $(A, \leq)$ be a lattice. An element $g \in A$ is called the greatest element of A if $a \leq g$ for all $a \in A$. Similarly, an element $s \in A$ is called the smallest (least) element of A if $s \leq a$ for all $a \in A$.

Example 11.5

   (i)   Consider N = the set of all natural numbers. Define $a \leq b \Leftrightarrow a$ divides b, for all $a, b \in N$. Then $(N, \leq)$ is a Poset. For any $x, y \in N$, we write $x \wedge y = \gcd\{x, y\}$ and $x \vee y = \mathrm{lcm}\{x, y\}$. Then $(N, \leq)$ is a lattice. Here 1 is the zero element. The greatest element does not exist.

   (ii) Let A be a set. Consider $\wp(A)$ = the power set of A. $(\wp(A), \subseteq)$ is a Poset (where $\subseteq$ is the set inclusion). For any $X, Y \in \wp(A)$, we write $X \wedge Y = X \cap Y$ and $X \vee Y = X \cup Y$. Then $(\wp(A), \subseteq)$ is a lattice. In this lattice, ϕ is the smallest element and A is the greatest element.

11.2.1 Properties of Lattices

Let $(L, \wedge, \vee)$ be an algebraic lattice and $x \in L$.

   1. $x \wedge x = x, x \vee x = x$ (idempotent laws)
   2. $x \vee y = y \vee x, x \vee y = y \vee x$ (commutative laws)
   3. $x \vee (y \vee z) = (x \vee y) \vee z, x \wedge (y \wedge z) = (x \wedge y) \wedge z$ (associative laws)
   4. $x \vee (x \wedge y) = x, x \wedge (x \vee y) = a$ (absorption laws)

Theorem 1

Let $(L, \leq)$ be a lattice. For $a, b, \in L$

   (i) $a \leq b \Leftrightarrow a \wedge b = a$
   (ii) $a \leq b \Leftrightarrow a \vee b = b$

Proof: Assume that $a \leq b$.

Since $a \leq a$, we have that a is a lower bound of a and b. Therefore $a \leq a \wedge b$ and $a \wedge b$ is the glb of a and b.

By definition of $a \wedge b$, we have $a \wedge b \leq a$. Therefore by antisymmetric property, we have $a \wedge b = a$.

Conserve: Suppose that $a \wedge b = a$. Then by definition of $a \wedge b$, $a = a \wedge b \leq b$.

Thus we have $a \wedge b = a \Rightarrow a \leq b$.

In a similar way we can prove (ii).

Example 11.6

Let a and b be two elements in a lattice $(L, \leq)$. Show that $a \wedge b = b$ if and only if $a \vee b = a$.

Solution:

Part (i): Suppose $a \wedge b = b$. Now

$$a = a \vee (a \wedge b) \qquad \text{(by absorption law)}$$
$$= a \vee b \qquad \text{(supposition law)}$$

Part (ii): Suppose $a \vee b = a$. Now

$$b = b \wedge (b \vee a) \qquad \text{(by absorption law)}$$
$$= b \wedge (a \vee b) \qquad \text{(by commutative law)}$$
$$= b \wedge a \qquad \text{(supposition law)}$$
$$= a \wedge b \qquad \text{(by commutative law)}$$

11.3   Algebraic Systems with One Binary Operation

In this section we discuss functions from a set $S \times S$ to S or more generally functions from S^n to S where $n = 1, 2, \ldots$

Definition

Let S be a set and g be a mapping from $S \times S \rightarrow S$. Then g is called a binary operation on S.

In general a mapping $g: S^n \rightarrow S$ is called an n – ary operation and n is called the order of the operation.

A mapping $g: S \rightarrow S$ is called a unary operation.

Note:

It is customary to denote a binary operation by a symbol such as $+, -, *, ., \cap, \cup, \wedge, \vee, \Delta$ etc.

For example $g(a, b)$ may be written as 'agb' or 'a*b'.

11.4   Properties of Binary Operations

Definitions

   (i)   A binary operation $g: S \times S \to S$ is said to be commutative if for every $a, b \in S$, $g(a, b) = g(b, a)$.

  (ii)   A binary operation $g: S \times S \to S$ is said to be associative if for every $a, b, c \in S$, $g(g(a, b), c) = g(a, g(b, c))$.

 (iii)   A binary operation $g: S \times S \to S$ is said to be distributive over the operation $h: S \times S \to S$ if for every $a, b, c \in S$, $g(a, h(a, b)) = h(g(a, b), g(x, z))$.

Example 11.7

   (i)   The operations addition and multiplication over the set of real numbers commutative and associative.

  (ii)   Union and intersection over the power set of any set are commutative and associative.

Example 11.8

How many commutative binary operations can be defined on a set A with 'n' elements? If A is a finite set with $|A| = 4$, determine how many binary operations are commutative?

[JNTUH, June 2010, Set No. 3]

Solution:

***Part* 1:** We have to define a binary operation on $A = \{a_1, a_2, ..., a_n\}$, an n element set. We have to define $a_i a_j$ to be an element of A. $\{a_i a_j \mid 1 \leq i \leq n, 1 \leq j \leq n\}$ contains n^2 elements. Each $a_i a_j$ must be in A. So far $a_i a_j$ there are n possibilities. Hence to define a binary operation, there are $(n)^{(n^2)}$ ways. Thus there exist $(n)^{(n^2)}$ binary operations.

***Part* 2:** To make the binary operation commutative, we have to have that $a_i a_j = a_j a_i$. So if we define $a_i a_j$ then automatically $a_j a_i$ will be defined.

	a_1	a_2	...	a_n
a_1				
a_2				
...				
a_n				

We have to define $a_i a_j$ for $1 \leq i \leq j$ for all $1 \leq j \leq n$.

Non diagonal entries are $(n^2 - n)$ in number we have to define $\dfrac{n^2 - n}{2}$ places $+ n$ for diagonal entries.

So we have to define $\dfrac{n^2 - n}{2} + n = \dfrac{n(n+1)}{2}$ entries and each entry will have n possibilities.

Hence the number of commutative binary operations are $n^{\frac{n(n+1)}{2}}$

Part 3: With n = 4 the number of binary operations is

$$n^{(n^2)} = 4^{(4^2)} = 4^{(16)}$$

Part 4: With n = 4 the number of commutative binary operations is

$$n^{\left(\frac{n(n+1)}{2}\right)} = 4^{10}$$

11.4.1 Result

Fill in the following table so that the binary operation * is commutative and has the idempotent property.

[JNTUH, June 2010, Set No. 1]

*	a	b	c
a	–	c	–
b	–	–	–
c	c	a	–

Proof: Given that the binary operation * satisfying the properties commutative and idempotent.

From the given table we have

    $a * b = c \Rightarrow b * a = c$ (by commutative property)

    $c * a = c \Rightarrow a * c = c$ (by commutative property)

    $c * b = a \Rightarrow b * c = a$ (by commutative property)

Since the binary operation * satisfying idempotent law we have

    $a * a = a;$

    $b * b = b$ and

    $c * c = c$

Now the given table is

*	a	b	c
a	a	c	c
b	c	b	a
c	c	a	c

Example 11.9

A binary composition * in R is defined by $a * b = a \cdot b^2$ for all a, b $\in$ R. Determine whether * is associative or not?

[JNTUH, June 2010, Set No. 3]

Solution: Given binary composition * in R is

$$a * b = a \cdot b^2 \qquad \text{for all a, } b \in R$$

Now we have to verify * is associative or not.

Let a, b, c $\in$ R

Consider $(a * b) * c = (a * b) \cdot c^2$

$$= (a \cdot b^2) \cdot c^2$$

$$= a \cdot b^2 \cdot c^2$$

Consider $a * (b * c) = a * (b * c^2)$

$$= a \cdot (b \cdot c^2)$$

$$= a \cdot b^2 \cdot c^4$$

$$(a * b) * c \neq a * (b * c)$$

Given binary composition * is not associative.

11.5   Semi Groups and Monoids

We begin our study of algebraic structures by investigating sets associated with single operations that satisfy certain axioms; that is, we wish to define an operation on a set in a way that will generalize such familiar structures as the integers Z together with the single operations of addition, matrix multiplication. We consider some algebraic systems with one binary operation on the set. These systems have useful applications in the theory of finite state machines, coding theory, and sequential mechanics.

Recall that a binary operation on a set S where S is non-empty, is a function from S × S into S. An n-ary operation on a set S is a function from S × S × ... × S (n times) into S. A unary operation is a function from S into S. If f is a binary operation on S, then for any two elements a, b in S the image of (a, b) under f is denoted by afb.

A non empty set together with a number of operations (one or more) operations defined on the set is called an algebraic system.

Definition

Let S be a non empty set. Then the operation * on S is said to be associative if $(a * b) * c = a * (b * c)$ for all, a, b, c $\in$ S.

Example 11.10

(i)   Take Z^+ = the set of positive integers. The binary operation '+' (usual addition) on Z^+ is an associative operation.

(ii)   Define * Z^+ as a * b = a^2 + b, where '+' is usual addition.

For any 2, 3, 4 ∈ Z^+, 2 * 3 = 2^2 + 3 = 4 + 3 = 7, (2 * 3) * 4 = 7 * 4 = 49 + 4 = 53.

Whereas 2 * (3 * 4) = 17. Therefore '*' on Z^+ is not associative.

Example 11.11

Take Z^+ = the set of positive integers. The binary operation '+' (usual addition) on Z+ is an associative operation.

Definition

Let (A, *) be an algebraic system where * is a binary operation on A. (A, *) is called a semigroup if the following conditions are satisfied:

(i)   '*' is a closure operation. That is., a * b ∈ A for all a, b ∈ A.

(ii)   '*' is an associative operation. That is., a * (b * c) = (a * b) * c, for all a, b, c ∈ A.

Example 11.12

(i)   Take E = {2, 4, 6, …}, the set of even positive integers. Define '+' on E as usual addition. Then (E, +) is a semi group.

(ii)   Take A = {a_1, a_2, …, a_n} be a non empty set. Let A^* be the set of all finite sequences of elements of A. That is, A^* consists of all words that can be formed from the set A. Let α, β be elements of A*. The operation catenation is a binary operation on A^*. For any two strings α = a_1 a_2 … a_n and β = b_1 b_2 … b_k, then α.β = $a_1a_2…a_n$ $b_1b_2…b_k$. It can be verified that for any α, β and γ of A^*, α. (β. γ) = (α. β). γ. Therefore (A^*.,) is a semi group.

(iii)   Let S be any set and P(S) the power set of S. Then (P(S), ∪) is a semi group, where ∪ is the set union.

(iv)   The set Z (the set of integers) with the binary operation subtraction is not a semigroup, since subtraction is not associative.

Example 11.13

Test whether the set Z (the set of integers), with binary operation * such that x * y = x^y is a semigroup.

Verification: Consider (2 * 2) * 3 = 2^2 * 3 = 4 * 3 = 4^3 = 64 and 2 * (2 * 3) = 2 * 2^3 = 2 * 8 = 2^8 = 256. Therefore (Z, *) is not a semigroup.

Definition

(i)   Let (S, *) be a semigroup and let T be a subset of S. If T is closed under the operation * (That is., a * b ∈ T whenever a and b are elements of T), then (T, *) is called a sub semigroup of (S, *).

(ii) Let (S, *) be a semigroup. For a ∈ S, we define $a^1 = a$ and $a^n = a^{n-1} * a$, n ≥ 2. For non-negative integers m, n we have $a^m * a^n = a^{m+n}$.

Example 11.14

Let (S, *) be a semigroup and T = $\{a^i \mid a \in S \text{ and } i \in Z^+\}$

Then for $a^i, a^j \in T$, we have $a^i * a^j = a^{i+j} \in T$ (since a ∈ S and $i + j \in Z^+$)

Therefore T is closed with respect to the operation *.

Hence (T, *) is a subsemigroup of (S, *).

Definition

Let (S, *) be a semigroup. An element a ∈ S is called a left-cancelable element if

a*x = a*y ⇒ x = y, for all x, y ∈ S.

Example 11.15

Show that if a and b are left-cancelable elements of a semigroup (S, *), then a*b is also a left cancelable element.

Solution: Take x, y ∈ S.

Now (a * b)* x = (a * b)* y ⇒ a * (b * x) = a * (b * y)   (by associative property)

⇒ b * x = b * y   (since a is left cancelable)

⇒ x = y   (since b is left cancelable)

11.5.1 Observation

We can define right cancelable element in a semigroup and the problem is true for right cancelable elements also.

Definition

Let (A, *) be an algebraic system where * is a binary operation on A. An element e in A is said to be a left identity (respectively, right identity) if for all x ∈ A, e * x = x (respectively, x * e = x) holds.

Example 11.16

(i)   Define '*' on A = {a, b, c, d} as follows:

*	a	b	c	d
a	d	a	b	c
b	a	b	c	d
c	a	b	c	c
d	a	b	c	d

Here both b and d are left identities.

(ii) Define 'o' on A = {a, b, c, d} as follows:

o	a	b	c	d
a	a	b	c	d
b	b	a	c	d
c	c	d	a	b
d	d	d	b	c

Here a is a right identity.

Definition

An element in an algebraic system is said to be an identity if it is both a left identity and a right identity. It can be observed that if e is a left identity, then either e is also a right identity or there is no right identity at all.

Note:

Observe that if e is a left identity, then either e is also a right identity or there is no right identity at all.

Definition

Let (M, *) be an algebraic system, where * is a binary operation on A. (M, *) is called a monoid if the following conditions are satisfied:

(i)   * is a closed operation

(ii) * is an associative operation

(iii) existence of identity

*	x	y
x	x	y
y	y	y

11.5.2   Result

For any binary operation * on a set M if identity element exists then it is unique.

Proof: Suppose e_1, e_2 are two identity elements in M.

Then $e_1 = e_1 * e_2$ (since e_2 is an identity)

$\qquad\quad = e_2$ (since e_1 is an identity)

Hence the identity element if it exists is unique.

Example 11.17

Let X be a non empty set. Write $X^X = \{f \mid f: X \to X\}$.

Let 'o' denotes the operation of composition of mappings.

That is, $(f o g)(x) = f(g(x))$ for all f, $g \in X^X$ and $x \in X$.

Now 'o' is a binary operation on X^X.

Also $f(x) = x$ for all $x \in X$ is the identity, as $(g o f)(x) = g(f(x)) = g(x) = f(g(x)) = (f o g)(x)$ for all $g \in X^X$. Therefore (X^X, o) is a monoid.

Example 11.18

Show that the set N of natural numbers is a semigroup under the operation *, where $x * y = \max \{x, y\}$. Is it a monoid?

Solution: Now

$\qquad (x * y) * z = \max \{\max\{x, y\}, z\} = \max \{x, y, z\}$

$x * (y * z) = \max \{x, \max\{y, z\}\} = \max \{x, y, z\}$. Hence * is associative. Thus (N, *) is a semi-group. Also $x*0 = \max\{x, 0\} = \max \{0, x\} = 0*x = x$. Therefore (N, *) is a monoid.

Example 11.19

   (i)   For any set S, $(\wp(S), \cup)$ where $\wp(S)$ is a power set of S, is a commutative semigroup. It is also a monoid with the empty set ϕ as the identity element.

  (ii)   The set (Z, +) is a monoid with identity 0.

 (iii)   Let (M, *) be a monoid with identity 'e' and T a non empty subset of M. If T is closed under the same operation '*' and $e \in T$, then (T, *) is called submonoid of (S, *).

11.5.3   Observations

   (i)   The associative property holds in any subset of a semigroup.

  (ii)   A submonoid of a monoid is itself a monoid.

Example 11.20

Let T be the set of even integers. Then (T, .) is a subsemigroup of the monoid (Z, .) where "." is usual multiplication. But (T, .) is not a submonoid, since the identity $1 \notin T$.

Example 11.21

(i) Suppose (S, *) is a semigroup, and let a ∈ S. For any n ∈ Z^+, we define the integral powers of a^n recursively as follows:

$a^1 = a$, $a^n = a^{n-1} * a$, $n \geq 2$. Write $T = \{a^n \mid n \in Z^+\}$.

Then (T, *) is a subsemigroup of (S, *).

(ii) Let (S, *) be a monoid and a ∈ S.

Define $a^0 = e$, $a^1 = a$, $a^n = a^{n-1} * a$, $n \geq 2$ (as in (i))

Write $T^1 = \{a^n \mid n \in Z^+, \cup \{0\}\}$. Then $(T^1, *)$ is a submonoid of (S, *).

Definition

Let (M, *) be a monoid. An element a ∈ M is called an idempotent element if $a^2 = a$.

Theorem 2

For any commutative monoid (M, *), show that the set of idempotent elements of M forms a sub-monoid.

Proof: Let (M, *) be any commutative monoid with identity e.

Write T = {x | x is an idempotent elements of M}. Now e * e = e and so e ∈ T. Therefore T is non-empty.

Take x, y ∈ T.

Now (x * y) * (x * y) = (x * y) * (y * x)   (since M is commutative)

$= x * (y * y) * x$ (since M is associative)

$= x * (y * x)$ (since y ∈ T is an idempotent)

$= x * (x * y)$ (since M is commutative)

$= (x * x) * y$ (since M is associative)

$= x * y$ (since x ∈ T is an idempotent)

Therefore x * y ∈ T. Hence (T, *) is a sub-monoid of (M, *).

Theorem 3

Let (S, .) and (T, *) be two monoids with identities e and e^1 respectively and f: S → T be an isomorphism. Then $f(e) = e^1$.

Proof: Let b ∈ T. Since f is onto, there exists a ∈ S such that f(a) = b.

Now b = f(a) = f(a . e) = f(a) * f(e) = b * f(e).

Similarly, b = f(e) * b.

Hence f(e) is an identity in T.

Since identity is unique, we have $f(e) = e^1$.

Example 11.22

Prove that the intersection of two submonoids of a monoid is a monoid.

[JNTUH, Nov 2008, Set No. 2]

Solution: Let M be a monoid under a binary operation '*' say, with identity 'e'.

Let M_1 and M_2 be two submonoids of M. Since M_1 and M_2 are submonoids, so these are monoids.

(Since a submonoid of a monoid is itself a monoid under the binary operation in monoid)

Therefore $e \in m_1$ and $e \in m_2 \Rightarrow e \in m_1 \cap m_2$.

Since $m_1 \cap m_2$ is a subset of m, the associative law holds in $m_1 \cap m_2$, because it holds in m.

This shows that, $m_1 \cap m_2$ forms a monoid with 'e' as the identity.

Example 11.23

Let $(S_1, *_1)$, $(S_2, *_2)$ and $(S_3, *_3)$ be semi-groups and f: $S_1 \rightarrow S_2$ and g: $S_2 \rightarrow S_3$ be homomorphisms. Prove that the mapping of *gof:* $S_1 \rightarrow S_3$ homomorphism

[JNTUH, Nov 2008, Set No. 4]

Solution: Let f: $S_1 \rightarrow S_2$ be a homomorphism of semigroups S_1 and S_2.

Let g: $S_2 \rightarrow S_3$ be a homomorphism of semigroups S_2 and S_3.

First we note that gof is a function from S_1 to S_3. Take any a, b $\in S_1$.

Then from the hypothesis of the theorem, we have

$$(gof)(a *_1 b) = g(f(a*_1 b)) = g(f(a)*_2 f(b))$$

$$= g(f(a)) *_3 g(f(b))$$

$$= (gof)(a) *_3 (gof)(b)$$

This shows that (gof) is a homomorphism from S_1 to S_3.

11.6   Homomorphism of Semi Groups and Monoids

11.6.1   Definition

Let (S, *) and (S^1, o) be two semigroups. A function f: $S \rightarrow S^1$ is called an isomorphism from (S, *) to (S^1, o) if

   (i)   f is one-to-one (that is, one-one and onto)

   (ii)   f(a * b) = f(a) o f(b) for all, a, b $\in$ S (homomorphism condition)

A homomorphism of a semigroup into itself is called a semigroup endomorphism.

Example 11.24

Consider (N, +) and $(Z_m, +_m)$. Define g: N → Z_m by g(a) = [i] where i is the remainder obtained when a is divided by m, for a ∈ N.

For a, b ∈ N, let g(a) = [i] and g(b) = [j]. Then

g(a + b) = [(a + b) (mod m)] = [(i + j) (mod m)] = [i] $+_m$ [j] = g(a) $+_m$ g(b).

Hence g is a homomorphism. Further g(0) = [0]. Hence g preserves the identity.

11.6.2   Result

If f is an isomorphism from (S, *) to (S^1, o), then f^1 is an isomorphism from (S^1, o) to (S, *).

Proof: Let $a^1, b^1 ∈ S^1$. Since f is onto, there exist a, b, ∈ S such that $f(a) = a^1$, $f(b) = b^1$. Then
$f^{-1} (a^1 o b^1) = f^1(f(a) o f(b)) = f^1(f(a * b))$ (Since f is homomorphism)

$$= (f^1 o f)(a * b)$$

$$= a * b$$

$$= f^1(a^1) * f^1(b^1)$$

Therefore f^1 is an isomorphism.

Example 11.25

Show that the semigroups (Z, +) and (E, +) where E is the set of all even integers, are isomorphic.

Solution: Define f: Z → E by f(n) = 2n

To show f is one-one: Suppose $f(n_1) = f(n_2) ⇒ 2n_1 = 2n_2 ⇒ n_1 = n_2$.

To show f is onto: Suppose b ∈ E. Then b is an even integer. Write $a = \dfrac{b}{2} ∈ Z$

Now $f(a) = f\left(\dfrac{b}{2}\right) = 2\left(\dfrac{b}{2}\right) = b$. Therefore f is one-one and onto.

To show f is homomorphism. Let m, n ∈ Z. Then

$$f(m + n) = 2(m + n) = 2m + 2n = f(m) + f(n)$$

Therefore f is a homomorphism and hence (Z, +) and (E, +) are isomorphic.

Definition

An equivalence relation 'R' on the semigroup (S, *) is called a congruence relation if aRa^1 and bRb^1 imply $(a * b) R (a^1 * b^1)$.

11.6.3   Observation

$a \equiv b \pmod n \Rightarrow a = qn + r$ and $b = tn + r$ for some $q, t, r \in Z \Rightarrow a - b$ is a multiple of n. That is., $n \mid a - b$.

Example 11.26

Semigroup $(Z, +)$ and the equivalence relation R on Z defined by aRb if and only if $a \equiv b \pmod 2$. If $a \equiv b \pmod 2$, then $2 \mid a - b$.

Now $a \equiv b \pmod 2$ and $c \equiv d \pmod 2 \Rightarrow 2 \mid a - b$ and $2 \mid c - d$.

$\Rightarrow \quad a - b = 2m, c - d = 2n$, where $m, n \in Z$

Adding $(a - b) + (c - d) = 2(m + n) \Rightarrow (a + c) - (b + d) = 2(m + n)$.

Therefore $(a + c) \equiv (b + d) \pmod 2$.

This shows that the relation is a congruence relation.

Example 11.27

Consider the semigroup $(Z, +)$ where '+' is the ordinary addition. Let $f(x) = x^2 - x - 2$. Define a relation R on Z by $a \ R \ b \Leftrightarrow f(a) = f(b)$.

Clearly aRa. So R is a reflexive relation.

Since $aRb \Leftrightarrow f(a) = f(b) \Leftrightarrow bRa$, we have that R is a symmetric relation.

Now aRb and $bRc \Leftrightarrow f(a) = f(b)$ and $f(b) = f(c)$

$\Leftrightarrow f(a) = f(c) \Leftrightarrow aRc$. Therefore R is a Transitive relation.

Hence 'R' is an equivalence relation.

Now we see R is not a congruence relation;

Since $f(-1) = f(2) = 0$, we have $-1R2$.

Since $f(-2) = f(3) = 4$, we have $-2R3$, but "not $(-1 + (-2)) \ R \ (2 + 3)$" (because $f(-3) = 10$ and $f(5) = 8$).

Definition

If $(S, *)$ and (T, o) are semigroups, then $(S \times T, \Delta)$ is a semigroup, where Δ defined by $(s_1, t_1) \ \Delta \ (s_2, t_2) = (s_1 * s_2, t_1 \ o \ t_2)$. This will become a semigroup, called product semigroup. If e_s and e_r are the identities of S and T then (e_s, e_r) is the identity element in $S \times T$.

Example 11.28

Consider the semigroup (Z, +) and the equivalence relation R on Z defined by aRb if and only if a ≡ b (mod 2). That is, aRb ⇔ a − b is divisible by 2. It is easy to verify that R is an equivalence relation on Z.

To verify that R congruence relation, suppose aRb and cRd.

Then a ≡ b (mod 2) and c ≡ d (mod 2) ⇒ a − b = 2m and c − d = 2n, for some integers m and n are integers. Adding we get (a − b) + (c − d) = 2m + 2n, so (a + c) − (b + d) = 2(m + n). This means that (a + c) ≡ (b + d) (mod 2).

Theorem 4

Let R be a congruence relation on the semigroup (S, *). Consider the relation ⊗ from S/R × S/R to S/R in which the ordered pair ([a], [b]) is for a and b in S, related to [a * b].

 (i)   ⊗ is a function from S/R × S/R to S/R.

  ⊗([a], [b]) = [a] ⊗ [b] = [a ⊗ b]

 (ii) (S/R, ⊗) is a semigroup.

***Proof*:** To verify that ⊗ is a function, suppose ([a], [b]) = ([a′], [b′]).

Then aR a′ and bR b′. Since R is a congruence relation on S, we have a*bR a′ * b′

 ⇒   [a*b] = [a′ * b′]

 ⇒   [a] ⊗ [b] = [a′] ⊗ [b′].

That is., ⊗([a], [b]) = ⊗([a′], [b′]).

This shows that ⊗ is a binary operation on S/R.

Next we verify that ⊗ is associative.

Now [a] ⊗ ([b] ⊗ [c]) = [a] ⊗ [b*c]

$$= [a*(b*c)] = [(a*b)*c] \quad \text{(by associativity of *)}$$

$$= [a * b] ⊗ [c] = ([a] ⊗ [b]) ⊗ [c].$$

Therefore ⊗ satisfies associative property. Hence S/R is a semigroup.

Definition

The semigroup S/R verified above is called the quotient semigroup or factor semigroup.

Example 11.29

Take a semigroup (Z, +). Define a relation 'R' on Z as follows:

Let n be a positive integer, aRb ⇔ a ≡ b (mod n).

We verify that 'R' is an equivalence relation. Since n divides $0 = a - a$, we have that

$a \equiv a \pmod n$ and so aRa. Suppose aRb, then $a \equiv b \pmod n$

$\Leftrightarrow n|\, a - b$

$\Leftrightarrow n|\, -(a - b)$

$\Leftrightarrow n|\, b - a$

$\Leftrightarrow b \equiv a \pmod n$. Therefore aRb $\Rightarrow$ bRa.

Suppose $a \equiv b \pmod n$ and $b \equiv c \pmod n$.

Then $n|\, a - b$ and $n|\, b - c \Rightarrow n|\, (a - b) + (b - c) \Rightarrow n|\, a - c$

Therefore $a \equiv c \pmod n$

Therefore aRb, bRc $\Rightarrow$ aRc. So R is an equivalence relation.

Take $n = 4$. The equivalence classes determined by the congruence relation $\equiv \pmod 4$ on Z. (It is denoted by Z_4).

$[0] = \{\dots -8, -4, 0, 4, 8, 12, \dots\} = [4] = [8] = \dots$

$[1] = \{\dots -7, -3, 1, 5, 9, 13, \dots\} = [5] = [9] = \dots$

$[2] = \{\dots -6, -2, 2, 6, 10, 14, \dots\} = [6] = [10] = \dots$

$[3] = \{\dots -5, -1, 3, 7, 11, 15, \dots\} = [7] = [11] = \dots$

Define $\oplus$ on Z_4 as follows:

$\oplus$	[0]	[1]	[2]	[3]
[0]	[0]	[1]	[2]	[3]
[1]	[1]	[2]	[3]	[0]
[2]	[2]	[3]	[0]	[1]
[3]	[3]	[0]	[1]	[2]

In general, $[a] \oplus [b] = [a + b]$. Thus Z_n has the 'n' equivalence classes [0], [1], [2], ..., [n − 1] and that $[a] \oplus [b] = [r]$, where r is the remainder when $a + b$ is divided by n. The following theorem establishes a relation between the structure of a semigroup (S, *) and the quotient semigroup (S/R, $\oplus$), where R is a congruence relation on (S, *).

Theorem 5

Let R be a congruence relation on a semigroup (S, *) and let (S/R, $\otimes$) be the corresponding quotient semigroup. Then the function f_R: S $\to$ S/R defined by $f_R(a) = [a]$ is an onto homomorphism.

Proof: Take $[a] \in$ S/R. Then $f_R(a) = [a]$, so f_R is an onto function. Let a, b $\in$ S, then

$f_R(a * b) = [a * b] = [a] \otimes [b] = f_R(a) \otimes f_R(b)$. Therefore f_R is a homomorphism.

11.6.4 Fundamental Theorem of Homomorphism

Let f: S → T be a homomorphism of the semigroup (S, *) onto the semigroup (T, o). Let R be the relation on S defined by a R b ⇔ f(a) = f(b) for a and b in S. Then

(i) R is a congruence relation;

(ii) (T, o) and the quotient semigroup (S/R, ⊗) are isomorphic.

Theorem 6

Let f: S → T be a homomorphism of the semigroup (S, .) onto the semigroup (T, *). Let R be a relation defined on S by aRb ⇔ f(a) = f(b) for all a, b ∈ S. Then

(i) R is a congruence relation;

(ii) (S/R, Θ) is isomorphic to (T, *)

Proof:

(i) Since f(a) = f(a), we have that aRa. Therefore R is a reflexive relation.

aRb ⇒ f(a) = f(b) ⇒ f(b) = f(a) ⇒ bRa. Therefore R is a symmetric relation.

aRb and bRc ⇒ f(a) = f(b) = f(c) ⇒ aRc. Therefore R is a transitive relation.

Hence R is an equivalence relation.

To verify that R is congruence relation let aR a′ and bR b′.

This means that f(a) = f(a′) and f(b) = f(b′). This implies that f(a) * f(b) = f(a′) * f(b′). That is, f(a.b) = f(a′.b′).

Therefore (a.b)R(a′.b′). Hence R is a congruence relation.

(ii) Consider the quotient semigroup (S/R, Θ). Define h: S/R → T as

h([a]) = f(a)

To show h is well defined, suppose [a] and [b]. Then aRb. This implies that f(a) = f(b).

Take b ∈ T. Since f is onto there exists a ∈ S such that f(a) = b. This means that h([a]) = f(a) = b.

To show that h is one one, suppose h([a]) = h([b]). This implies f(a) = f(b) ⇒ aRb ⇒ [a] = [b].

Also h([a] Θ [b]) = h([a.b]) = f(a.b) = f(a)*f(b). Therefore h is a homomorphism and hence an isomorphism.

Example 11.30

Show that the function 'f' from (N, +) to (N, *) where N is the set of all natural numbers, defined by f(x) = 2^x ∀x ∈ N is a homomorphism.

[JNTUH, Nov 2008, Set No. 2]

Solution: Given that f is a function from (N, +) to (N, *) defined by $f(x) = 2^x$ for all $x \in N$, N is the set of all natural numbers.

Let $x, y \in N$.

Then $f(x + y) = 2^{x+y} = 2^x.2^y = f(x).f(y)$

   (i) $f(x) = f(y) \Rightarrow 2^x = 2^y \Rightarrow x = y$

          so f is well defined.

   (ii)   If we take any y in (N, *) then $y = 2^x$ for some $x \in (N, +)$

          $\Rightarrow f(x) = 2^x$

          Therefore f is onto.

So f is homomorphism from (N, +) to (N, *).

Example 11.31

Explain in detail, the algebraic systems: Endomorphism and Automorphism with suitable examples.

[JNTUH, Nov 2008, Set No. 1]

Solution:

Endomorphism: Let (S, o) and (T, *) be two algebraic systems such that $T \subseteq S$. A homomorphism f from (S, o) to (T, *) in such a case is called endomorphism.

Example: Let G be a group and let 'e' be the identity element of G. Then show that the mapping f: $G \rightarrow G$ defined by $f(a) = e$, for all $a \in G$ is an endomorphism of G.

Solution: Let a, b be any two elements of G. Then $f(a) = e$, $f(b) = e$.

   Now $f(ab) = e$ (since a, b $\in G \Rightarrow ab \in G$ and $f(ab) = e$)

          $= e.e = f(a) f(b)$

   Thus f is a homomorphism from G into G. Therefore f is an endomorphism of G.

Automorphism: Let (S, o) and (T, *) be two algebraic systems. If S = T, then an isomorphism from (S, o) to (T, *) is called an Automorphism.

Example: Let G be an abelian group and f: $G \rightarrow G$ be defined such that $f(x) = \dfrac{1}{x}$ for all x $\in$ G. Then show that f is an Automorphism.

Solution: Let G be an abelian group.

Define a function f: G $\rightarrow$ G as

$$f(x) = \frac{1}{x} \text{ for all } x \in G$$

Now x, y $\in$ G

$$f(x) = f(y) \Rightarrow \frac{1}{x} = \frac{1}{y}$$

$\Rightarrow$ x = y $\Rightarrow$ f is one-one.

Let x $\in$ G $\Rightarrow \frac{1}{x} \in$ G such that $f(x) = \frac{1}{x}$

$\Rightarrow$ f is onto.

Now $f(xy) = \frac{1}{xy} = \frac{1}{x}.\frac{1}{y} = f(x).f(y)$

Therefore f: G $\rightarrow$ G is an Automorphism.

Exercises

Properties

1. Let R be the set of all real numbers with '≤' as the partial order. Also, let B be the open interval (1, 2). Find

    (i) All upper bounds of B;

    (ii) All lower bounds of B.

    Ans: (i) Every real number ≥ 2 is an upper bound of B.

    (ii) Every real number ≤ 1 is a lower bound of B.

2. Show that every partially ordered set has atmost one greatest element and atmost one least element.

3. For the poset shown in the following Hasse diagram find

    (i) All maximal elements;

    (ii) All minimal elements.

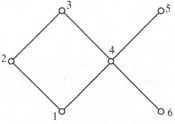

Ans: (i) maximal elements 3, 5, (ii) minimal elements 1, 6

Lattices as Algebraic System

1. Show that the poset {1, 3, 6, 12, 24} is a Lattice.

2. Find the maximal and minimal elements of the poset {2, 4, 5, 10, 12, 20, 25}

   Ans: min. elements 2, 5, max. elements 12, 20, 25

3. Show that sublattice of a modular lattice is modular.

Algebraic Systems with One Binary Operation

1. Indicate whether the usual division is a binary operation.

   Ans: (i) No

2. Let '*' denote a binary operation on the set of natural numbers, given by a * b = a. Then show that '*' is not commutative, but it is associative.

3. Is the operation '*' defined by $a * b = a^b$ a binary operation on the set of all positive integers? If it is so, is it commutative? Is it associative?

   Ans: yes, not commutative, not associative.

Properties of Binary Operations

1. Let X = R and '*' is defined by a * b = a + b + 1.

   Then show that

   (i) '*' is commutative

   (ii) '*' is associative

   (iii) "*" is having identity '–1'

Semi Groups and Monoids

1. Let Z be the set of positive integers and '*' be an operation on 'S' = Z × Z defined by

   (i) (a, b) * (c, d) = (ac, bd)

   (ii) (a, b) * (c, d) = (a + c, b + d)

   Then show that S is a semi-group.

2. Show that

   (i) (N, .) is a monide

   (ii) (N, +) is not a monoid

   where N is the set of Natural numbers.

3. In each of the following cases, a binary operation '*' on the set A = {1, 2} is defined through a multiplication table. Determine whether < A, *> is a semigroup or a monoid.

(i)

*	a	B
a	a	B
b	b	b

(ii)

*	a	B
a	B	b
b	a	a

Ans: (i) Monoid (ii) neither a semigroup nor a monoid

Homomorphism of Semi-groups and Monoids

1. Show that the function f from $< Z, + >$ to $< E, + >$ defined by $f(x) = x^2$ for all $x \in Z$ is not a homomorphism.

2. On the set $A = N \times N$, where N is the set of all natural numbers, the operation '*' is defined by

   $(a, b) * (c, d) = (a + c, b + d)$

   where '+' is the usual addition. Prove that $<A, *>$ is a semigroup and the function $f: A \rightarrow Z$ is defined by $f(a, b) = a - b$ is a homomorphism.

3. Let $S = N \otimes N$. Let '*' be the operation on S defined by $(a, b) * (c, d) = (ac, bd)$.

   Show that a mapping $f: S$ Q, defined as $f(a, b) = \dfrac{a}{b}$ is a homomorphism.

CHAPTER - 12

Algebraic Structures
(Groups and Rings)

We begin our study of algebraic structures by investigating sets associated with single operations that satisfy certain reasonable axioms; that is, we wish to define an operation on a set in a way that will generalize such familiar structures as the integers Z together with the single operations of addition, matrix multiplication. We study the important algebraic object known as group which serve as one of the fundamental building blocks for the abstract algebra. In fact group theory has several applications in every area where symmetry occurs. Applications of groups also can be found in physics, chemistry. Some of exciting applications of group theory have arisen in fields such as particle physics, and binary codes.

> **LEARNING OBJECTIVES**
>
> ♦ *to understand the fundamental Concepts of Group Theory*
> ♦ *to know the concept of Coset*
> ♦ *to understand the concept: Ring and examples*

12.1 Fundamental Concepts in Group Theory

Definition

(i) We recollect that for a non empty set G, a binary operation on G is mapping from G × G to G. In general, binary operations are denoted by *, ., o etc.

(ii) A non empty set G together with a binary operation * is called a group if the algebraic system (G, *) satisfies the following four axioms:

    (a) *Closure axiom:* a, b are elements of G, implies a*b is an element of G.

    (b) *Associative axiom:* (a*b)*c = a*(b*c) for all elements a, b, c in G.

(c) *Identity axiom:* There exists an element 'e' in G such that

    a*e = e*a = a for all a ∈ G.

(d) *Inverse axiom:* For any element a in G there corresponds an element b in G such that a*b = e = b*a.

Note:

The element e of G (given in identity axiom) is called an identity element. The element b (given in the inverse axiom) is called an inverse of a in G.

Definition

Let (G, *) be a group. Then (G, *) is said to be a commutative group (or Abelian group) if it satisfies the commutative property: a*b = b*a for all a, b in G.

Example 12.1

Take G = {–1, 1}. Then (G, .) is a commutative group w.r.t. the usual multiplication of numbers.

Solution:

Closure axiom: Clearly a, b is in G for all a, b in G.

Associative axiom: Since 1, –1 are real numbers, this axiom holds.

.	–1	+1
–1	1	–1
1	–1	1

Identity axiom: 1.a = a = a.1 for all elements a ∈ G.

   Hence 1 is the identity element.

Inverse axiom: The element 1 is the inverse of 1, and –1 is the inverse of –1.

Commutative axiom: (–1).1 = 1.(–1). Therefore commutative law holds in (G, .)

   Hence (G, .) is a commutative group.

Example 12.2

(i)   (Z, +) where Z is the set of all integers is an abelian group. 0 is the identity element and –a is the inverse of a.

(ii) The set of all n×n matrices under matrix addition is an abelian group with 0 matrix (null matrix) as the identity element and –A is the inverse of A.

(iii) (N, +) where N is the set of natural numbers, is not a group (since no element has an inverse in N with respect to addition).

      (for example, 3 ∈ N but the additive inverse of 3 is –3 ∉ N).

(iv) The set of all non-regular n×n matrices forms a group under matrix multiplication with I_n, the identity matrix of order n, as the identity element, and A^{-1} as the inverse of the matrix A. Since for any two matrices A and B we have AB ≠ BA, in general we can conclude that this is not an abelian group.

Example 12.3

Let Z_m be the set of all equivalence classes for the relation congruent modulo m and $+_m$ is the modulo m addition. Take m = 5 and verify that $(Z_5, +_5)$ is an abelian group.

Solution: We give the table of values with respect to the operation $+_5$ on Z_5.

Clearly $Z_5 = \{[0], [1], [2], [3], [4]\}$.

$+_5$	[0]	[1]	[2]	[3]	[4]
[0]	[0]	[1]	[2]	[3]	[4]
[1]	[1]	[2]	[3]	[4]	[0]
[2]	[2]	[3]	[4]	[0]	[1]
[3]	[3]	[4]	[0]	[1]	[2]
[4]	[4]	[0]	[1]	[2]	[3]

By observing the table we can conclude that $(Z_5, +_5)$ is an abelian group.

Example 12.4

Let p be a prime number. Consider the set $(Z_p - \{[0]\}, \times_p)$ where $\times_p$ is multiplication modulo p. Verify that for p = 5, Z_5 forms a commutative group with respect to multiplication modulo 5. Write the inverses of each element.

Solution: The multiplication table for $(Z_5 - \{[0]\}, \times_5)$ is given below.

Clearly $Z_5 - \{[0]\} = \{[1], [2], [3], [4]\}$.

$\times_5$	[1]	[2]	[3]	[4]
[1]	[1]	[2]	[3]	[4]
[2]	[2]	[4]	[1]	[3]
[3]	[3]	[1]	[4]	[2]
[4]	[4]	[3]	[2]	[1]

From the table, the identity element is [1]; the inverse of [2] is [3], the inverse of [3] is [2], the inverse of [4] is [4].

Theorem 1

Given that G is a group and a, b ∈ G. Then

(i)   The equations a.x = b and y.a = b have unique solutions for x and y in G.

(ii)   The cancellation laws: a.u = a.w $\Rightarrow$ u = w; and u.a = w.a $\Rightarrow$ u = w for all u, w in G hold.

Proof:

(i)   Let a, b in G.

Since G is a group, we have a^{-1} is in $G \Rightarrow a^{-1}.b, b.a^{-1}$ are in G (by closure property)
Write $x = a^{-1}.b$ and $y = b.a^{-1}$

Consider $a.x = a.(a^{-1}.b) = (a.a^{-1}).b = e.b = b$. Similarly $y.a = b$

Therefore $x = a^{-1}.b$ and $y = b.a^{-1}$ are required solutions

Next we show that these solutions are unique

Suppose x_1, x_2 are two solutions of $a.x = b$

Then $a.x_1 = b$ and $a.x_2 = b$. So $a.x_1 = b = a.x_2$

Now $x_1 = e.x_1 = (a^{-1}.a)x_1 = a^{-1}.(a.x_1) = a^{-1}.(a.x_2) = (a^{-1}.a)x_2 = e.x_2 = x_2$. Therefore $x_1 = x_2$. Hence $a.x = b$ has unique solution. Similarly, we show that $y.a = b$ has unique solution.

(ii) Now we will prove that the cancellation laws holds good in G.

Suppose $a.u = a.w$. Consider u

Now $u = e.u = (a^{-1}.a).u = a^{-1}.(a.u) = a^{-1}.(a.w) = (a^{-1}.a).w = e.w = w$

Now suppose $u.a = w.a$. Consider u

Now $u = u.e. = u.(a.a^{-1}) = (u.a).a^{-1} = (w.a).a^{-1} = w.(a.a^{-1}) = w.e = w$

Hence both the left and right cancellation laws holds in G.

12.1.1   Result

Let G be a non-empty set and '.' be a binary operation on G which is associative. If for all $a \in G$ there exists $e \in G$ such that $e.a = a$ and for all $a \in G$, there exists $b \in G$ such that $b.a = e$ then

(i) $a.m = a.n \Rightarrow m = n$

(ii) for all $a \in G$, $a.e = a$ (e is right identity)

(iii) for all $a \in G$ there exists $b \in G$ such that $a.b = e$.

Proof:

(i)   Suppose $a.m = a.n$. Consider $m = e.m = (b.a)m = b(a\ m) = b(a\ n) = (ba)n = en = n$. Therefore $m = n$

(ii) Given that $e.a = a$ for all $a \in G$. Now $e.e = e \Rightarrow (ba).e = ba$ (since $ba = e$)
$\Rightarrow b(ae) = ba \Rightarrow ae = a$ ((by (i)).

(iii) Let $a \in G$. By our assumption $ba = e$. Consider $b(ab) = (ba)b = eb = b = be$
((by (ii)) $\Rightarrow b(ab) = be \Rightarrow ab = e$.

The proof is complete

Theorem 2

If G is a group, then

    (i)   The identity element of G is unique.

    (ii)   Every element in G has unique inverse in G.

    (iii) For any a in G, we have $(a^{-1})^{-1} = a$.

    (iv) For all a, b in G, we have $(a.b)^{-1} = b^{-1}.a^{-1}$.

Proof:

    (i)   Let e, f be two identity elements in G. Since e is the identity we have $e.f = f$. Since f is the identity we have $e.f = e$. Therefore $e = e.f = f$. Hence the identity element is unique.

    (ii)   Let a be in G and a_1, a_2 are two inverses of a in G.

$$
\begin{aligned}
\text{Now} \quad a_1 &= a_1.e && \text{(since e is the identity)}\\
&= a_1.(a.a_2) && \text{(since } a_2 \text{ is the inverse of a)}\\
&= (a_1.a)a_2 && \text{(by associativity)}\\
&= e.a_2 && \text{(since } a_1 \text{ is the inverse of a)}\\
&= a_2.
\end{aligned}
$$

        Hence the inverse of an element in G is unique.

    (iii) Let $a \in G$. Since $a.a^{-1} = e = a^{-1}.a$, we have that a is the inverse of a^{-1}. Hence $(a^{-1})^{-1} = a$.

    (iv) Let a, b $\in$ G. Consider $(b^{-1}.a^{-1})(a.b) = b^{-1}.(a^{-1}.a).b = b^{-1}.e.b = b^{-1}.b = e$. Similarly $e = (a.b)(b^{-1}.a^{-1})$. This shows that $(a.b)^{-1} = b^{-1}.a^{-1}$.

Theorem 3

Show that G = {1, 2, 3, 4, 5} is not a group under addition modulo 6 and multiplication modulo 6.

[JNTUH, Nov 2010, Set No. 3]

Proof: Given that G = {1, 2, 3, 4, 5}.

We have to show that $(G, +_6)$ and $(G, ._6)$ are not groups.

Consider the addition and multiplication tables modulo 6.

$+_6$	1	2	3	4	5
1	1	3	4	5	0
2	3	4	5	0	1
3	4	5	0	1	2
4	5	0	1	2	3
5	0	1	2	3	4

for elements 1, 5 in G = {1, 2, 3, 4, 5}

$1 +_6 5 = o \notin G$

with respect to addition modulo 6 fails closer property.

Therefore $(G, +_6)$ is not a group.

Consider the table of multiplication modulo 6 on G = {1, 2, 3, 4, 5}

·6	1	2	3	4	5
1	1	2	3	4	5
2	2	4	0	2	4
3	3	0	3	0	3
4	4	2	0	4	2
5	5	4	3	2	1

By observing the table of multiplication modulo 6 on G = {1, 2, 3, 4, 5}

For 2, 3 ∈ G. $2 ._6 3 \notin G$

Therefore $(G, ·6)$ is not a group under multiplication modulo 6.

Theorem 4

If G is the set of all matrices of the form $\begin{pmatrix} a & 0 \\ 0 & a^{-1} \end{pmatrix}$ where a ≠ e, prove that G is an abelian

group under matrix multiplication.

[JNTUH, Nov 2010, Set No. 3]

Proof: Let G be the set of all matrices of the form $\begin{pmatrix} a & 0 \\ 0 & a^{-1} \end{pmatrix}$ where a ≠ e.

We have to show that G is an Abelian group under matrix multiplication.

Let x, y ∈ G. The $x = \begin{pmatrix} a & 0 \\ 0 & a^{-1} \end{pmatrix}$ and $y = \begin{pmatrix} b & 0 \\ 0 & b^{-1} \end{pmatrix}$ where a, b are elements of matrices

with a ≠ e ≠ b.

$$\text{Now } xy = \begin{pmatrix} a & 0 \\ 0 & a^{-1} \end{pmatrix} . \begin{pmatrix} b & 0 \\ 0 & b^{-1} \end{pmatrix} = \begin{pmatrix} ab + 0.0 & a.0 + 0.b^{-1} \\ 0.b + a^{-1}.0 & 0.0 + a^{-1}.b^{-1} \end{pmatrix}$$

$$= \begin{pmatrix} ab + 0 & 0 + 0 \\ 0 + 0 & 0 + a^{-1}b^{-1} \end{pmatrix}$$

$$= \begin{pmatrix} ab & 0 \\ 0 & (ab)^{-1} \end{pmatrix} \text{ is in the given matrix form.}$$

Therefore x y $\in$ G.

Thus G is closed under matrix multiplication.

Let x, y and z in G.

Then $\quad x = \begin{pmatrix} a & 0 \\ 0 & a^{-1} \end{pmatrix} \quad y = \begin{pmatrix} b & 0 \\ 0 & b^{-1} \end{pmatrix} \quad$ and $\quad z = \begin{pmatrix} c & 0 \\ 0 & c^{-1} \end{pmatrix}$

Now consider $xy = \begin{pmatrix} a & 0 \\ 0 & a^{-1} \end{pmatrix} \begin{pmatrix} b & 0 \\ 0 & b^{-1} \end{pmatrix}$

$$= \begin{pmatrix} a.b+0.0 & a.0+0.b^{-1} \\ 0.b+a^{-1}.0 & 0.0+a^{-1}.b^{-1} \end{pmatrix}$$

$$= \begin{pmatrix} ab+0 & 0+0 \\ 0+0 & 0+a^{-1}b^{-1} \end{pmatrix}$$

$$= \begin{pmatrix} ab & 0 \\ 0 & (ab)^{-1} \end{pmatrix}$$

Again consider $(xy)z = \begin{pmatrix} ab & 0 \\ 0 & (ab)^{-1} \end{pmatrix} . \begin{pmatrix} c & 0 \\ 0 & c^{-1} \end{pmatrix}$

$$= \begin{pmatrix} ab.c+0.0 & ab.0+0.c^{-1} \\ 0.c+(ab)^{-1}.0 & 0.0+(ab)^{-1}.c^{-1} \end{pmatrix}$$

$$= \begin{pmatrix} abc+0 & 0+0 \\ 0+0 & 0+(abc)^{-1} \end{pmatrix}$$

$$= \begin{pmatrix} abc & 0 \\ 0 & (abc)^{-1} \end{pmatrix}$$

Similarly we can show $x(yz) = \begin{pmatrix} abc & 0 \\ 0 & (abc)^{-1} \end{pmatrix}$

Therefore matrix multiplication satisfying the associative property.

Let $x \in G$ such that $xy = x$. where y is a matrix in the form $\begin{pmatrix} z & 0 \\ 0 & z^{-1} \end{pmatrix}$

Consider $xy = \begin{pmatrix} a & 0 \\ 0 & a^{-1} \end{pmatrix} \begin{pmatrix} z & 0 \\ 0 & z^{-1} \end{pmatrix} = \begin{pmatrix} az+0.0 & a.0+o.z^{-1} \\ 0.z+a^{-1}.0 & 0.0+a^{-1}z^{-1} \end{pmatrix} = \begin{pmatrix} a & 0 \\ 0 & a^{-1} \end{pmatrix}$

$\Rightarrow \begin{pmatrix} az+0 & 0+0 \\ 0+0 & 0+a^{-1}z^{-1} \end{pmatrix} = \begin{pmatrix} a & 0 \\ 0 & a^{-1} \end{pmatrix}$

$\Rightarrow \quad az = a$ and $(a.z)^{-1} = a^{-1}$

$\Rightarrow \quad z = 1.$

Therefore $y = \begin{pmatrix} 1 & 0 \\ 0 & 1 \end{pmatrix}$ is the multiplication identity on G.

Take x, y $\in$ G such that $xy = I_2$ where $I_2 = \begin{pmatrix} 1 & 0 \\ 0 & 1 \end{pmatrix}$.

Since x and y are in G. Then they are in the form $\begin{pmatrix} a & 0 \\ 0 & a^{-1} \end{pmatrix}$.

Consider $xy = I_2$

$\Rightarrow \begin{pmatrix} a & 0 \\ 0 & a^{-1} \end{pmatrix} \begin{pmatrix} b & 0 \\ 0 & b^{-1} \end{pmatrix} = \begin{pmatrix} 1 & 0 \\ 0 & 1 \end{pmatrix}$

$\Rightarrow \begin{pmatrix} a.b+0.0 & a.0+0.b^{-1} \\ 0.b+a^{-1}.0 & 0.b+a^{-1}.b^{-1} \end{pmatrix} = \begin{pmatrix} 1 & 0 \\ 0 & 1 \end{pmatrix}$

$\Rightarrow \begin{pmatrix} ab+0 & 0+0 \\ 0+0 & 0+a^{-1}b^{-1} \end{pmatrix} = \begin{pmatrix} 1 & 0 \\ 0 & 1 \end{pmatrix}$

By equality of matrices

$ab = 1$ and $(ab)^{-1} = 1$

$b = \dfrac{1}{a}$

Therefore $\begin{pmatrix} 1/a & 0 \\ 0 & 1 \end{pmatrix}$ is the multiplicative inverse of $x = \begin{pmatrix} a & 0 \\ 0 & a^{-1} \end{pmatrix}$. This is true for all x in G.

Therefore G is a group with respect to matrix multiplication.

Example 12.5

If 0 is an operation on Z defined by x 0 y = x + y + 1, then prove that (Z, 0) is an Abelian group.

[JNTUH, June 2010, Set No. 1]

Proof: Let Z be the set of all integers.

Define x 0 y = x + y + 1 where x, y ∈ z.

We have to show that (Z, 0) is an Abelian group.

Let x, y ∈ z. Then x + y is also an integer.

Now x + y + 1 is also an integer.

⇒   for any x, y ∈ Z, x 0 y = x + y + 1 ∈ Z.

⇒   "0" satisfies the closure property on Z.

Let x, y, z ∈ Z.

$$
\begin{aligned}
\text{Now } (x\ 0\ y)\ 0\ z &= (x + y + 1)\ 0\ z && \text{[by definition of ``0'']} \\
&= (x + y + 1) + z + 1 && \text{[by definition of ``0'']} \\
&= x + y + 1 + z + 1 \\
&= x + (y + z + 1) + 1 && \text{[since Z satisfies associativity]} \\
&= x + (y\ 0\ z) + 1 \\
&= x\ 0\ (y\ 0\ z)
\end{aligned}
$$

"0" satisfies the associative law.

Existence of Identity: Let e ∈ Z such that e 0 a = a = a 0 e.

$$
\begin{aligned}
\text{Now e 0 a = a } &\Rightarrow & e + a + 1 &= a \\
&\Rightarrow & e + 1 &= 0 \\
&\Rightarrow & e &= -1
\end{aligned}
$$

Similarly a 0 e = a ⇒ e = −1, **−1** ∈ Z such that

−1 0 a = a = a 0 −1

⇒ −1 is the identity element in (Z, 0)

Existence of Inverse: Let a ∈ Z, and b ∈ Z, be the inverse of a ∈ (Z, 0).

Therefore a 0 b = −1 ⇒ a + b + 1 ⇒ b = − 2 − a

Similarly b 0 a = −1 ⇒ b + a + 1 = −1 ⇒ b = − a − 2

Therefore for each element a ∈ Z, has an inverse in (Z, 0)

Commutative law: Let a, b ∈ Z, then a 0 b = a + b + 1

$$= b + a + 1 = b\ 0\ a$$

Therefore, a 0 b = b 0 a for all a, b ∈ Z

Hence (Z, 0) is an Abelian group.

Example 12.6

Consider the algebraic system (Q, *), where Q is the set of rational numbers and '*' is a multiplication operation defined by

$$a * b = a + b - ab \qquad \forall\, a, b \in Q$$

Determine whether (Q, +) is a group.

[JNTUH, June 2010, Set No. 2]

Solution: Let Q be the set of rational numbers.

Define an algebraic system (Q, *) as

$$a * b = a + b - ab \qquad \forall\, a, b \in Q$$

where '*' is a multiplication operation.

Since a + b − ab is a rational number for all rational numbers a, b. So the given operation '*' is a binary operation on Q.

(i) For all a, b, c ∈ Q

$$(a * b)*c = (a + b - ab) * c$$
$$= (a + b - ab) + c - (a + b - ab)c$$
$$= a + b - ab + c - ac - bc + abc$$
$$= a + (b + c - bc) - a(b + c - bc)$$
$$= a * (b + c - bc) = a * (b * c)$$

Hence '*' is associative.

(ii) For any a ∈ Q

$$a * 0 = a + 0 - a \,.\, 0 = a$$

and $0 * a = 0 + a - 0 \,.\, a = a$

So '0' is the identity element under the operation '*'.

(iii) Clearly by the definition of '*', it is commutative.

(iv) Since there is a/a−1, a ≠ 1 inverse element for the definition of *,

i.e., $a * \left[\dfrac{a}{a-1} \right] = a + \dfrac{a}{a-1} - a\left(\dfrac{a}{a-1} \right) = 0, a \neq 1$

Thus (Q, *) is a group under '*'.

Example 12.7

Show that the identity element in a group is unique?

[JNTUH, June 2010, Set No. 2]

Solution: Suppose that e_1 and e_2 are identity elements in a group G.

Since e_1 is an identity element in G,

we have $ae_1 = e_1a = e_1$ for all $a \in G$.

Taking $a = e_2$ in this we get

$$e_2\ e_1 = e_1\ e_2 = e_1$$

Since e_2 is an identity element in G, a similar argument shows that

$$e_1\ e_2 = e_2\ e_1 = e_2$$

This shows that $e_1 = e_2$, that is the identity is unique in every group G.

Example 12.8

Prove that the set Z of all integers with the binary operation $a * b = a + b + 1 \ \forall \ a, b \in Z$ is an abelian group.

[JNTUH, Nov 2008, Set No. 1]

Solution: If $a, b \in Z$, then $a + b + 1 \in Z$ that is, $a * b \in Z$.

This verifies that Z is closed under the given operation '*'.

Now we find that, for any $a, b, c \in Z$

$$a * (b * c) = a * (b + c + 1) = \{a + (b + c + 1)\} + 1$$
$$= \{(a + b + 1) + c\} + 1$$
$$= (a * b) * c$$

This shows that * is associative in Z.

For any $a \in Z$,

$$a * (-1) = (a - 1) + 1 = a, \text{ and}$$
$$(-1) * (a) = -1 + a + 1 = a$$

Thus -1 is the identity element in Z under '*'.

Also for any $a \in Z$,

$$a * \{-(a + 2)\} = a - (a + 2) + 1 = -1 \text{ and}$$
$$-(a + 2) * a = -(a + 2) + a + 1 = -1$$

This shows that every a ∈ Z has the inverse

$$- (a + 2) \in Z \text{ under } *.$$

We note that, for any a, b ∈ Z,

$$a * b = a + b + 1 = b + a + 1 = b * a$$

Thus '*' is commutative.

The above facts prove that {Z, *} is an abelian group.

12.2   Subgroups

Definition

Let (G, o) be a group. A non-empty subset H of G is said to be a subgroup of G if H itself forms a group under the product in G.

Theorem 5

A non-empty subset H of a group G is a subgroup of G if and only if

(i)   a, b ∈ H ⇒ ab ∈ H and

(ii)   a ∈ H ⇒ a^{-1} ∈ H

Proof: Suppose that H is a subgroup of G

⇒ H itself is a group under the product in G. Therefore (i), (ii) holds.

Converse: Suppose H satisfies (i) and (ii). By (i), H satisfies the closure property.

For any a, b, c ∈ H, we have that a, b, c ∈ G implies that a(bc) = (ab)c.

Therefore (H, .) is a subgroup of (G, .).

Example 12.9

If H is a non-empty finite subset of a group G and H is closed under multiplication, then H is a subgroup of G.

Proof: Suppose H is a non-empty finite subset of a group G and H is closed under multiplication. Now we have to show that H is a subgroup of G.

It is enough to show that a ∈ H ⇒ a^{-1} ∈ H

Since H is a non-empty, there exists a ∈ H. Now a, a ∈ H ⇒ a^2 ∈ H.

Similarly a^3 ∈ H, ..., a^m ∈ H, ...

Therefore H ⊇ {a, a^2, ...}. Since H is finite, we have that there must be repetitions in a, a^2, ... Therefore there exists integers r, s with r > s > 0 such that a^r = a^s

$\Rightarrow \quad a^r . a^s = a^0$

$\Rightarrow \quad a^{r-s} = e \Rightarrow e \in H$ (since $r - s > 0$ and $a \in H \Rightarrow a^{r-s} \in H$).

Since $r - s - 1 \geq 0$, we have $a^{r-s-1} \in H$ and $a . a^{r-s-1} = a^{r-s} = e \in H$.

Hence a^{r-s-1} acts as the inverse of a in H. Hence H is a subgroup.

Example 12.10

(i) The set of even integers with respect to addition (E, +) is a subgroup of the group (Z, +).

(ii) If k is a positive integer then (kZ, +) is a subgroup of (Z, +).

Example 12.11

Consider G = Z, the group of integers with respect to addition. Write H = $\{5x \mid x \in G\}$.
Suppose a, b $\in$ H $\Rightarrow$ a = 5x, b = 5y for some x, y $\in$ G

$\Rightarrow \quad a + b = 5x + 5y = 5(x + y) \in H$. Also $- a = - 5x = 5(- x) \in H$.

Therefore H is a subgroup of G.

Example 12.12

Let (G, .) be a group. Let H = $\{x \mid x \in G$ and $x . y = y . x$ for all $y \in G\}$.

Prove that H is a subgroup of G.

Solution: Since $e . y = y . e$ for all $y \in G$, we have $e \in H$.

Take $x_1, x_2 \in H$. Then $x_1 . y = y . x_1$ and $x_2 . y = y . x_2$ for all $y \in G$.

Now $(x_1 . x_2).y = x_1 . (x_2 . y)$

$$= x_1 . (y. x_2)$$
$$= (x_1 . y) . x_2$$
$$= (y . x_1) . x_2$$
$$= y . (x_1 . x_2)$$

Therefore $x_1 . x_2 \in H$.

Take $x \in H$. Now $x . y = y . x$ for all $y \in G$.

Now $x^{-1} . y = (y^{-1} . x)^{-1} = (x . y^{-1})^{-1}$ (since $x \in H \Rightarrow x . y^{-1} = y^{-1} . x$)

$$= y. x^{-1} \text{ for any } y \in G.$$

This shows that $x^{-1} \in H$. Hence (H, .) is a subgroup of (G, .).

Theorem 6

Let (G, *) be a group. H be any non-empty subset of G. Then H is a subgroup of G if and only if $a * b^{-1} \in H$ whenever a, b $\in$ H.

Definition

Let G be a group. If G contains only a finite number of elements then G is called a finite group. If G contains infinite number of elements then G is called an infinite group. If G is a finite group then the order of G is the number of elements in G. If G is infinite group, then we say that order of G is infinite. The order of G is denoted by O(G). If G is a group and $a \in G$, then the order of 'a' is defined as the least positive integer m such that $a^m = e$. If there is no positive integer n such that $a^n = e$ then 'a' is said to be of infinite order.

Example 12.13

(i) Let G be the set of all integers and + be the usual addition of numbers. Then (G, +) is an Abelian group. Here '0' is the additive identity and – x is the additive inverse of x, for any x in G. This (G, +) is an infinite group and so O(G) is infinite.

(ii) Consider Q, the set of rational numbers, and R the set of all real numbers. Clearly these two are finite Abelian groups w. r. t. usual addition.

(iii) From the above, it is clear that the set G consisting of –1 and 1 is a group w. r. t. usual multiplication. This group is a finite group and O(G) = 2.

Example 12.14

Let G be a group, $a \in G$. Then $(a) = \{a^i \mid i = 0, \pm 1, ...\}$ is a subgroup of G.

Solution: Let $x, y \in (a) \Rightarrow x = a^i$ and $y = a^j$ for some $i, j \in Z$. Now $x. y = a^i. a^j = a^{i+j} \in (a)$ (since $i + j \in Z$).

Also $x^{-1} = (a^i)^{-1} = a^{-i} \in (a)$ (since $a^i. a^{-i} = a^{i-i} = a^0 = e \Rightarrow (a^i)^{-1} = a^{-i}$).

Therefore $x, y \in (a) \Rightarrow x. y \in (a)$ and $x^{-1} \in (a)$. Hence (a) is a subgroup of G.

Definition

Let G be a group and $a \in G$. Then $(a) = \{a^i \mid i = 0, \pm 1, ...\}$ is called the cyclic subgroup generated by the element $a \in G$. In other words, G is said to be a cyclic group if there exists an element $a \in G$ such that G = (a).

Example 12.15

(i) The set of integers with respect to addition, (Z, +), is a cyclic group with generator 1.

(ii) The multiplicative group, the cube roots of unity, $\{1, \omega, \omega^2\}$ is a cyclic group with generators ω and ω^2.

(**Verification:** $\omega^0 = 1, \omega^1 = \omega, \omega^2 = \omega^2, \omega^3 = 1$ and $(\omega^2)^0 = 1, (\omega^2)^1 = \omega^2, (\omega^2)^2 = \omega^4 = \omega$).

12.2.1   Notation

If ~ is an equivalence relation on S, then [a], the class of a, is defined by

$$[a] = \{b \in S \mid b \sim a\}.$$

Definition

Let G be a group, H be a subgroup of G, a b $\in$ G. We say that a is congruent to b (mod H), written as a $\equiv$ b (mod H) if a b^{-1} $\in$ H.

Theorem 7

The relation a $\equiv$ b (mod H) is an equivalence relation.

Proof:

(i)   *Reflexive:* Since H is a subgroup of G, we have that a a^{-1} = e $\in$ H for

$$a \in G \Rightarrow a \equiv a \text{ (mod H)}.$$

(ii) *Symmetric:* Suppose a $\equiv$ b (mod H) $\Rightarrow$ ab^{-1} $\in$ H $\Rightarrow$ (ab^{-1})$^{-1}$ $\in$ H (since H is a subgroup of G) $\Rightarrow$ (b^{-1})$^{-1}$ a^{-1} $\in$ H $\Rightarrow$ ba^{-1} $\in$ H $\Rightarrow$ b $\equiv$ a (mod H).

(iii) *Transitive:* Suppose a $\equiv$ b (mod H), b $\equiv$ c (mod H) $\Rightarrow$ ab^{-1} $\in$ H, bc^{-1} $\in$ H

$$\Rightarrow (ab^{-1})(bc^{-1}) \in H \text{ (since H is a subgroup of G)} \Rightarrow a(b^{-1}b)c^{-1} \in H \Rightarrow aec^{-1} \in H$$

$$\Rightarrow ac^{-1} \in H \Rightarrow a \equiv c \text{ (mod H)}.$$

Therefore the relation a $\equiv$ b (mod H) satisfies (i) reflexive, (ii) symmetric, (iii) transitive properties. Thus the relation is an equivalence relation.

Definition

Let G be a group and H be a subgroup of G. Then H is called a normal subgroup of G whenever aH = Ha for all a $\in$ G.

Example 12.16

If (G, *) and (H, Δ) are two groups and f: G $\rightarrow$ H is a homomorphism then prove that the Kernal of 'f' is a normal subgroup.   [JNTUH, Nov 2010, Set No. 2]

Proof: Let (G, *) and (H, Δ) be two groups.

Suppose f: G $\rightarrow$ H is a homomorphism.

We have to show kerf = {x $\in$ G | f(x) = e^1, where e^1 is identity in H} is a normal subgroup.

Since f(e) = e^1 $\Rightarrow$ e $\in$ kerf.

Therefore kerf $\neq$ ϕ.

Let x, y $\in$ kerf. Then f(x) = e^1 and f(y) = e^1.

Now f(xy^{-1}) = f(x) . f(y^{-1})   [since f is a homomorphism]

$$= f(x) . (f(y))^{-1}$$

$$= e^1 . (e)^{-1}$$

$$= e^1$$

$\Rightarrow xy^{-1} \in$ kerf. This is true for all x, y $\in$ kerf.

Let g $\in$ G and x $\in$ kerf

Then g $\in$ G and f(x) = e^1 where e^1 is identity in H.

Consider $f(g^{-1} x g) = f(g)^{-1} f(x) . f(g)$

$$= (f(g))^{-1} f(x) f(g)$$
$$= (f(g))^{-1} . e^1 f(g)$$
$$= (f(g))^{-1} . f(g)$$
$$= e^1$$

$\Rightarrow g^{-1} x g \in$ kerf. This is true for all x $\in$ kerf, g $\in$ G.

Therefore kerf is a normal subgroup.

Example 12.17

Prove that H = {0, 2, 4} forms a sub group of $<Z_6, +_6>$.

[JNTUH, Nov 2008, Set No. 4] [JNTUH, Nov 2010, Set No. 4]

Proof: Take $<Z_6, +_6>$. Write H = {0, 2, 4}

We have to show that H = {0, 2, 4} is a sub group of $(Z_6, +_6)$. It is enough to show that $(H, +_6)$ is a group.

Since for any a, b $\in$ H a, b $\in Z_6$ and Z_6 is a group with respect to '$+_6$' implies that H satisfies closure, associative, identity and inverse with respect to $+_6$.

Therefore $(H, +_6)$ is a group.

Example 12.18

Consider the group G = {1, 2, 4, 7, 8, 11, 13, 14} under multiplication modulo 15. Construct the multiplication table of G and verify whether G is cyclic or not.

[JNTUH, Nov 2010, Set No. 4]

Proof: Given G = {1, 2, 4, 7, 8, 11, 13, 14}

First we construct the multiplication table of modulo 15.

15	1	2	4	7	8	11	13	14
1	1	2	4	7	8	11	13	14
2	2	4	8	14	1	7	11	13
4	4	8	1	13	2	14	7	11
7	7	14	13	4	11	2	1	8
8	8	1	2	11	4	13	14	7
11	11	7	14	2	13	1	8	4
13	13	11	7	1	14	8	4	2
14	14	13	11	8	7	4	2	1

Given G is a finite group with multiplication modulo 15 as binary operation.

From the above table it is easy to show that G is a cyclic group.

Example 12.19

Let G be a group of order P, where P is a prime. Find all subgroups of G.

[JNTUH, June 2010, Set No. 4]

Solution: Given that $O(G) = P$, a prime number. We know that $\{e\}$ is a subgroup of G when 'e' is the identity in G.

Let H be a subgroup of G with $H \neq \{e\}$. Then there exists $e \neq a \in H$.

Since $o(a)$ divides $O(G) = P$.

We have that $o(a) = 1$ or $o(a) = P$.

If $o(a) = 1$ then $a^1 = e$ and so

$$a = e, \text{ a contradiction to the selection of 'a'.}$$

Hence $o(a) = P$ which implies that

$$\{e, a, a^2, \dots, a^{P-1}\} = G$$

So $G = \{e, a, a^2, \dots, a^{P-1}\} \subseteq H$

This shows that $H = G$.

Hence there are only two subgroups namely $\{e\}$ and G if $O(G) = P$, a prime number.

Example 12.20

Consider $Z_{20} = \{0, 1, 2, \dots, 19\}$ under addition modulo 20. Let H be the subgroup generated by 5.

   (i)   Find the elements and order of H.

   (ii) Find the cosets of H in Z_{20}.

[JNTUH, June 2010, Set No. 4]

Solution: Given that H is the subgroup generated by 5 then $H = \{5, 10, 15, 0\}$

So $0(H) = 4$

$\{0 + H, 1 + H, 2 + H, 3 + H, 4 + H\}$ is the set of all cosets of H in Z_{20}.

Note that $a + H = b + H \Leftrightarrow a - b \in H$

So

$$0 + H = 5 + H = 10 + H = 15 + H$$
$$1 + H = 6 + H = 11 + H = 16 + H$$

$$2 + H = 7 + H = 12 + H = 17 + H$$
$$3 + H = 8 + H = 13 + H = 18 + H$$
$$4 + H = 9 + H = 14 + H = 19 + H$$

Example 12.21

Consider the group $G = \{1, 2, 4, 7, 8, 11, 13, 14\}$ under multiplication modulo 15:

   (a)   Construct the multiplication table of G.

   (b)   Find the values of 2^{-1}, 7^{-1} and 11^{-1}.

   (c)   Find the orders and subgroups generated by 2, 7 and 11.

   (d)   Is G cyclic.

[JNTUH, Nov 2008, Set No. 3]

Solution:

   (a)

X_{15}	1	2	4	7	8	11	13	14
1	1	2	4	7	8	11	13	14
2	2	4	8	14	1	7	11	13
4	4	8	1	13	2	14	7	11
7	7	14	13	4	11	2	1	8
8	8	1	2	11	4	13	14	7
11	11	7	14	2	13	1	8	4
13	13	11	7	1	14	8	4	2
14	14	13	11	8	7	4	2	1

   (b)   The value of $2^{-1} = 8$

          The value of $7^{-1} = 13$

          The value of $11^{-1} = 11$

   (c)   Left to the student

   (d)   No

12.3   Cosets

Definition

If H is a subgroup of G and a ∈ G, then write Ha = {ha | h ∈ H} is called the right coset of H in G. aH = {ah | h ∈ H} is called the left coset (or the left coset of H determined by H).

Note:

   (i)   We denote the set of left cosets of H in G by G/H is the quotient set with respect to the equivalence relation ≡ (mod H).

        It is clear that if H is a normal subgroup, then the coset relation is a congruence relation.

Verification: Let a ≡ p (mod H) and b ≡ q (mod H). Then by definition, $p^{-1}a \in H$ and $q^{-1}b \in H$. Let $p^{-1}a = h_1$ and $q^{-1}b = h_2$ for some $h_1, h_2 \in H$.

Now $(pq)^{-1}(ab) = (q^{-1}p^{-1})(ab) = q^{-1}(p^{-1}a)b = q^{-1}(h_1\,b) = q^{-1}(b\,h_3)$ (since bH = Hb)

$= (q^{-1}b)h_3 = h_2h_3 \in H$. Therefore (ab) = (pq) (mod H). Thus ≡ (mod H) is a congruent relation on G.

(ii) Consider the quotient set G/H. Define the operations on G/H as aH*bH = (ab)H. Then (G/H, *) is a group. We call this as the quotient group.

Example 12.22

Consider the group $(Z_4, +_4)$ given in the following table.

Clearly Z_4 = {[0], [1], [2], [3]}.

$+_4$	[0]	[1]	[2]	[3]
[0]	[0]	[1]	[2]	[3]
[1]	[1]	[2]	[3]	[0]
[2]	[2]	[3]	[0]	[1]
[3]	[3]	[0]	[1]	[2]

Then H = {[0], [2]} is a subgroup of G. Now we will list the left cosets determined by the elements of Z_5 as follows.

Left coset determined by [0] is {[0], [2]}.

Left coset determined by [1] is {[1], [3]}.

Left coset determined by [2] is {[0], [2]}.

Left coset determined by [3] is {[1], [3]}.

These are the only two distinct left cosets of H in G.

12.3.1   Remark

(i) If a ∈ H, then aH = H.

(ii) Right coset can be defined in the same manner.

Example 12.23

There is a one-to-one correspondence between any two right cosets of H in G.

Proof: Let H be a subgroup of G and Ha, Hb be two right cosets of H in G (for some a, b ∈ G).

Define φ: Ha → Hb by φ(ha) = hb for all ha ∈ Ha.

To show φ is one-one, take $h_1a, h_2a \in$ Ha such that φ(h_1a) = φ(h_2a)

$\Rightarrow h_1 b = h_2 b$

$\Rightarrow h_1 = h_2$ (by cancellation law)

$\Rightarrow h_1 a = h_2 a$

Therefore ϕ is one-one.

To show ϕ is onto, let $hb \in Hb \Rightarrow h \in H$. Now $ha \in Ha$ and ϕ (ha) = hb. Therefore ϕ is onto.

Note:

Since H = He we have that H is also a right coset of H in G and by the Problem 12.3.5, any right coset of H in G have O(H) elements.

12.3.2   Theorem (Lagranges Theorem)

If G is a finite group and H is a sub group of G, then O(H) is a divisor of O(G).

Proof: Let G be a finite group and H is a subgroup of G with O(G) = n, O(H) = m, (since G is finite, H is also finite).

We know that any two right cosets are either disjoint or identical.

Now suppose Ha_1, Ha_2, ..., Ha_k are only distinct right coset of H in G

$\Rightarrow G = Ha_1 \cup Ha_2 \cup ... \cup Ha_k$

$\Rightarrow O(G) = O(Ha_1) + O(Ha_2) + ... + O(Ha_k)$

$\Rightarrow O(H) + O(H) + ... + O(H)$ (k times)   (since every right coset has O(H) elements)

$\Rightarrow O(G) = k. O(H)$

$\Rightarrow n = k.m \Rightarrow (n/m) = k$

Hence O(H) divides O(G).

12.3.3   Remark

Converse of the Lagranges theorem is not true: that is, "if G is a finite group and k | O(G) then there exists a subgroup H of G such that O(H) = k" is not true. For this fact, observe the following example.

Example 12.24

Consider the symmetric group S_4. We know that S_4 = {f: A → A | f is a bijection and A = {1, 2, 3, 4}}. Clearly $|S_4|$ = 24 = (= 4!). Now A_4 = the set of all even permutations in S_4. Then $|A_4|$ = 12. It can be verified that any six elements of A_4 cannot form a subgroup.

Therefore $6 \mid O(A_4)$ but A_4 contains no subgroup of order 6. (refer the section: Permutation Groups).

12.4   Rings

Definition

(i) A non empty set R is said to be a ring (or an associative ring) if there exists two operations "+" and "." on R such that

   (a) (R, +) is an abelian group;

   (b) (R, .) is a semi-group; and

   (c) for any a, b, c $\in$ R we have $a(b + c) = ab + ac$, $(a + b)c = ac + bc$.

(ii) Let (R, +, .) be a ring. If $1 \in R$ such that $a.1 = 1.a = a$ for every a $\in$ R, then we say that R is a ring with identity (or unit) element. If $a.b = b.a$ for all a, b $\in$ R, then R is said to be a commutative ring.

Example 12.25

(i)   (Z, +, .) is a commutative ring with identity, where Z is the set of integers.

(ii) (2Z, +, .) is a commutative ring without identity, where 2Z is the set of even integers.

(iii) (Q, +, .) is a commutative ring with identity, where Q is the set of rationals.

(iv) $(Z_n, +, .)$ is a commutative ring with identity, where Z_n is the set of integers modulo n

Definition

(i)   If R is a commutative ring then $0 \neq a \in R$ is said to be a zero divisor if there exists $0 \neq b \in R$ such that $ab = 0$.

(ii) A commutative ring is said to be an integral domain if it has no zero divisors.

(iii) A ring R is said to be a Boolean ring if $x^2 = x$ for all x $\in$ R (in other words, each element of R is an idempotent).

Example 12.26

Let R be the set of all formal square arrays $\begin{pmatrix} a & b \\ c & d \end{pmatrix}$ where a, b, c, d are any real numbers.

Define $\begin{pmatrix} a_1 & b_1 \\ c_1 & d_1 \end{pmatrix} + \begin{pmatrix} a_2 & b_2 \\ c_2 & d_2 \end{pmatrix} = \begin{pmatrix} a_1 + a_2 & b_1 + b_2 \\ c_1 + c_2 & d_1 + d_2 \end{pmatrix}$. It is easy to see that R forms

an abelian group under addition with $\begin{pmatrix} 0 & 0 \\ 0 & 0 \end{pmatrix}$ as the zero element and $\begin{pmatrix} -a & -b \\ -c & -d \end{pmatrix}$ is the inverse　　　of　　$\begin{pmatrix} a & b \\ c & d \end{pmatrix}$.　　We　　define　　the　　multiplication　　in　　R　　by

$$\begin{pmatrix} a & b \\ c & d \end{pmatrix} \cdot \begin{pmatrix} r & s \\ t & u \end{pmatrix} = \begin{pmatrix} ar+bt & as+bu \\ cr+dt & cs+du \end{pmatrix}.$$

The element $\begin{pmatrix} 1 & 0 \\ 0 & 1 \end{pmatrix}$ acting as multiplicative unit element. Clearly R is a ring.

Since $\begin{pmatrix} 1 & 0 \\ 0 & 0 \end{pmatrix} \cdot \begin{pmatrix} 0 & 0 \\ 1 & 0 \end{pmatrix} = \begin{pmatrix} 0 & 0 \\ 0 & 0 \end{pmatrix}$, we have that R is not an integral domain.

Since $\begin{pmatrix} 1 & 0 \\ 0 & 0 \end{pmatrix} \cdot \begin{pmatrix} 0 & 0 \\ 1 & 0 \end{pmatrix} = \begin{pmatrix} 0 & 0 \\ 0 & 0 \end{pmatrix} \neq \begin{pmatrix} 0 & 0 \\ 1 & 0 \end{pmatrix} = \begin{pmatrix} 0 & 0 \\ 1 & 0 \end{pmatrix} \begin{pmatrix} 1 & 0 \\ 0 & 0 \end{pmatrix}$, we have that R is not commutative.

Definition

A ring R is said to be a division ring if $(R^*, .)$ is a group (here $R^* = R - \{0\}$). A division ring is said to be a field if it is commutative (we will learn this concept in the next section).

Example 12.27

If R is a ring then for all a, b $\in$ R we have

- (i)　　$a0 = 0 = 0a$
- (ii)　　$a(-b) = (-a)b = -ab$
- (iii)　　$(-a)(-b) = ab$. If in addition if R has identity 1, then
- (iv)　　$(-1)a = -a, (-1)(-1) = 1$

Solution:

- (i)　　$a0 = a(0+0) = a0 + a0$. Now

  $0 + a0 = a0 = a0 + a0 \Rightarrow a0 = 0$　　(by right cancellation law).

  Similarly, we can prove that $0 = 0a$.

- (ii)　　$0 = a0 = a(b + (-b)) = ab + a(-b)$

  $\Rightarrow -(ab) = a(-b)$ and $0 = 0b = (a + (-a))b = ab + (-a)b$

  $\Rightarrow -(ab) = (-a)b$

- (iii)　　$(-a)(-b) = -(a(-b)) = -(-(ab)) = ab$
- (iv)　　from (ii), $(-1)a = -(1a) = -a$

  from (iii), $(-1)(-1) = 1.1 = 1$

Example 12.28

Consider the ring $(Z_m, +_m, \times_m)$ for all $m \in Z$.

(i)   For $m = 6$, we have $[2] \times_m [3] = [0]$ but $0 \neq [2] \neq [3] \neq 0$. So Z_6 is not an integral domain.

(ii)   For $m = 7$, $(Z_m, +_m, \times_m)$ is an integral domain.

(iii) For any prime number 'p', the set of integers modulo p, that is Z_p is a field.

12.4.1   The Pigeon Hole Principle

If a objects are distributed over m places and if $a > m$, then some place receives at least two objects.

Theorem 8

A finite integral domain is a field.

Proof: We know that in an integral domain we have $ab = 0 \Rightarrow a = 0$ or $b = 0$. Now it suffices to show that every non-zero element has multiplicative inverse. Let D be an integral domain.

Now we show

(i)   there exists $1 \in D$ such that $a \cdot 1 = a$ for all $a \in D$,

(ii) $0 \neq a \in D \Rightarrow$ there exists $b \in D$ such that $ab = 1$.

Let $D = \{x_1, x_2, ..., x_n\}$ and $0 \neq a \in D$. Now $x_1a, x_2a, ..., x_na$ are all distinct

(If $x_ia = x_ja$, then $(x_i - x_j)a = 0 \Rightarrow x_i - x_j = 0 \Rightarrow x_i = x_j$ (since $a \neq 0$)).

Therefore $D = \{x_1a, x_2a, ..., x_na\}$. Since $a \in D$, $a = x_ka$ for some $1 \leq k \leq n$.

Again since D is commutative, we have $x_ka = a = ax_k$.

We show x_k is the identity element. For this, let $y \in D$, then $y = x_ia$ for some i.

Now consider $y.x_k = (x_ia) x_k = x_i (ax_k) = x_ia = y$

Thus $yx_k = y$ for all $y \in D$. Therefore x_k is the identity element.

For $x_k \in D = \{x_1a, x_2a, ..., x_na\} \Rightarrow x_k = x_ja$ for some $1 \leq j \leq n$.

Therefore x_j is the multiplicative inverse of a. Hence D is a field.

Definition

A subset T of a ring (S, +, .) is called a subring if (T, +, .) is itself a ring.

Example 12.29

The set of all even integers is a subring of (Z, +, .).

Definition

(i) Let $(R, +, .)$, $(R^1, +, .)$ be two rings. A mapping $\phi: R \to R^1$ is said to be a homomorphism   (or a ring-homomorphism) if

(a) $\phi(a + b) = \phi(a) + \phi(b)$,

(b) $\phi(ab) = \phi(a) \phi(b)$ for all a, b $\in$ R.

(ii) Let $\phi: R \to R^1$ be a homomorphism. Then the set $\{x \in R \mid \phi(x) = 0\}$ is called the kernal of ϕ and is denoted by kerϕ.

(iii) A homomorphism $\phi: R \to R^1$ is said to be an isomorphism if ϕ is one one and, onto.

(iv) R and R^1 are said to be isomorphic, if there exist an isomorphism $\phi: R \to R^1$.

Example 12.30

Let $\phi: R \to R^1$ be a homomorphism.

Then

(i) ker$\phi = \{0\} \Leftrightarrow \phi$ is one one.

(ii) If ϕ is onto then ϕ is an isomorphism if and only if ker$\phi = \{0\}$, where 0 is the additive identity if R.

Solution: (i) Suppose ker $\phi = \{0\}$. To show ϕ is 1-1.

Suppose x, y $\in$ R such that $\phi(x) = \phi(y)$.

Then $\phi(x) - \phi(y) = 0$

$\Rightarrow$ $\phi(x) + \phi(-y) = 0$

$\Rightarrow$ $\phi(x - y) = 0 \Rightarrow x - y \in$ ker $\phi = \{0\}$

$\Rightarrow$ $x - y = 0$

$\Rightarrow$ x = y. Therefore ϕ is one one.

Conserve: Suppose ϕ is one one.

Since $\phi(0) = 0$, we have $0 \in$ ker ϕ

$\Rightarrow$ $\{0\} \subseteq$ ker ϕ. Now let y $\in$ ker ϕ

$\Rightarrow$ $\phi(y) = 0 = \phi(0)$ (since $\phi(0) = 0$)

$\Rightarrow$ y = 0   (since ϕ is one one)

$\Rightarrow$ ker $\phi \subseteq \{0\}$. Therefore ker $\phi = \{0\}$.

(ii) Suppose ϕ is an isomorphism.

Since ϕ is one one, we have ker $\phi = \{0\}$ (by (i)).

Converse: Suppose ker$\phi = \{0\} \Rightarrow \phi$ is 1-1   (by (i)).

Since ϕ is onto, we have ϕ is a bijection.

Hence ϕ is an isomorphism.

Exercises

Groups

1. Show that A = {1, –1, i, –i}, the set of all fourth roots of unity forms an abelian group under multiplication.

2. Let G be the set of all non-zero real numbers and let $a * b = \frac{1}{2}ab$. Show that < G,*> is an abelian group.

3. Let S = {R – {0}} × R. Define the binary operation '*' on S by (u, v) * (x, y) = (ux, vx + y). Prove that <S, *> is a non abelian group.

Subgroups

1. Show that the set of all even integers forms a subgroup of the group of all integers under usual addition.

2. Prove that H = {x ∈ Z / x = 3y, for some integer y} forms a subgroup of < Z, +>.

3. Let G be a group. For a fixed element a ∈ G, define H_a = {x ∈ G | ax = xa} forms a subgroup of G.

Cosets

1. Prove that the only left coset of a subgroup H of a group G which is also a subgroup of G is H itself.

2. Find all the left cosets of the subgroup formed by {1, –1} in the group A = {1, –1, i, –i} under multiplication. Also obtain a coset decomposition of A.

    Ans: {1, –1} and {i, –i}

3. Find the right cosets of the following:

    (i) S_1 = {[0], [3]} in <Z_6, +>

    (ii) S_2 = {[0], [3], [6], [9]}

    Ans: (i) {[0], [3]}, {[1], [4]}, {[2], [5]}

    (ii) {[0], [3], [6], [9]}, {[1], [4], [7], [10]}, {[2], [5], [8], [11]}

Rings

1. Show that the set A = {0, 1, 2, 3, 4, 5} is a commutative ring with respect to $+_6$ and $\times_6$.

2. If R is a ring such that a^2 = a for all a ∈ R, then prove that

    (i) a + a = 0 for all a ∈R (ii) a + b = 0 ⇒ a = b (iii) R is a commutative ring

3. If addition and multiplication modulo 10 is defined on the set of even integers R = {0, 2, 4, 6, 8} then prove that the resulting system is a ring with unity.

CHAPTER - 13

Permutations and Combinations

For most applications of computer problems, one normally needs to know, at least approximately, how much storage will be required and about how many operations are necessary. The basic idea of this lesson is to give the brief idea about the concepts of basic counting principles, permutations and combinations.

13.1 Basic Counting Principles

For a set A, the number of elements in A is denoted by $|A|$.

If $A = \{1, 2, 3, 4\}$, then $|A|$ = (the number of elements in A) = 4. There are mainly two elementary or basic principles in counting problems. They are

   (i)   Sum Rule   (ii)   Product Rule

13.1.1 Sum Rule

If A is any set which is the union of disjoint non-empty subsets, say $A_1, A_2, \ldots A_n$, then $|A| = |A_1| + |A_2| + |A_3| + \ldots |A_n|$.

Observe the above sum rule. There is no element common in the subsets $A_1, A_2, \ldots, A_n$ of A. Since $A = A_1 \cup A_2 \cup \ldots \cup A_n$. So each element of A appears in exactly one of the subsets $A_1, A_2, \ldots, A_n$. In other words $A_1, A_2, \ldots, A_n$ is a partition of A.

Example 13.1

Suppose there are 15 boys and 10 girls in a class and we wish to select one of these students (either boy or girl) as a class representative. The number of ways selecting a boy is 15 and the number of ways selecting a girl is 10. Therefore the number of ways of selecting a student either boy or girl is $15 + 10 = 25$.

LEARNING OBJECTIVES

- to know the Basic Counting Principles
- to understand the concepts: Permutations, Combinations
- to identify different types: Circular Permutations
- to understand Restricted Permutations, Restricted Combinations
- to develop the problem solving skills related to Permutations, Combinations

Example 13.2

Find the number of ways to choose a book from a library which has 12 mathematics, 10 physics and 16 computer science books.

Solution: $12 + 10 + 16 = 38$

13.1.2   General Rule for Counting Event (or Sum Rule)

If $E_1, E_2, ..., E_n$ are mutually exclusive events and E_1 can happen m_1 ways, E_2 can happen m_2 ways, ..., E_n can happens m_n ways, then E_1 or E_2 or ... or E_n happens in $m_1 + m_2 + ... + m_n$ ways.

Example 13.3

In how many ways can we draw

  (i)   A heart or a spade from an ordinary deck of playing cards?

  (ii)   A heart or an ace?

  (iii) An ace or a king?

  (iv)   A card numbered 2 through 10?

  (v)   A numbered card or a king?

Solution:

  (i)   Since there are 13 hearts and 13 spades we may draw a heart or a spade in

         $13 + 13 = 26$ ways;

  (ii)   We may draw a heart or an ace in $13 + 3 = 16$ ways since there are only 3 aces that are not hearts.

  (iii) We may draw an ace or a king in $4 + 4 = 8$ ways.

  (iv)   There are 9 cards numbered 2 through 10 in each of suits, clubs, diamonds, hearts, or spades, so we may choose a numbered card in 36 ways.

  (v)   (Note that we are counting aces as distinct from numbered cards).

Therefore, we may choose a numbered card or a king in $36 + 4 = 40$ ways.

Example 13.4

  (i)   How many ways can we get a sum of 4 or of 8 when two distinguishable dice (say one die is red and the other is white) are rolled?

  (ii)   How many ways can we get an even sum?

Solution: Label the outcome of a 1 on the red die and a 3 on the white die as the ordered pair $(1, 3)$.

  (i)   Then we see that the outcomes $(1, 3)$, $(2, 2)$, and $(3, 1)$ are the only ones whose sum is 4. Thus there are 3 ways to obtain the sum 4.

Similarly, we obtain the sum 8 from the outcomes (2, 6), (3, 5), (4, 4), (5, 3), and (6, 2). Thus, there are $3 + 5 = 8$ outcomes whose sum is 4 or 8.

(ii)   The number of ways to obtain an even sum is the same as the number of ways to obtain either the sum 2, 4, 6, 8, 10, or 12. There is 1 way to obtain the sum 2, 3 ways to obtain the sum 4, 5 ways to obtain 6, 5 ways to obtain an 8, 3 ways to obtain a 10, and 1 way to obtain a 12.

Therefore, there is $1 + 3 + 5 + 5 + 3 + 1 = 18$ ways to obtain an even sum.

13.1.3   Product Rule

If $S_1, \ldots, S_n$ are nonempty sets, then the number of elements in the Cartesian product $S_1 \times S_2 \times \ldots \times S_n$, is the product $\prod_{i=1}^{n}|S_i|$. That is,

$$|S_1 \times S_2 \times \ldots \times S_n| = \prod_{i=1}^{n}|S_i|.$$

Example 13.5

Let us illustrate $S \times S$ by a tree diagram (see Figure) where and

$$S_1 = \{a_1, a_2, a_3, a_4, a_5\} \text{ and } S_2 = \{b_1, b_2, b_3\}.$$

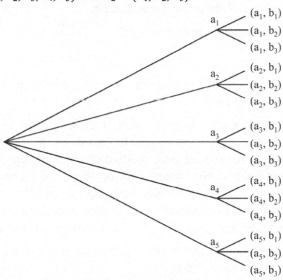

Example 13.6

Find the number of elements in $A \times B$ where $A = \{1, 2\}$ and $B = \{5, 7\}$.

Solution: Given that $A = \{1, 2\}$, $B = \{5, 7\}$.

By the definition, $A \times B = \{(a, b) \mid a \in A \text{ and } b \in B\}$.

Now $A \times B = (1, 2) \times (5, 7) = \{(1, 5), (1, 7), (2, 5), (2, 7)\}$.

Therefore the number of elements in $A \times B$ is $|A \times B| = 4$.

Note:

In general, if $a_1, a_2, \ldots, a_n$ are n distinct elements of a set A and $b_1, b_2, \ldots, b_m$ are m distinct elements of a set B then $A \times B = \bigcup\limits_{i=1}^{n}(a_i \times B)$

Example 13.7

(i)   How many 3-digit numbers can be formed using the digits 1, 3, 4, 5, 6, 8, and 9?

(ii) How many can be formed if no digit can be repeated?

Solution:

(i)   Since each of the 3 digits can be filled with 7 possibilities, we have that there are 7^3 such 3-digit numbers.

(ii) Since there are 7 possibilities for the hundreds digit but once one digit is used it is not available for the tens digit (since no digit can be repeated in this example). Thus there are only 6 possibilities for the tens digit, and then for the same reason there are only 5 possibilities for the units digit.

Therefore the required number is 7 . 6 . 5

Example 13.8

How many three-digit numbers are there which are even and have no repeated digits? (Here we may use all digits 0 through 9).

Solution: For a number to be even it must end in 0, 2, 4, 6, or 8. There are two cases to consider. First, suppose that the number ends in 0; then there are 9 possibilities for the first digit and 8 possibilities for the second since no digit can be repeated. Hence there are 9 . 8 three-digit numbers that end in 0. Now suppose the number does not end in 0. Then there are 4 choices for the last digit (2, 4, 6, or 8); when this digit is specified, then there are only 8 possibilities for first digit, since the number cannot begin with 0. Finally, there are 8 choices for the second digit and therefore there are 8 . 8 . 4 numbers that do not end in 0.

Since these two cases are mutually exclusive, the sum rule gives 9 . 8 + 8 . 8 . 4 even three-digit numbers with no repeated digits.

13.1.4 Indirect Counting

Sometimes it is beneficial to solve some combinatorial problems by counting indirectly, that is, by counting the complement of a set.

Example 13.9

Suppose that we draw a card from a deck of 52 cards and replace it before the next draw. In how many ways can 10 cards be drawn so that the tenth card is a repetition of previous draw?

Solution: First we count the number of ways we can draw 10 cards so that the 10th card is not a repetition.

To analyze this as follows:

First, choose the 10th card will be. This can be done in 52 ways.

If the first 9 draws are different from this, then each of the 9 draws can be chosen from 51 cards.

Thus there are 51^9 ways to draw the first 9 cards different from the 10th card.

Hence there are $(51^9)(52)$ ways choose 10 cards with the 10th card different from any of previous 9 draws.

So that there are $52^{10} - (51^9)(52)$ ways to draw 10 cards where the 10th card is the repetition since there are 52^{10} ways to draw 10 cards with replacements.

13.1.5   One to One Correspondence

A one to one correspondence between two sets A and B is a one to one function from A to B.

13.1.6   Applications to Computer Science

A 2-valued Boolean function of n variables is defined by assigning a value either 0 or 1 to each of the 2^n n-digit binary numbers.

Example 13.10

How many Boolean functions are there of "n" variables?

Solution: Since there are 2 ways to assign a value to each of the 2^n binary n-tuples, by the rule of product there are $2.2 \dots 2 = 2^{2^n}$ ways to assign all the values, and therefore 2^{2^n} different Boolean functions (of n variables) are possible.

Note:

A 2-valued Boolean function can also be represented as a tabular form which is also known as truth tables.

Example 13.11

The following is the truth table for a 2-valued Boolean function for 3 variables:

x	y	z	f(x, y, z)
0	0	0	1
0	0	1	1
0	1	0	0
0	1	1	0
1	0	0	1
1	0	1	0
1	1	0	1
1	1	1	0

Example 13.12

Write the truth table of 2-valued Boolean function of four variables.

Solution:

x	y	z	ω	f(x, y, z)
0	0	0	0	1
0	0	0	1	1
0	0	1	0	0
0	0	1	1	0
0	1	0	0	1
0	1	1	0	0
0	1	0	1	1
0	1	1	1	1
1	0	0	0	0
1	0	0	1	1
1	0	1	0	0
1	1	0	0	0
1	1	0	1	1
1	0	1	1	1
1	1	1	0	0
1	1	1	1	0

Definition

For each positive integer 'n' we define n! = n.(n – 1).(n – 2) ... 3.2.1 = the product of all integers from 1 to n.

Also define 0! = 1 and 1! = 1

For example,

(i)   3! = 3.2.1 = 6 and

(ii) 6! = 6.5.4.3.2.1 = 720.

Example 13.13

Find the factorials for (i) 7; (ii) 9; (iii) 11.

Solution: (i) 7! = 5040; (ii) 9! = 362880; (iii) 11! = 39916800

Example 13.14

How many ways are there to roll two distinguishable dice to yield a sum that is divisible by 3?

[JNTUH, Nov 2008, Set No. 3]

Solution: The two dice are distinguishable. So we get a sum which is divisible by 3, in the following cases:

(1, 2), (1, 5), (2, 1), (2, 4), (3, 3), (3, 6), (4, 2), (4, 5), (5, 1), (5, 4), (6, 3), (6, 6)

Therefore, there are 12 ways where we get the sum that is divisible by 3.

Example 13.15

How many six digit numbers are there with exactly one 5?

[JNTUH, Nov 2008, Set No. 2]

We have to find the number of six digit numbers with exactly one 5.

Suppose the places are as in the figure

1	2	3	4	5	6

Case (i): Suppose that 5 is in place – 1.

The remaining places can be filled with the nine numbers (0, 1, 2, 3, 4, 6, 7, 8, 9).

Therefore the number of such numbers can be formed in $9 \times 9 \times 9 \times 9 \times 9 (= 9^5)$ ways.

Case (ii): Suppose 5 is in place – 2.

The first place can be filled with 8 numbers (1, 2, 3, 4, 6, 7, 8, 9).

Therefore the number of such numbers that can be formed is $8 \times 9 \times 9 \times 9 \times 9 = 8 \times 9^4$.

Case (iii): Suppose 5 is in place – 3.

As in case (ii), the number of such numbers is 8×9^4.

Case (iv): If 5 is in place – 4 then also the required number is 8×9^4.

Case (v): If 5 is in place – 5 then the required number is 8×9^4.

Case (vi): If 5 is in place – 6 then the required number is 8×9^4.

Hence, the six digit numbers (with exactly one 5) that can be formed is

$$9^5 + 6\,(8 \times 9^4)$$
$$= 9^4\,[1 + 48] = (49)\,(9^4)$$

Example 13.16

One type of automobile license plate number in Massachusetts consists of one letter and five digits. Compute the number of such license plate numbers possible.

[JNTUH, Nov 2010, Set No. 3]

Solution: The condition that "letter is in the beginning of the license plate number" is not given.

There are six places. The place of the letter can be selected in 6 ways. The letter place can be filled in 26 ways.

The other five places can be filled in 10^5 ways by using 0 or 1 or 2 or 3 or 4 or 5 or 6 or 7 or 8 or 9 (repetition is allowed).

Hence the required number is

$$6 \times 26 \times 10^5$$

13.2   Permutations

Combinatorics is a study of arrangements of objects, is an important part of discrete mathematics. In this section, we shall study the permutations, with some illustrations. An experiment means a physical process that has a number of observable outcomes. Simple examples are tossing of a coin which has two possible outcomes HEAD and TAIL, rolling a die which has six possible outcomes 1, 2, ..., 6. We would like to know how many possible outcomes are there in selecting 10 student representatives from 3000 students.

When we consider the outcomes of such experiments we shall follow the following rules.

Let us recollect the counting principles from the last section.

13.2.1   Fundamental Principles of Counting

(i)   *Rule of Sum:* If the object A may be chosen in 'm' ways, and B in 'n' ways, then "either A or B" (exactly one) may be chosen in m + n ways. This can be generalized for any 'p' objects.

(ii)   *Rule of Product:* If the object A may be selected in m ways and the object B in n ways, and both the selections can be made in 'mn' ways. This can be generalized for any 'p' objects.

Example 13.17

If there are 42 ways to select a representative for class A and 50 ways to select a representative for the class B, then to select a pair one from A and other from B, by the rule of product, there are 42 × 50 ways to select the representative for both the class A and B.

13.2.2   Permutations of Distinct Things

We can choose first thing from the given n distinct things in n ways. The second can be selected in (n – 1) ways, ..., The rth thing in n – (r – 1) ways.

So by the repeated application of product rule, the number required is n(n – 1) ... (n–(r – 1)) ways, n ≥ r., it is denoted by p(n, r).

If r = n, then p(n, n) = n(n – 1) ... (n – n + 1) = n(n – 1) ... 2.1 = n!.

Therefore $p(n, r) = \dfrac{n(n-1)(n-2)...(n-(r-1))...2.1}{(n-r)...2.1}$

$$= \frac{n!}{(n-r)!}$$

$$= \frac{p(n, n)}{p(n-r, n-r)}$$

or   p(n, n) = p(n, r). p(n – r, n – r).

p(n, r) is also denoted by $^{n}P_{r}$.

So $^{n}P_{r} = \dfrac{n!}{(n-r)!} = n(n-1)...(n-r+1)$

Example 13.18

Prove that $^{n}P_{r} = {}^{n-1}P_{r} + r.\,{}^{n-1}P_{r-1}$

(In other notation, p(n, r) = p(n – 1, r) + r p (n – 1, r – 1) for 1 ≤ r ≤ (n – 1)

Solution:

$$RHS = {}^{n-1}P_{r} + r.\,{}^{(n-1)}P_{(r-1)}$$

$$= \frac{(n-1)!}{(n-1-r)!} + r\frac{(n-1)!}{\left[(n-1)-(r-1)\right]!}$$

$$= (n-1)! \left\{ \frac{1}{(n-1-r)!} + \frac{r}{(n-r)!} \right\}$$

$$= (n-1)! \left\{ \frac{1}{(n-1-r)!} + \frac{r}{(n-r)\ (n-1-r)!} \right\}$$

$$= \frac{(n-1)!}{(n-1-r)!} \left\{ 1 + \frac{r}{n-r} \right\}$$

$$= \frac{(n-1)!}{(n-r-1)!} \left\{ \frac{n-r+r}{n-r} \right\}$$

$$= \frac{(n-1)!}{(n-r-1)!} \left\{ \frac{n}{(n-r)} \right\}$$

$$= \frac{n(n-1)!}{(n-r-1)!\ (n-r)}$$

$$= \frac{n!}{(n-r)!} = {}^nP_r = \text{LHS}$$

13.2.3 Permutations with Repetitions

The number of permutations of n objects taken 'r' at a time with repetition [which is same as the number of ways of filling r blank spaces with n objects].

After selecting the first object in n ways, the next object can be selected in n ways and so on. Therefore by the rule of product, we get that

$$n \times n \times \ldots \times n = n^r \text{ ways}$$

Therefore the number of permutations of n things taken r at a time with repetition allowed is n^r.

Example 13.19

A computer password consists of a letter of the alphabet followed by 3 or 4 digits. Find

  (i)   The total number of passwords that can be formed, and

  (ii) The number of passwords in which no digit repeats.

Solution:

  (i)   Note that there are 26 alphabets and 10 digits.

        By product rule, the number of 4-character passwords is 26.10.10.10. = 26000. Similarly the number of 5-character passwords is 26.10.10.10.10. = 260000. Hence the total number of passwords is 26000 + 260000 = 286000.

(ii) Given that the digits cannot be repeated. The first digit after alphabet can be taken from any one out of 10, the second digit from remaining 9 digits and so on. Thus the number of 4-character passwords is 26.10.9.8 = 18720 and the number of 5-character passwords is 26.10.9.8.7 = 131040 (by the product rule). Hence, the total number of passwords is 149760.

Example 13.20

Determine the number of 5-digit decimal numbers that contain no repeated digits and does not have a leading 0.

Solution: There are 10 digits 0, 1, 2, 3, 4, 5,6, 7, 8, 9. Here n = 10. We can form 5 digit numbers with no repeated digits in P(10, 5) = $10 \times 9 \times 8 \times 7 \times 6 = 30240$ ways.

Among these 30240 numbers there are $9 \times 8 \times 7 \times 6 = 3024$ numbers with leading 0. Thus there are 30240 − 3024 = 27216, 5-digit numbers with no repetition and without leading zero.

Example 13.21

Suppose there are 6 boys and 5 girls.

(i) In how many ways can they sit in a row?

(ii) In how many ways can they sit in a row if the boys and girls are each to sit together?

(iii) In how many ways can they sit in a row if the girls are to sit together and the boys do not sit together?

(iv) How many seating arrangements are there with no two girls sitting together?

Solution:

(i) There are 6 + 5 = 11 persons and they can sit in P(11, 11) = 11! ways.

(ii) The boys among themselves can sit in 6! ways and the girls among themselves can sit in 5! ways. They can be considered as 2-units and can be permuted in 2! ways. Thus the required seating arrangements can be in 2! 6! 5! ways.

(iii) The boys can sit in 6! ways and the girls in 5! ways. Since girls have to sit together they are considered as one unit. Among the 6 boys either 0 or 1 or 2 or 3 or 4 or 5 or 6 have to sit to the left of the girls unit. Of these seven ways 0 and 6 cases have to be omitted as the boys do not sit together. Thus the required number of arrangements = $5 \times 6! \times 5!$.

(iv) The boys can sit in 6! ways. There are seven places where the girls can be placed. Thus total arrangements are $P(7, 5) \times 6!$.

Example 13.22

In how many ways, the letters of the following given words can be arranged?

(i) RAM

    (ii)   COMBINE

    (iii) FRACTION

    (iv) POWER

Solution:

    (i)   The word 'RAM' contains three letters and all are different. So the number of required arrangements $= {}^3P_3 = 3! = 3 \times 2 \times 1 = 6$.

    (ii)   The word 'COMBINE' contains seven different letters. So the required number is ${}^7P_7 = 7! = 5040$.

    (iii) The word 'FRACTION' contains eight different letters. Hence the number of arrangements $= {}^8P_8 = 8! = 40320$.

    (iv) As the number of letters in 'POWER' is 5 and all distinct, the required number is ${}^5P_5 = 5! = 120$.

13.2.4   Dearrangements

If n things are arranged in a row, the number of ways in which they can be dearranged so that no one of them occupies its original place is

$$n!\left(1 - \frac{1}{1!} + \frac{1}{2!} + \frac{1}{3!} + ... + (-1)^n . \frac{1}{n!}\right)$$

If r things go to wrong place out of n things then $(n - r)$ things go to original place

(Here $r < n$).

If D_n = No. of ways, if all n things go to wrong place

and D_r = No. of ways, if r things go to wrong place

Then $D_n = {}^nC_{n-r} D_r$

Where $D_r = r!\left(1 - \frac{1}{1!} + \frac{1}{2!} + \frac{1}{3!} + ... + (-1)^n . \frac{1}{r!}\right)$

Example 13.23

A person writes greeting card to six friends and addresses the corresponding envelopes. In how many ways can the greeting cards be placed in the envelopes so that

    (i)   at least two of them are in the wrong envelopes

    (ii) all the cards are in the wrong envelopes

Solution:

    (i)   The number of ways in which at least two of them are in the wrong envelopes

$$= \sum_{r=2}^{6} {}^nC_{n-r} D_r$$

$$= {}^nC_{n-2} D_2 + {}^nC_{n-3} D_3 + {}^nC_{n-4} D_4 + {}^nC_{n-5} D_5 + {}^nC_{n-6} D_6$$

Here n = 6

$$= {}^6C_4.2!\left(1-\frac{1}{1!}+\frac{1}{2!}\right) + {}^6C_3.3!\left(1-\frac{1}{1!}+\frac{1}{2!}+\frac{1}{3!}\right) + {}^6C_2.4!$$

$$\left(1-\frac{1}{1!}+\frac{1}{2!}-\frac{1}{3!}+\frac{1}{4!}\right) + {}^6C_1.5!\left(1-\frac{1}{1!}+\frac{1}{2!}-\frac{1}{3!}+\frac{1}{4!}-\frac{1}{5!}\right)$$

$$= 15 + 40 + 135 + 264 + 265$$

$$= 719.$$

(ii) The number of ways in which all letters be placed in wrong envelopes

$$= 6!\left(1-\frac{1}{1!}+\frac{1}{2!}-\frac{1}{3!}+\frac{1}{4!}-\frac{1}{5!}+\frac{1}{6!}\right)$$

$$= 720\left(\frac{1}{2}-\frac{1}{6}+\frac{1}{24}-\frac{1}{120}+\frac{1}{720}\right)$$

$$= 360 - 120 + 30 - 6 + 1 = 265$$

Theorem 1

Show that the number of r – permutations of a set of 'n' (distinct) elements is given by

$$p(n, r) = \frac{n!}{(n-r)!}$$

[JNTUH, Nov 2010, Set No. 4]

Proof: Since there are 'n' distinct objects, the first position of an r – permutation may be filled in n ways.

Once this was done, the second position can be filled in (n – 1) ways, since no repetitions are allowed and there are n – 1 objects left to choose from.

The third can be filled in n – 2 ways and so on until the rth position is filled in n – r + 1 ways.

By applying the product rule, we conclude that

$$p(n, r) = n (n-1) (n-2) \ldots (n-r+1)$$

$$= \frac{n!}{(n-r)!}$$

(From the definition of factorials)

13.3 Circular Permutations, Restricted Permutations

In this section, we consider the circular permutations and restricted permutations. We illustrate these concepts

13.3.1 Circular Permutations

The permutations considered so far are called linear permutations as the objects are being arranged in a row (line). Suppose we arrange them in a circle, see the figures.

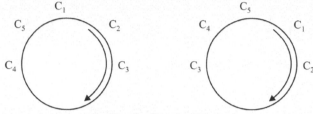

Circular Permutation

The arrangements are considered to be the same if the objects are in the same order clockwise. Therefore keeping c_1 in a fixed position there are $(n - 1)!$ arrangements for the remaining objects. Thus the following statement is true.

There are $(n - 1)!$ permutations of n distinct objects in a circle.

Example 13.24

(i)   How many ways are there to seat 10 boys and 10 girls around a circular table.

(ii) If boys and girls sit alternate how many ways are there

Solution:

(i)   There are in total 20 members to sit around the table. So n = 20

The number of circular permutations is $(n - 1)! = (20 - 1)! = 19$

(ii) Suppose 10 boys were first arranged around the table. The number of such arrangements is 9!. There are 10 places in between boys (for girls to sit alternatively). So the number of ways = 10!.

Hence the required number of circular permutations is $(9!) \times (10!)$

13.3.2 Clockwise and Anticlockwise Permutations

Consider the two figures given here

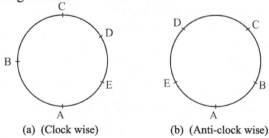

(a) (Clock wise)   (b) (Anti-clock wise)

The Fig. (a) shows a circular permutation of the letters A, B, C, D, E. We can observe that it is clockwise.

The Fig. (b) shows a circular permutation of letters A, B, C, D, E, We can observe that it is anticlockwise.

If A, B, C, D, E are persons sitting around a round table, then figure (a) and figure (b) represents two different circular permutations.

If A, B, C, D, E are five flowers arranged on a ring then figure (a) and figure (b) represents the same arrangement. In this case, there is no distinction between clockwise and anticlock wise permutations.

If there is no difference between the clockwise and anticlock wise permutations, then the number of circular permutations of n things is $\dfrac{(n-1)!}{2}$

Similarly, forming a necklace with n different beads, can be done in $\dfrac{(n-1)!}{2}$

Example 13.25

Find the number of ways in which 7 dissimilar flowers can be arranged in a

   (i)   line

   (ii)   circular

Solution: Take 7 dissimilar flowers.

   (i)   Now we can arrange these 7 dissimilar flowers in a line as 7! ways. That is the number of ways of arranging these flowers in a line

$$= 7! = 5040.$$

   (ii)   Now the number of ways of arranging 7 dissimilar flowers in a circle

$$= (7-1)! = 6! = 720$$

Example 13.26

There are six gents and four ladies to dine at a round table. In how many ways can be they seat themselves so that no two ladies are together?

Solution: Given   number of gents = 6

                 number of ladies = 4

We have to arrange them to seat around the table so that no two of ladies sit together.

First arrange all the 6 gents and then arrange ladies between them.

Number of ways arranging 6 gents = (6 – 1)! = 120.

Now, number of ways of arranging ladies

   $= 6\,P_4 = 360$

Therefore, the required number of arrangements

   $= 120 \times 360 = 43200$

Example 13.27

15 males and 10 females members are seated in a round table meeting. How many ways they can seated if all the females seated together.

[JNTUH, June 2010, Set No. 1]

Solution: Number of males $= 15$

   Number of females $= 10$

Since all female seated together. Take them as 1 unit.

Then number of persons $= 15 + 1 = 16$

Now the number of ways 16 can be arranged in round table is $= (16 - 1)! = 15!$

Since all females sit together and they can be arranged 10! ways.

Therefore number of ways that 15 males and 10 females can be seated in round table is $15! \times 10!$

13.3.3   Restricted Permutations

The number of distinguishable permutations of n objects in which the first object appears in m_1 times, second object in m_2 ways, ... so on, is $\dfrac{n!}{m_1! m_2!...m_k!}$, where m_k is the kth object appears in m_k times.

Example 13.28

Find the number of different letter arrangements that can be formed using the letters in the word "MATHEMATICS".

Solution: Total number of letters n $= 11$ (with repetitions)

   Number of M's $= 2$

   Number of T's $= 2$

   Number of A's $= 2$

and the letters H, C, S, E, appears once.

Therefore the required number of permutations is $\dfrac{11!}{2!2!2!1!1!1!1!} = 6652800$

Example 13.29

Find the number of ways in which the letters of the word "ARRANGEMENT" can be arranged so that two R's and two A's do not occur together.

Solution: There are n = 11 letters in "ARRANGEMENT", there are two A's, two R's, two N's and two E's.

Therefore the number of arrangement that can be formed is

$$\frac{11!}{(2!)(2!)(2!)(2!)} = 495(7!)$$

when two A's occur together, then two A's forms a unit.

So the number of arrangements when two A's occur together is

$$\frac{10!}{2!\ 2!\ 2!} = (90).(7!) \qquad\qquad .....(13.1)$$

Similarly when two R's occur together, then the number of arrangements is

$$\frac{10!}{(2!)\ (2!)\ (2!)} = (90)\ (7!) \qquad\qquad .....(13.2)$$

The number of arrangements in which two A's and two R's occur is

$$\frac{9!}{(2!)\ (2!)} = (18).7! \qquad\qquad .....(13.3)$$

Therefore the number of arrangements in which two A's and two R's do not occur together is

$$(479)\ (7!) - 90.\ (7!) - 90\ (7!) + 18.(7!)$$
$$(333)(7!) = 16,78,320$$

Example 13.30

How many anagrams (arrangements of letters) are there of

{7 . a, 5 . c, 1 . d, 5 . e, 1 . g, 1 . h, 7 . i, 3 . m, 9 . n, 4 . o, 5 . t}

[JNTUH, Nov 2008, Set No 1]
[JNTUH, June 2010, Set No. 2]

Solution: Given that

Total number of a's = 7

Total number of c's = 5

Total number of d's = 1

Total number of e's = 5

Total number of g's = 1

Total number of h's = 1

Total number of i's = 7

Total number of m's = 3

Total number of n's = 9

Total number of o's = 4

Total number of t's = 5

Therefore the total number of letters = 48

The number of arrangements of these letters

$$= \frac{48!}{(7!)(5!)(1!)(5!)(1!)(1!)(7!)(3!)(9!)(4!)(5!)}$$

13.4   Combinations and Restricted Combinations

The selection (group) of a number of things taken (some or all) of the given things at a time is called a combination.

The number of combinations of n distinct things taken r $(1 \leq r \leq n)$ at a time is denoted by $^{n}C_{r}$ (or) $C(n, r)$.

Theorem 2

$$C(n, r) = \frac{n!}{(n - r!)}$$

Proof: Any r permutations of n objects without repetition can be obtained by selecting r objects and then arranging the r objects in all possible orders.

Selection can be made in $C(n, r)$ ways and arrangements can be made in r! ways.

Thus, $P(n, r) = r! \, C(n, r)$.

This implies that $C(n, r) = \dfrac{n!}{(n-r)!r!} = \dbinom{n}{r}$

Theorem 3

If n and r are natural numbers such that $1 \le r \le n$ then

$$^n C_r + {}^n C_{r-1} = {}^{(n+1)} C_r$$

[In other notation $C(n, r) + C(n, r - 1) = C(n + 1, r)$]

Proof: Take L.H.S

$$\text{LHS} = {}^n C_r + {}^n C_{r-1}$$

$$= \frac{n!}{r!(n-r)!} + \frac{n!}{(r-1)!(n-r+1)!} \qquad \text{(by def)}$$

$$= \frac{n!}{r(r-1)!(n-r)!} + \frac{n!}{(r-1)!(n-r+1)(n-r)!}$$

$$= \frac{n!}{(r-1)!\ (n-r)!} \left[\frac{1}{r} + \frac{1}{n-r+1} \right]$$

$$= \frac{n!}{(r-1)!\ (n-r)!} \left[\frac{n-r+1+r}{r(n-r+1)} \right]$$

$$= \frac{(n+1)\ n!}{r(r-1)!\ (n-r+1)(n-r)!}$$

$$= \frac{(n+1)!}{r!\ (n-r+1)}$$

$$= {}^{n+1} C_r = \text{RHS}$$

Therefore ${}^n C_r + {}^n C_{r-1} = {}^{n+1} C_r, \qquad (1 \le r \le n)$

Example 13.31

Evaluate the following

(i) $\quad {}^9 C_4$

(ii) $\quad {}^{100} C_{96}$

(iii) $\quad {}^{51} C_{49}$

Solution:

(i)   By the definition we know

$$^{n}C_{r} = \frac{n!}{r!(n-r)!}$$

So $^{9}C_{4} = \dfrac{9!}{4!(9-4)!}$

$$= \frac{9!}{4!\ 5!}$$

$$= \frac{9 \times 8 \times 7 \times 6 \times 5!}{4 \times 3 \times 2 \times 1 \times 5!} = \frac{9 \times 8 \times 7 \times 6}{4 \times 3 \times 2 \times 1} = 126$$

Therefore $^{9}C_{4} = 126$

(ii)   By the known formula, we have

$$^{n}C_{r} = {}^{n}C_{n-r}$$

So $^{100}C_{96} = {}^{100}C_{100-96} = {}^{100}C_{4}$

$$= \frac{100!}{4!\ (100-4)!}$$

$$= \frac{100 \times 99 \times 98 \times 97 \times 96!}{4 \times 3 \times 2 \times 1 \times 96!}$$

$$= \frac{100 \times 99 \times 98 \times 97}{4 \times 3 \times 2 \times 1} = 3921225$$

Therefore $^{100}C_{96} = 3921225$

(iii)   By the definition

$$^{51}C_{49} = \frac{51!}{49!(51-49)!}$$

$$= \frac{51!}{49!\ 2!}$$

$$= \frac{51 \times 50 \times 49!}{2 \times 1 \times 49!} = \frac{51 \times 50}{2 \times 1} = 1275$$

Therefore $^{51}C_{49} = 1275$

Theorem 4

For $0 \leq r \leq n$ show that ${}^{n}C_{n-r} = {}^{n}C_{r}$

Proof: ${}^{n}C_{n-r} = \dfrac{n!}{(n-r)! \left[n-(n-r) \right]!}$

$$= \dfrac{n!}{(n-r)! \left[r! \right]}$$

$$= {}^{n}C_{r}$$

Now we provide some problems related to restricted combinations.

Example 13.32

How many different committees can be formed consisting of 4 men and 3 women out of 7 men and 5 women?

Solution: No. of ways of selecting 4 men out of 7 men is

$$= {}^{7}C_{4} = \dfrac{7!}{4! \ (7-4)!}$$

$$= \dfrac{7!}{4! \ 3!} = \dfrac{7 \times 6 \times 5 \times 4!}{3 \times 2 \times 1 \times 4!} = 35$$

No. of ways selecting 3 women out of 5 women is

$$= {}^{5}C_{3} = \dfrac{5!}{3! \ 2!} = \dfrac{5 \times 4 \times 3}{3 \times 2 \times 1} = 10$$

Therefore the number of committees can be formed consisting of 4 men and 3 women is

$$= 35 \times 10 = 350$$

Example 13.33

In how many ways 4 white balls and 6 black balls be arranged in a row so that no two white balls are together.

Solution: Total number of white balls = 4

          Total number of black balls = 6

We first arrange 6 black balls in a row and this can be done in one way.

     □ B □ B □ B □ B □ B □ B □

If 4 white balls are arranged in the space marked by □, where no two white balls would be together. Out of 7 possible spaces, 4 spaces for white balls can be choosen in $^7C_4$ ways.

Therefore total number of ways of arranging all balls with the given condition is

$$1 \times {}^7C_4 = {}^7C_4$$

$$= \frac{7!}{4! \, (7-4)!}$$

$$= \frac{7!}{4! \, 3!}$$

$$= \frac{7 \times 6 \times 5 \times 4!}{3 \times 2 \times 1 \times 4!}$$

$$= \frac{7 \times 6 \times 5}{3 \times 2 \times 1} = 35$$

Example 13.34

(a) In how many ways a committee of 3 be formed chosen from 10 people.

(b) How many committees of 3 or more can be chosen from 10 people?

Solution:

(a) C(10, 3) ways

(b) C(10, 3) + C(10, 4) + C(10, 5) + … + C(10, 10), which is also equal to $- 2^{10} -$ C(10, 1) C(10, 2).

Example 13.35

From a class of 25 students, 10 are to be chosen for an excursion party. There are 3 students who decide that either all them will join or none of them will join. In how many ways can they be chosen?

Solution: When three particular students join the excursion party:

In this case, we have to choose 7 students from the remaining 22 students. This can be done in $^{22}C_7$ ways.

When three particular students do not join the excursion party:

In this case, we have to choose 10 students from the remaining 22 students. This can be done in $^{22}C_{10}$ ways.

Hence, the required number of ways = $^{22}C_7 + {}^{22}C_{10} = 817190$.

Example 13.36

Applying the multiplication principle, show that a set S with n elements has 2^n subsets.

[JNTUH, Nov 2010, Set No. 3]

Solution: Let $0 \leq r \leq n$.

To form a subset T of S such that $|T| = r$, the first element can be selected in n ways, the second element can be selected in $(n - 1)$ ways and so on, finally the rth element can be selected in $(n - r - 1)$ ways. Therefore by using multiplication principle, the subset T with $|T| = r$ can be selected in $n(n - 1) \ldots (n - r + 1)$ ways. Therefore $n(n-1) \ldots (n-r+1) = {}^nC_r$ number of subsets T can be formed with $|T| = r$.

Then a subset of 'S' can have no element, one element, two element, ..., $n - 1$ elements and 'n' elements.

Therefore, the total number of subsets of 'S' is

$$1 + 1 + {}^nC_1 + {}^nC_2 + {}^nC_3 + \ldots + {}^nC_{n-1}$$

$$= 1 + {}^nC_1 + {}^nC_2 + \ldots + {}^nC_n \qquad (\text{since } {}^nC_1 = n, \ {}^nC_n = 1)$$

$$= 1 + {}^nC_1(1) + {}^nC_2(1)^2 + \ldots + {}^nC_{n-1}(1)^{n-1} + {}^nC_n(1)^n$$

$$= (1 + 1)^n \qquad \qquad (\text{by binomial theorem})$$

$$= 2^n.$$

Example 13.37

15 males and 10 females are seated in a round table meeting. How many ways they can be seated if all the females seated together?

[JNTUH, June 2010, Set No. 1]

Solution: Total number of males = 15

Total number of females = 10

10 females can be seated together in 10! ways.

In a round table meeting, the number of ways they can be seated if all the females seated together = 15! 10

Example 13.38

In how many ways can we draw a heart or spade from ordinary deck of playing cards? a heart or an ace? an ace or a king? A card numbered 2 through 10?

[JNTUH, Nov 2008, Set No. 3]

Solution: Total number of playing cards in a deck = 52

(i)   Total number of hearts = 13

Total number of spades = 13

The number of ways of drawing a heart = $^{13}C_1 = 13$

The number of ways of drawing a spade = $^{13}C_1 = 13$

Therefore the number of ways we draw a heart or spade = 13 + 13 = 26

(ii) A heart or an ace = $^{13}C_1 + {}^3C_1 = 13 + 3 = 16$

(iii) An ace or a king = $^4C_1 + {}^4C_1 = 4 + 4 = 8$

(iv) There are 9 cards numbered 2 through 10 in each of 4 suits, clubs, diamonds, hearts or spades. So we may choose a numbered card in 36 ways.

Example 13.39

A Group of 8 scientists is composed of 5 psychologists and 3 sociologists.

(a)   In how many ways can a committee of 5 be formed?

(b)   In how many ways can a committee of 5 be formed that has 3 psychologists and 2 sociologists?

[JNTUH, Nov 2008, Set No. 4]

Solution: Total number of scientists = 8

Psychologists = 5

Sociologists = 3

(i)   The total number of ways a committee formed with 5 members

$$= {}^8C_5 = \frac{8!}{5!\ 3!}$$

$$= \frac{6 \times 7 \times 8}{3 \times 2 \times 1} = 56$$

(ii) 3 psychologists selected from 5 is in $^5C_3$ ways

2 sociologists selected from 3 is in $^3C_2$ ways

So the number of ways committee formed of 5 members with 3 psychologists and 2 sociologists

$$= {}^5C_3 . {}^3C_2 = \frac{5!}{3!\ 2!} \times \frac{3!}{2!\ 1!} = \frac{120 \times 6}{12 \times 2} = 30$$

Example 13.40

How many arrangements are there of {8 . a, 6 . b, 7 . c} in which each 'a' is on atleast one side of another 'a'?

[JNTUH, Nov 2008, Set No. 1]; [JNTUH, June 2010, Set No. 2]

Solution: Given that

Total number of a's = 8

Total number of b's = 6

Total number of c's = 7

b's and c's (all put together $6 + 7 = 13$) can be arranged in $\dfrac{13!}{6! \, 7!}$ ways.

To place a's there are 14 places

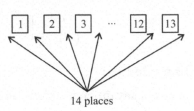

14 places

Case (i): All a's put together.

In this case, the number of arrangements is

$$\frac{13!}{(6!)(7!)}\left({}^{14}C_1 \right)$$

Case (ii): All a's divided into 2 parts.

In this case, the possibilities are

$$a\,a\,a\,a\,a\,a\,,\,a\,a$$
$$a\,a\,a\,a\,a\,,\,a\,a\,a$$
$$a\,a\,a\,a\,,\,a\,a\,a\,a$$
$$a\,a\,a\,,\,a\,a\,a\,a\,a$$
$$a\,a\,,\,a\,a\,a\,a\,a\,a$$

So, in this case, the number of possible arrangements is

$$\frac{13!}{(6!)(7!)}\left[5.\left({}^{14}C_2 \right) \right]$$

Case (iii): All a's divided into 3 parts.

In this case, the possibilities are

$$a\,a\,a\,,\,a\,a\,a\,,\,a\,a$$
$$a\,a\,a\,,\,a\,a\,,\,a\,a\,a$$
$$a\,a\,,\,a\,a\,a\,,\,a\,a\,a$$
$$a\,a\,a\,a\,,\,a\,a\,,\,a\,a$$
$$a\,a\,,\,a\,a\,a\,a\,,\,a\,a$$
$$a\,a\,,\,a\,a\,,\,a\,a\,a\,a$$

So, in this case, the number of possible arrangements is

$$\frac{13!}{(6!)(7!)}\left[6.\left(^{14}C_3\right)\right]$$

Case (iv): All a's divided into 4 parts.

The only possibility is a a, a a , a a , a a

So, in this case, the number of possible arrangements is

$$\frac{13!}{(6!)(7!)}\left[^{14}C_4\right]$$

Hence, the required number is

$$\frac{13!}{(6!)(7!)}\left[^{14}C_4+6\left(^{14}C_3\right)+5\left(^{14}C_2\right)+^{14}C_1\right]$$

Example 13.41

In how many different orders can 3 men and 3 women be seated in a row of 6 seats if:

  (i)   any one may sit in any of the seats

  (ii) the first and last seats must be filled by men

 (iii) men and women are seated alternatively

 (iv) all members of the same sex seated in adjacent seats?

[JNTUH, Nov 2008, Set No. 2]

Solution: There are totally 3 men + 3 women = 6

  (i)   Any one may sit in any of the seats

        $6! = 720$

  (ii) Two males can be selected in $^3C_2\,(=3)$ ways

      Arrangement of two males in two seats can be done in 2! (= 2) ways.

      In the remaining 4 seats 3 seats for women can be selected in $^4C_3\,(=4)$ ways.

      3 women can be sit in 3 places in 3! (= 6) ways.

      Hence the required number is

        $2 \times 3 \times 4 \times 6 = 144$ ways

 (iii) Now we have to find the number of ways when men and women are seated alternatively.

The 3 men can be seated in 3! ways in even places, then the 3 women can be seated in 3! ways in odd places.

Then the total number of arrangements in this ways is

$$3! \times 3! = 6 \times 6 = 36$$

Similarly, 3 men can be seated in 3! ways in odd places, then 3 women can be seated in 3! ways in even places.

So the total number of arrangements are

$$3! \times 3! = 6 \times 6 = 36$$

Therefore the required number of arrangements are

$$36 + 36 = 72$$

(iv)   3 men were seated in adjacent seats in 3! ways.

3 women were seated in adjacent seats in 3! ways.

The total number of ways = $3! \times 3! = 6 \times 6 = 36$

Therefore all members of the same sex seated in adjacent seats = $2 \times 36 = 72$

Example 13.42

Eight people enter an elevator at the first floor. The elevator discharges a passenger on each successive floor, until it empties on the fifth floor. How many different ways can this happen?

[JNTUH, June 2010, Set No. 1]

Solution: Total number of people entered the elevator at the first floor = 8 (say 'n')

Number of floors = 5 (say 'r')

By the given hypothesis, on each successive floor discharges at least one passenger until it completes in 5th floor.

Suppose

n_1 = the number of people came out in 2nd floor

n_2 = the number of people came out in 3rd floor

n_3 = the number of people came out in 4th floor

n_4 = the number of people came out in 5th floor

Now $n_1 + n_2 + n_3 + n_4 = 8$ and $n_1 \geq 1, n_2 \geq 1, n_3 \geq 1, n_4 \geq 1$

$\Rightarrow$ $n_1 > 0, n_2 > 0, n_3 > 0, n_4 > 0$

So $n_1 + n_2 + n_3 + n_4 = 8$

Therefore the number of possible ways is

$$C(n + r - 1, r) = C(4 + 8 - 1, 8) = C(11, 8)$$

Example 13.43

Compute the number of 10-digit numbers which contain only the digits 1, 2 and 3 with the digit 2 appearing in each number twice?

[JNTUH, June 2010, Set No. 4]

Solution:

1	2	3	4	5	6	7	8	9	10

The digit 2 can be kept at 2 places in $^{10}C_2$ ways. The remaining 8 places may contain either 1 or 3. So the remaining 8 places can be filled in

$$2 \times 2 \times \ldots (8 \text{ times}) = 2^8 \text{ ways.}$$

Thus the number of such numbers is $\left(^{10}C_2 \right) \times \left(2^8 \right)$

13.5   The Pigeonhole Principle and its Applications

If n pigeons are assigned to m pigeonholes, and m < n, then atleast one pigeonhole contains two or more pigeons.

For discussion, suppose each pigeonhole contains at most 1 pigeon. Then at most m pigeons have been assigned. But since m < n, not all pigeons have been assigned pigeonholes. This is a contradiction. Hence, at least one pigeonhole contains two or more pigeons.

13.5.1   The Extended Pigeonhole Principle

If n pigeonholes are occupied by kn + 1 or more pigeons, where k is a positive integer then at least one pigeon hole is occupied by k + 1 or more pigeons.

Example 13.44

Show that if any five numbers from '1' to '8' are chosen, then two of them will add to 9.

Solution: Let us construct four different two element sets {a, b} such that a + b = 9.

$$A_1 = \{1, 8\}$$
$$A_2 = \{2, 7\}$$
$$A_3 = \{3, 6\}$$
$$A_4 = \{4, 5\}$$

If any five numbers chosen it must contain atleast one of these sets.

Since there are only four sets, then according to pigeonhole principle, two of the chosen numbers belong to same set. These two numbers when we add up we get 9.

Example 13.45

Assume that there are 10 distinct pairs of gloves in a drawer. Show that if you choose 11 single gloves at random from the drawer, you are certain to have a pair.

Solution: The 10 distinct pairs constitute 10 pigeonholes. The 11 single gloves correspond to 11 pigeons. Therefore, there must be atleast one pigeonhole with two gloves and thus you will certainly have draw at least one pair of globes.

Exercises

Basics of Counting

(i)   How many different license plates are available if each plate contains a sequence of two letters followed by four digits (no sequence of letters are prohibited)

Ans: $26 \times 26 \times 10 \times 10 \times 10 \times 10 = 6760000$

(ii) How many ways are there to roll two distinguishable dice to yield a sum is divisible by 3?

Ans: 12

(iii) Find the number of binary sequences of length n.

Ans: 2^n

(iv) Find the number of 3-digit even numbers with no repeated digits.

Ans: 328

Permutations, Dearrangements Permutations with Repetition of Objects

(i)   How many different strings (sequences) of length 4 can be formed using the letters of the word FLOWER?

Ans: 360

(ii) In how many ways can 7 women and 3 men be arranged in a row if the 3 men must always stand next to eachother?

Ans: 3! 8!

(iii) Find 'n' if $2P(n, 2) + 50 = P(2n, 2)$

Ans: 5

(iv)   How many 8-digit telephone numbers have one or more repeated digits?

Ans: $10^8 - P(10, 8)$

Circular Permutations, Restricted Permutations

(i)   In how many ways can 5 children arrange themselves in a ring?

Ans: 4!

(ii)   Find the number of permutations of the letters of the word MISSISSIPPI. How many of these begin with an I? How many of these begin and end with an S?

Ans: 34650, 12600, 7560

(iii)   In how many ways can 3-men and 3-women be seated at a round table if

    (a)   No restriction is imposed?

    (b)   Each women is to be between two men?

Ans: (a) 120 (b) 12

Combinations and Restricted Combinations

(i)   From seven consonants and five vowels, how many sets consisting of four different consonants and three different vowels can be formed?

Ans: $C(7, 4) . C(5, 3) . 7! = 17,64,000$

(ii)   How many committees of 5 or more can be chosen from '9' people?

Ans: $C(9, 5) + C(9, 6) + C(9, 7) + C(9, 8) + C(9, 9)$

(iii)   Determine the number of diagonals formed by joining the angular points of an octagon.

Ans: 20

Pigeonhole Principle and its Applications

(i)   Show that at least in any set of seven classes there must be two that meet on the same day assuming that no class holds on Sunday?

(ii)   Show that if the integer b is such that $b - 1$ or $b - 2$ is divisible by 3, then $b^2 - 1$ is divisible by 3.

(iii)   How many persons must be choosen in order that atleast five of them will have birth days in the same calendar month?

Ans: 49 (at the least)

CHAPTER - 14

Binomial Theorem

If x and a are real numbers then for all $n \in N$ we have that

$$\left(x+a\right)^n = {}^nC_0 x^n a^0 + {}^nC_1 x^{n-1}a^1 + {}^nC_2 x^{n-2}a^2 +$$
$$\dots + {}^nC_r x^{n-r}a^r + \dots +$$
$$ {}^nC_{n-1}x^1 a^{n-1} + {}^nC_n x^0 a^n$$

That is, $\left(x+a\right)^n = \sum_{r=0}^{n} {}^nC_r x^{n-r}a^r$ and it is called as

Binomial Theorem.

14.1 Binomial and Multinomial Coefficients

14.1.1 Pascal's Triangle

We know that $(x + a)^0 = 1$

$(x + a)^1 = x + a = 1 \cdot x + 1 \cdot a$

$(x + a)^2 = x^2 + 2ax + a^2 = 1 \cdot x^2 + 2 \cdot ax + 1 \cdot a^2$

$(x + a)^3 = x^3 + 3x^2a + 3xa^2 + a^3$

$\qquad = 1 \cdot x^3 + 3 \cdot x^2a + 3 \cdot xa^2 + 1 \cdot a^3$

We observe that the coefficients in the above expansions follow a particular pattern as shown in the Table 14.1.

Table 14.1

Index of the binomial	Coefficients of various terms				
0			1		
1		1		1	
2		1	2	1	
3	1		3	3		1

We observe that each row is bounded by 1 on both sides. Any entry, except the first and last, in a row is the sum of two entries in the preceding row, one on the immediate left and the other on immediate right. The above pattern is known as Pascal's triangle. It has been checked that the above pattern also holds good for the coefficients in the expansions of the binomial expressions having index (exponent) greater than 3 as shown in the Table 14.2.

Table 14.2

Index of the binomial	Coefficients of various terms
0	1
1	1 1
2	1 2 1
3	1 3 3 1
4	1 4 6 4 1
5	1 5 10 10 5 1

Using the above Pascal's triangle, we may express $(x + a)^n$ for $n = 1, 2, 3, \ldots$

$$(x + a)^1 = x + a$$

$$(x + a)^2 = x^2 + 2ax + a^2$$

$$(x + a)^3 = x^3 + 3ax^2 + 3a^2x + a^3$$

$$(x + a)^4 = x^4 + 4ax^3 + 6a^2x^2 + 4a^3x + a^4$$

$$(x + a)^5 = x^5 + 5ax^4 + 10a^2x^3 + 10a^3x^2 + 5a^4x + a^5$$

and so on.

14.1.2   Properties of Binomial Coefficients or Combinatorial Identities

An identity that results from some counting process is called a combinatorial identity. Some identities involving binomial coefficients are given below:

1. In the expansion of $(1 + x)^n$ the coefficients of terms equidistant from the beginning and end are equal. [That is., $^nC_r = \, ^nC_{n-r}$].

2. The sum of the binomial coefficients in the expansion of $(1 + x)^n$ is 2^n.
    That is., $C_0 + C_1 + C_2 + \ldots + C_n = 2^n$.

    or $\displaystyle\sum_{r=0}^{n} {}^nC_r = 2^n$

    $$\left[(1+1)^n = \, ^nC_0 + \, ^nC_1 + \ldots ^nC_n = C_0 + C_1 + \ldots + C_n\right]$$

3. $0 = (1-1) = (1-1)^n = C_0 - C_1 + C_2 - C_3 + C_4 + \ldots + (-1)^n C_n$

$$\Rightarrow 0 = C_0 - C_1 + C_2 - C_3 + C_4 + ...$$

$$\Rightarrow C_0 + C_2 + C_4 + ... = C_1 + C_3 + C_5 + ...$$

4. The sum of the coefficients of the odd terms in the expansion of $(1 + x)^n$ is equal to the sum of the coefficients of the even terms and each is equal to 2^{n-1}.

$C_0 + C_1 + ... + C_n = 2^n$. Since $C_0 + C_2 + C_4 + ... = C_1 + C_3 + ...$ we have that $C_0 + C_2 + C_4 + ... = C_1 + C_3 + C_5 + ... = 2^{n-1}$

5. The number of ways (or combinations) of n different things selecting at least one of them is $^nC_1 + {}^nC_2 + {}^nC_3 + ... + {}^nC_n = 2^n - 1$

6. Newton's Identity:

$$^nC_r \cdot {}^rC_k = {}^nC_k \cdot {}^{(n-1)}C_{(r-k)} \qquad \text{for } n \geq r \geq k \geq o$$

7. Pascal Identity:

$$^{(n+1)}C_r = {}^nC_r + {}^nC_{(r-1)}$$

8. Vandermonde's Identity:

$$^{(n+m)}C_r = {}^nC_0 \cdot {}^mC_r + {}^nC_1 \cdot {}^mC_{(r-1)} + ... + {}^nC_r \cdot {}^mC_0$$

$$= \sum_{k=0}^{r} {}^mC_{(r-k)} \cdot {}^nC_k$$

for $n \geq r \geq 0$ and $m \geq r \geq 0$.

Theorem 1

Prove the identity $^nC_r = {}^nC_{(n-r)}$

Proof (combinational version): If r objects are chosen from n objects there are n – r objects are left. Thus selection of r objects from n objects is same as to pick out n – r objects. Hence for every r-combination automatically there is an associated (n – r) combination and conversely. This proves the identity.

Theorem 2

Prove the Pascal Identity $^{(n+1)}C_r = {}^nC_r + {}^nC_{(r-1)}$ where n and r are positive integers such that $n \geq r$.

Proof (combinational version): A choice of 'r' of the 'n + 1' objects $x_1, x_2, ... x_n$ may or may not include x_{n+1}. If it does not contain x_{n+1} then r objects have to be chosen from $x_1, x_2, ... x_n$ and there are ^nC_r such choices.

If this choice contain x_{n+1} then $r - 1$ further objects have to be chosen. From $x_1, x_2, \ldots, x_n$ and there are $^nC_{(r-1)}$ such choices. So by the rule of sum, the total number of choices is $^nC_r + {}^nC_{(r-1)}$ which must be equal to $^{(n+1)}C_r$.

Therefore $^{(n+1)}C_r = {}^nC_r + {}^nC_{(r-1)}$

Pascal's formula: Pascal's formula gives a recurrence relation for the computation of Binomial coefficient, given the initial data $C(n, 0) = C(n, n) = 1$ for all n. Notice that no multiplication is needed for this computation. One can obtain the numbers by constructing a triangular array using very simple arithmetic. The triangular array is usually called Pascal's triangle. One can label the rows of the triangular array by n = 0, 1, 2 and the positions within the nth row as k = 0, 1, 2, …, n. The zero row of the triangle is the single entry 1 and the first row be a pair of entries each equal to 1. This gives the first two rows, the nth row of the triangle, which contains n + 1 numbers, can be formed from the preceding row by the following rules.

(a)   The first (k = 0) and the last (k = n) entries are both equal to 1.

(b)   For $1 \le k \le n - 1$, the kth entry in the nth row is the sum of the $(k - 1)^{th}$ and k^{th} entries in the (n − 1) rows.

The first eight rows of Pascal's triangle are shown in the following diagram.

```
                1
              1   1
            1   2   1
          1   3   3   1
        1   4   6   4   1
      1   5  10  10   5   1
    1   6  15  20  15   6   1
  1   7  21  35  35  21   7   1
```

$$\binom{0}{0}$$

$$\binom{1}{0}\binom{1}{1}$$

$$\binom{2}{0}\binom{2}{1}\binom{2}{2}$$

$$\binom{3}{0}\binom{3}{1}\binom{3}{2}\binom{3}{3}$$

$$\binom{4}{0}\binom{4}{1}\binom{4}{2}\binom{4}{3}\binom{4}{4}$$

$$\binom{5}{0}\binom{5}{1}\binom{5}{2}\binom{5}{3}\binom{5}{4}\binom{5}{5}$$

$$\binom{6}{0}\binom{6}{1}\binom{6}{2}\binom{6}{3}\binom{6}{4}\binom{6}{5}\binom{6}{6}$$

$$\binom{7}{0}\binom{7}{1}\binom{7}{2}\binom{7}{3}\binom{7}{4}\binom{7}{5}\binom{7}{6}\binom{7}{7}$$

(a)   (b)

Note that ^nC_r is also denoted by $\begin{pmatrix} n \\ r \end{pmatrix}$

A basic property of binomial coefficients is illustrated by Pascal's triangle. If we evaluate the numbers, we can find that we obtain the same numbers as in the first six rows of Pascal's triangle. Each number in the triangle is the sum of the two numbers above it.

For example, take n = 5 and k = 3, we have $\begin{pmatrix} 5 \\ 3 \end{pmatrix} = \begin{pmatrix} 4 \\ 2 \end{pmatrix} + \begin{pmatrix} 4 \\ 3 \end{pmatrix}$, which is the particular case of Pascal's identity.

14.1.3   Multinomial Coefficients

The expression in the form $x_1 + x_2$ is a binomial. A multinomial is an expression of the form $x_1 + x_2 + \ldots + x_n$, with $n \geq 3$. Just as binomial coefficients appear in the expansion of powers of a binomial, multinomial coefficients appear when a power of a multinomial is expanded.

14.1.4   Multinomial Theorem

For real numbers $a_1, \ldots, a_k$ and for $n \in N$, we have

$$\left(a_1 + \ldots + a_k\right)^n = \sum_{n_1 + \ldots + n_k} \begin{pmatrix} n \\ n_1 \ldots n_k \end{pmatrix} a_1^{n_1} \ldots a_k^{n_k}$$

Here $\begin{pmatrix} n \\ n_1 \ldots n_k \end{pmatrix}$ stands for $\dfrac{n!}{n_1! n_2! \ldots n_k!}$ is called the multinomial coefficient and the sum is over all possible ways to write n as $n_1 + n_2 + \ldots + n_k$

Example 14.1

Find the number of arrangements of the letters in the word ACCOUNTANT.

Solution: Total number of letters in the word ACCOUNTANT is 10. Out of which A occurs twice, C occurs twice, N occurs twice, T occurs twice and the rest are all different. Since some of the letters are repeated, we apply multinomial theorem.

Hence the number of arrangements is

$$\frac{10!}{2!2!2!2!} = 226800$$

Note:

Like the term "binomial coefficient", the term "multinomial coefficient" comes from considering algebraic expressions. Given real numbers $a_1, a_2, \ldots, a_k$, consider the power

$$(a_1 + \ldots + a_k)^n = (a_1 + \ldots + a_k)(a_1 + \ldots + a_k) \ldots (a_1 + \ldots + a_k)$$

After performing this product, but before collecting like terms, a typical term in this product has the form

$$a_1^{n_1} \ldots a_k^{n_k}$$

The coefficient of $a_1^{n_1} \ldots a_k^{n_k}$ after collecting like terms is equal to the number of ways of picking n_1 factors equal to a_1 and a_2 factors equal to a_2, and so on, as we multiply the n copies of $a_1 + a_2 + \ldots + a_k$. This is precisely the multinomial coefficient.

$$\binom{n}{n_1 n_2 \ldots n_k}$$

Example 14.2

Find the coefficient of $x^3 y^2 z^2$ in $(x + y + z)^9$?

Solution: This is the same as how many ways one can choose x from three brackets, y from 2 brackets and z from two brackets in the expansion.

$(x + y + z) (x + y + z) \ldots (x + y + z)$ (9 times)

This can be done in $\binom{9}{3\ 2\ 2} = \dfrac{9!}{3!2!2!} = 15120$

Example 14.3

A student is allowed to select at most n books from a collection of $(2n + 1)$ books. If the total number of ways in which he can select a book is 63, find the value of n.

Solution: Given that the student selects at most n books from a collection of $(2n + 1)$ books. It means that he selects one book or two books or three books or ... or n books. Hence by the given hypothesis.

$$^{2n+1}C_1 + {}^{2n+1}C_2 + {}^{2n+1}C_3 + \ldots + {}^{2n+1}C_n = 63 \qquad \ldots\ldots(14.1)$$

But we know that

$$^{2n+1}C_0 + {}^{2n+1}C_1 + {}^{2n+1}C_2 + \ldots + {}^{2n+1}C_{2n+1} = 2^{2n+1}$$

Now, $^{2n+1}C_0 = {}^{2n+1}C_{2n+1} = 1$

Then, $2 + \left({}^{2n+1}C_1 + {}^{2n+1}C_2 + {}^{2n+1}C_3 + \ldots + {}^{2n+1}C_n \right)$

$$+ \left({}^{2n+1}C_{n+1} + {}^{2n+1}C_{n+2} + {}^{2n+1}C_{n+3} + \ldots + {}^{2n+1}C_{2n-1} + {}^{2n+1}C_{2n} \right)$$

$$\Rightarrow 2 + \left({}^{2n+1}C_1 + {}^{2n+1}C_2 + {}^{2n+1}C_3 + \ldots + {}^{2n+1}C_n \right) + \left({}^{2n+1}C_{n-1} + {}^{2n+1}C_{n-2} + \ldots + {}^{2n+1}C_2 + {}^{2n+1}C_1 \right)$$

$$= 2^{2n+1} \quad \left[\because {}^{2n+1}C_r\right]$$

$$\Rightarrow \quad 2 + 2\left({}^{2n+1}C_1 + {}^{2n+1}C_2 + {}^{2n+1}C_3 + ... + {}^{2n+1}C_n\right)$$

$$= 2^{2n+1}$$

$$\Rightarrow \quad 2 + 2 \cdot 63 = 2^{2n+1} \qquad \text{[from (14.1)]}$$

$$\Rightarrow \quad 1 + 63 = 2^{2n}$$

$$\Rightarrow \quad 2^6 = 2^{2n}$$

$$\Rightarrow \quad 6 = 2n$$

Therefore n = 3

14.1.5 Multinomial Theorem – An Explanation

(i) If there are l objects of one kind, m objects of second kind, n objects of third kind and so on; then the number of ways choosing r objects out of these objects

(i.e., l + m + n + ...) is the coefficient of x^r in the expansion of

$(1 + x + x^2 + ... x^l)(1 + x + x^2 + ... x^m)(1 + x + x^2 + ... x^n)...$

Further if one object of each kind is to be included, then the number of ways of choosing r objects out of these l + m + n + ... objects is the coefficient of x^r in the expansion of $(x + x^2 + x^3 + ... x^l)(x + x^2 + x^3 + ... x^m)(x + x^2 + x^3 + ... x^n)...$

(ii) If there are l objects of one kind, m objects of second kind, n objects of third kind and so on, then the number of possible arrangement/permutations of r objects out of l + m + n + ... objects is the coefficient of x^r in the expansion of r!

$$\left(1 + \frac{x}{1!} + \frac{x^2}{2!} + ... \frac{x^l}{l!}\right)\left(1 + \frac{x}{1!} + \frac{x^2}{2!} + ... \frac{x^m}{m!}\right)\left(1 + \frac{x}{1!} + \frac{x^2}{2!} + ... \frac{x^m}{n!}\right)...$$

14.1.6 How to Find Number of Solutions of the Equation

Let the equation be

$$\alpha + 2\beta + 3\gamma + ... + q\theta = n \qquad \qquad .....(14.2)$$

(i) If zero included then number of solutions of equation (14.2)

= coefficient of x^n in $(1 + x + x^2 + ...)(1 + x^2 + x^4 + ...)(1 + x^3 + x^6 + ...)$

$(1 + x^q + x^{2q} + ...)$

= coefficient of x^n in $(1 + x)^{-1}(1 + x^2)^{-1}(1 + x^3)^{-1} ... (1 + x^q)^{-1}$

(ii) If zero excluded then the number of solutions of equation (14.2)

$\qquad$ = coefficient of x^n in $(x + x^2 + x^3 + ...) (x^2 + x^4 + x^6 + ...) (x^3 + x^6 + x^9 + ...)$

$\qquad ... (x^q + x^{2q} + ...)$

$\qquad$ = coefficient of x^n in $x^{1+2+3+ \, \cdots \, +q} (1 - x)^{-1} (1 - x^3)^{-1} ... (1 - x^q)^{-1}$

$\qquad$ = coefficient of $x^{n - \frac{q(q+1)}{2}}$ in $(1 - x)^{-1} (1 - x^2)^{-1} (1 - x^3)^{-1} ... (1 - x^n)^{-1}$

Example 14.4

Find the number of combinations and permutations of 4 letters taken from the word EXAMINATION.

Solution: There are 11 letters

$\qquad$ A, A; I, I; N, N; E, X, M, T, O.

The number of combinations

$\qquad$ = coefficient of x^4 in $(1 + x + x^2)^3 (1 + x)^5$

$\qquad$ [since 2A's, 2I's, 2N's, 1 E, 1 X, 1 M, 1 T and 1 O]

$\qquad$ = coefficient of x^4 in $\{(1 + x)^3 + x^6 + 3(1 + x)^2 x^2 + 3 (1 + x) x^4\} (1 + x)^5$

$\qquad$ = coefficient of x^4 in $\{(1 + x)^8 + x^6 . (1 + x)^5 + 3x^2 (1 + x)^7 + 3x^4 (1 + x)^6\}$

$\qquad$ = $^8C_4 + 0 + 3.^7C_2 + 3$

$\qquad$ = $\dfrac{8.7.6.5}{1.2.3.4} + 3.\dfrac{7.6}{1.2} + 3$

$\qquad$ = $70 + 63 + 3 = 136$

Number of permutations

$\qquad$ = coefficient of x^4 in $4! \left(1 + \dfrac{x}{1!} + \dfrac{x^2}{2!}\right)^2 . \left(1 + \dfrac{x}{1!}\right)^5$

$\qquad$ = coefficient of x^4 in $4! \left(1 + x + \dfrac{x^2}{2}\right)^3 (1 + x)^5$

$\qquad$ = coefficient of x^4 in $4!$

$\qquad \left\{(1 + x)^3 + \dfrac{x^6}{8} + \dfrac{3}{2}(1 + x)^2 x^2 + \dfrac{3}{4}x^4 (1 + x)\right\}(1 + x)^5$

$\qquad$ = coefficient of x^4 in $4!$

$$\left\{ \left(1+x\right)^8 + \frac{x^6}{8}\left(1+x\right)^5 + \frac{3}{2}x^2 + \left(1+x\right)^7 + \frac{3}{4}x^4\left(1+x\right)^6 \right\}$$

$$= 4! \left\{ {}^8C_4 + 0 + \frac{3}{2} \cdot {}^7C_2 + \frac{3}{4} \right\}$$

$$= 24 \left\{ \frac{8.7.6.5}{1.2.3.4} + \frac{3}{2} \cdot \frac{7.6}{1.2} + \frac{3}{4} \right\}$$

$$= 8.7.6.5 + 6\,(3.7.6) + 6.3$$

$$= 1680 + 756 + 18 = 2454$$

Example 14.5

Find the number of non-negative integral solutions of $x_1 + x_2 + x_3 + 4x_4 = 20$

Solution: Number of non-negative integral solutions of the given equation

$$= \text{coefficient of } x^{20} \text{ in } (1-x)^{-1}(1-x)^{-1}(1-x)^{-1}(1-x^4)^{-1}$$

$$= \text{coefficient of } x^{20} \text{ in } (1-x)^{-3}(1-x^4)^{-1}$$

$$= \text{coefficient of } x^{20} \text{ in } \left(1 - {}^3C_1 x + {}^4C_2 x^2 + {}^5C_3 x^3 + {}^6C_4 x^4 \right.$$

$$+ \ldots + {}^{10}C_8\, x^8 + \ldots {}^{14}C_{12}\, x^{12}$$

$$+ \ldots + {}^{18}C_{16}\, x^{16} + \ldots + {}^{22}C_{20}\, x^{20} + \ldots \Big)$$

$$\left(1 + x^4 + x^8 + x^{12} + x^{16} + x^{20} + \ldots \right)$$

$$= 1 + {}^6C_4 + {}^{10}C_8 + {}^{14}C_{12} + {}^{18}C_{16} + {}^{22}C_{20}$$

$$= 1 + {}^6C_2 + {}^{10}C_2 + {}^{14}C_2 + {}^{18}C_2 + {}^{22}C_2$$

$$= 1 + \left(\frac{6.5}{1.2}\right) + \left(\frac{10.9}{1.2}\right) + \left(\frac{14.13}{1.2}\right) + \left(\frac{18.17}{1.2}\right) + \left(\frac{22.21}{1.2}\right)$$

$$= 1 + 15 + 45 + 91 + 153 + 231$$

$$= 536$$

Example 14.6

Find the coefficient of $x^4 y^7$ in the expansion of $(x-y)^{11}$.

[JNTUH, Nov 2010, Set No. 1]

Solution: By the definition of binomial theorem, we have

$$(x+y)^n = {}^nC_0\, x^n + {}^nC_1\, x^{n-1}y + \ldots + {}^nC_r\, x^{n-r}y^r + \ldots + {}^nC_n\, y^n$$

Coefficient of $x^4 y^7$ in $(x-y)^{11}$ is

$${}^{11}C_7 = \frac{11!}{7!\,4!} = \frac{8 \times 9 \times 10 \times 11}{4 \times 3 \times 2 \times 1}$$

$$= 330$$

Example 14.7

Using the binomial theorem to prove that

$$3^n = \sum_{r=0}^{n} C(n,\, r)\, 2^r$$

[JNTUH, Nov 2010, Set No. 2]

Solution: By the binomial expansion for any positive integer n, we have

$$(x+y)^n = \sum_{r=0}^{n} \left({}^nC_r\right) x^r y^{n-r} \qquad\qquad \ldots.(14.3)$$

By taking $x = 2$ and $y = 1$ in (1), we get that

$$(2+1)^n = \sum_{r=0}^{n} \left({}^nC_r\right) 2^r (1)^{n-r}$$

$$\Rightarrow \quad 3^n = \sum_{r=0}^{n} \left({}^nC_r\right) 2^r$$

$$\Rightarrow \quad 3^n = \sum_{r=0}^{n} C(n,\, r)\, 2^r$$

Example 14.8

If $x > 2$, $y > 0$, $z > 0$, then find the number of solutions of $x + y + z + w = 21$

[JNTUH, Nov 2010, Set No. 2]

Solution: Given that

$$x + y + z + w = 21 \qquad\qquad \ldots.(14.4)$$

where $x > 2$, $y > 0$, $z > 0$ and $w > 0$

Let us set

$$y_1 = x - 2 \quad \Rightarrow \quad x = y_1 + 2$$

$$y_2 = y$$

$$y_3 = z$$

$$y_4 = w$$

Then y_1, y_2, y_3, y_4 are all non-negative, put these values in (14.4), then

$$(y_1 + 2) + y_2 + y_3 + y_4 = 21$$

$$\Rightarrow \quad y_1 + y_2 + y_3 + y_4 = 21 - 2 = 19$$

Therefore the number of non-negative integer solutions of the equation is

$$C(n + r - 1, \ r) = C[4 + 19 - 1, \ 19] = {}^{22}C_{19} = \frac{22!}{19! \ 3!}$$

14.2   Generating Functions of Permutations and Combinations

In this section, we study the concept 'generating function' and some problems involving the notion of permutations and combinations.

We provide some examples related to different generating functions so that the reader can understand how the generating functions are useful in counting process in different situations.

14.2.1   The Concept "Generating Function"

Consider a sequence of real numbers $b_0, b_1, \ldots$ Let us denote this sequence by $\{b_n\}$, $n = 0, 1, 5, 2, \ldots$ or just $\{b_n\}$

The function

$$f(x) = b_0 + b_1 x + b_2 x^2 + \ldots + b_n x^n + \ldots = \sum_{k=0}^{\infty} b_k x^k \qquad \ldots(14.5)$$

is called a generating function for the sequence $\{b_n\}$.

Note that b_k is the coefficient of x^k in the expansion of $f(x)$ is a series of powers of x. The term $b_0 = b_0 x^0$ is the constant term.

If $f(x)$ is the generating function of the sequence $\{b_n\}$, then we say that $f(x)$ generates the sequence $\{b_n\}$. The series on the RHS of the expansion eq. (14. 5) is known as the power series expansion (or) formal power series.

If all the coefficients are equal to zero after certain point, then $f(x)$ is just a polynomial. In particular, if $b_k \neq 0$ and $b_j = 0$ for all $j \geq (k + 1)$ then $f(x)$ is a polynomial of degree k.

Example 14.9

Consider the expansion

$$(1-x)^{-1} = 1 + x + x^2 + \ldots = \sum_{m=0}^{\infty} x^m \qquad \ldots(14.6)$$

Then

$$f(x) = (1-x)^{-1} = \sum_{m=0}^{\infty} x^m \text{ is the generating function for the sequence } 1, 1, \ldots$$

Example 14.10

Consider the expansion

$$(1+x)^{-1} = 1 - x + x^2 - x^3 + \ldots = \sum_{k=0}^{\infty} (-1)^k x^k \qquad \ldots(14.7)$$

Then

$$f(x) = (1+x)^{-1} = \sum_{k=0}^{\infty} (-1)^k x^k \text{ is the generating function for the sequence } 1, -1, 1, -1, \ldots$$

Note:

Consider the expansion for $(1+x)^n$

$$(1+x)^n = 1 + nx + \frac{n(n-1)}{1.2}x^2 + \frac{n(n-1)(n-2)}{1.2.3}x^3 + \ldots$$

$$= \sum_{r=0}^{\infty} \frac{n(n-1)(n-2)\ldots(n-r+1)}{r!} \qquad \ldots(14.8)$$

So for a real number 'n', we get that $f(x) = (1+x)^n$ is a generating function for the sequence

$$1, \frac{n}{1!}, \frac{n(n-1)}{2!}, \frac{n(n-1)(n-2)}{3!}, \ldots$$

In case, if 'n' is a positive integer, the expansion given in eq. (14.8) terminates with the term containing x^n. (Recall the binomial theorem for a positive integral index).

So if 'n' is a positive integer, then $(1+x)^n$ generates the sequence,

$$\binom{n}{0}, \binom{n}{1}, \ldots, \binom{n}{n}, \ldots$$

If 'n' is a real number (not necessarily a positive integer), let us define $\left({}^{n}C_r \right)$ as follows:

$$\binom{n}{0} = 1 \qquad\qquad .....(14.9)$$

and $\qquad \binom{n}{r} = {}^{n}C_r = \dfrac{n(n-1)(n-2)...(n-r+1)}{r!} \qquad \text{for all } r \geq 1 \qquad .....(14.10)$

Then the expansion in eq. (14.8) can be written as

$$(1+x)^n = \sum_{r=0}^{\infty} \binom{n}{r} x^r \qquad .....(14.11)$$

for a given real number 'n'.

Hence the function $f(x) = (1+x)^n = \sum_{r=0}^{\infty} \binom{n}{r} x^r$ is a generating function for the sequence

$${}^{n}C_0, {}^{n}C_1, {}^{n}C_2, ..., {}^{n}C_r, ...$$

If "n" is a positive integer, then all the terms of this sequence beyond the term $\left({}^{n}C_n \right)$ are identically "0" (as mentioned above).

Example 14.11

We know that, for any infinitely differentiable function f(x), we get the expansion:

$$f(x) = f(0) + \frac{x}{1!}f^1(0) + \frac{x^2}{2!}f^{11}(0) + ... \qquad .....(14.12)$$

$$= \sum_{r=0}^{\infty} \frac{f^{(r)}(0)}{r!} x^r$$

This is known as "Maclaurins" series.

Thus $\qquad f(x) = \sum_{r=0}^{\infty} \frac{f^{(r)}(0)}{r!} x^r$

is a generating function for the sequence $\{b_n\}$, where $b_k = \dfrac{f^{(k)}(0)}{k!}$

So for any infinitely differentiable function f(x), we can find a sequence $\{b_n\}$ such that f(x) is a generating function of $\{b_n\}$.

Example 14.12

Find the sequence generated by the function.

$$f(x) = (3 + x)^3$$

Solution: Consider the given function

$$f(x) = (3 + x)^3 \qquad\qquad\qquad .....(14.13)$$

$$(3+x)^3 = \left[3\left(1+\frac{x}{3}\right)\right]^3$$

$$= 3^3\left(1+\frac{x}{3}\right)^3$$

$$= 27 \times \left(1+\frac{x}{3}\right)^3$$

$$= 27 \times \left\{1 + \left(^3C_1\right)\left(\frac{x}{3}\right) + \left(^3C_2\right)\left(\frac{x}{3}\right)^2 + \left(^3C_3\right)\left(\frac{x}{3}\right)^3\right\}$$

(by Binomial Theorem)

$$= 27 \times \left\{1 + x + \frac{x^2}{3} + \frac{x^3}{27}\right\}$$

$$= 27 \times \left\{1 + x + \frac{x^2}{3} + \frac{x^3}{27}\right\}$$

$$= 27 + 27x + 9x^2 + x^3$$

$$= 27 + 27x + 9x^2 + x^3 + 0.x^4 + 0.x^5 + ...$$

This shows that the sequence generated by the generating function

$(3 + x)^3$ is 27, 27, 9, 1, 0, 0, ...

Example 14.13

Find the sequence generated by the function $\dfrac{1}{1-x} + 2x^3$

Solution: Consider the given function

$$f(x) = \frac{1}{1-x} + 2x^3 = (1-x)^{-1} + 2x^3$$

$$= (1 + x + x^2 + \ldots) + 2x^3$$

$$= 1 + x + x^2 + 3x^3 + x^4 + \ldots$$

Hence the sequence generated by the given generating function $\frac{1}{1-x} + 2x^3$ is

$$1, 1, 1, 3, 1, 1, \ldots$$

Example 14.14

Find the sequence generated by the function $(1+3x)^{-\frac{1}{3}}$.

Solution: We have by binomial expansion,

$$(1+3x)^{-1/3} = 1 + \sum_{r=1}^{\infty} \frac{(-1/3)(-1/3 - 1)(-1/3 - 2)\ldots(-1/3 - r + 1)}{r!}(3x)^r$$

$$= 1 + \sum_{r=1}^{\infty} \frac{(-1)(-4)(-7)\ldots(-3r + 2)}{r!} x^r$$

This shows that the sequence generated by the generating function $(1+3x)^{-1/3}$ is

$$1, \; -1, \; \frac{(-1)(-4)}{2!}, \; \frac{(-1)(-4)(-7)}{3!}, \ldots$$

Example 14.15

Find the sequence generated by the function $f(x) = 2x^2(1-x)^{-1}$

Solution: The given function is

$$f(x) = 2x^2(1-x)^{-1}$$

$$= 2x^2(1 + x + x^2 + \ldots)$$

$$= 2x^2 + 2x^3 + 2x^4 + \ldots$$

$$= 0 + 0.x + 2x^2 + 2x^3 + 2x^4 + \ldots$$

Hence the sequence generated by the generating function.

$2x^2(1-x)^{-1}$ is $0, 0, 2, 2, 2, \ldots$

14.2.2   Properties of Generating Functions with Respect to Sum and Derivative

(i)   If $f(x)$ is a generating function for a sequence $\{a_n\}$ and $g(x)$ is a generating function for a sequence $\{b_n\}$, then $d_1\ f(x) + d_2\ g(x)$ is a generating function for the sequence $\{d_1\ a_n + d_2\ b_n\}$, where d_1 and d_2 are two given real numbers.

(ii) If $f(x)$ is a generating function for a sequence $\{b_n\}$, then $x\ f'(x)$ (where $f'(x)$ is the derivative of $f(x)$) is a generating function for the sequence $\{n.\ b_n\}$

(iii) If $f(x)$ is a generating function for a sequence $\{b_n\}$ and c is a constant then $c.f(x)$ is a generating function for the sequence $\{cb_n\}$.

Example 14.16

Find the sequence generated by the function $3x^3 + e^{2x}$.

Solution: Write $f(x) = 3x^3$ and $g(x) = e^{2x}$

Given function is $f(x) + g(x) = 3x^3 + e^{2x}$ .....(14.14)

Since

$$e^x = 1 + \frac{x}{1!} + \frac{x^2}{2!} + \frac{x^3}{3!} + ...$$

We have that

$$g(x) = e^{2x} = 1 + \frac{(2x)}{1!} + \frac{(2x)^2}{2!} + ...$$

$$= 1 + \frac{2}{1!}x + \frac{2^2}{2!}x^2 + \frac{2^3}{3!}x^3 + ...$$

$$f(x) = 3x^3 = 0 + 0.x + 0.x^2 + 3.x^3 + 0.x^4 + ...$$

Note that $f(x)$ is generating function for the sequence $0, 0, 0, 3, 0, 0, ...$

$g(x)$ is generating function for

$$1, 2, \frac{2^2}{2!}, \frac{2^3}{3!}, ...$$

Hence the sequence generated by the function $f(x) + g(x) = 3x^3 + e^{2x}$ is given by

$$\left(0,\ 0,\ 0,\ 3,\ 0,\ 0,\ ...\right) + \left(1,\ 2,\ \frac{2^2}{2!},\ \frac{2^3}{3!},\ ...\right)$$

$$= 1,\ \frac{2}{1!},\ \frac{2^2}{2!},\ 3 + \frac{2^3}{3!},\ \frac{2^4}{4!},\ ...$$

Example 14.17

Find the generating function for the sequence 1, –2, 3, –4, ...

Solution: Given sequence is 1, –2, 3, –4, ...

We have to find the generating function related to the given sequence.

By the binomial expansion, we have that

$$(1+x)^{-2} = 1 - 2x + 3x^2 - 4x^3 + ... \qquad \qquad .....(14.15)$$

$$= \sum_{n=0}^{\infty} (-1)^n (n+1) x^n$$

Hence $f(x) = (1+x)^{-2} = \dfrac{1}{(1+x)^2}$ is the generating function for the sequence 1, –2, 3, 4.

Example 14.18

Find the generating function for the sequence 1, 2, 3, 4, 5, ...

Solution: Given sequence is 1, 2, 3, 4, ...

We have to find the related generating function.

We know the binomial expansion that

$$(1-x)^{-2} = 1 + 2x + 3x^2 + 4x^3 + ... \qquad \qquad .....(14.16)$$

$$= \sum_{n=0}^{\infty} (n+1) x^n$$

From (14.16), it follows that

$$f(x) = \dfrac{1}{(1-x)^2} = (1-x)^{-2}$$

is the generating function for the given sequence 1, 2, 3, 4, ...

Example 14.19

Suppose that f(x) is a generating function for the sequence $\{a_n\}$ and g(x) is the generating function for the sequence $\{b_n\}$. Express g(x) in terms of f(x) if

$$b_3 = 3, \, b_7 = 7, \, b_n = a_n \qquad \qquad \text{for } n \neq 3, 7$$

Solution: Given that $g(x) = \sum\limits_{n=0}^{\infty} b_n x^n$

$$= b_0 + b_1 x + b_2 x^2 + \ldots + b_n x^n + \ldots$$

Since $\quad\quad\quad b_3 = 3, \; b_7 = 7, \; b_n = a_n \;\text{ for } n \neq 3, 7$

We have that

$$g(x) = a_0 + a_1 x + a_2 x^2 + 3x^3 + a_4 x^4 + a_5 x^5 + a_6 x^6 + 7x^7 + a_8 x^8 + \ldots$$

$\Rightarrow \quad\quad\quad g(x) = \left\{ \sum\limits_{r=0}^{\infty} a_r x^r - a_3 x^3 - a_7 x^7 \right\} + 3x^3 + 7x^7$

$\Rightarrow \quad\quad\quad g(x) = f(x) + (3 - a_3) x^3 + (7 - a_7) x^7$

Example 14.20

Find the related generating function for the sequence 0, 1, –2, 3, –4, …

Solution: We know that

$$0 + x - 2x^2 + 3x^3 - 4x^4 + \ldots$$
$$= x\left(1 - 2x + 3x^2 - 4x^3 + \ldots\right) = x(1 + x)^{-2}$$

This shows that $f(x) = x(1+x)^{-2}$ is the generating function for the given sequence

0, 1, –2, 3, –4, …

Example 14.21

Find the related generating function for the sequence 0, 1, 2, 3, …

We know that

$$0 + x + 2x^2 + 3x^3 + \ldots = x(1 + 2x + 3x^2 + \ldots) = x(1 - x)^{-2} \quad\quad \ldots\ldots(14.17)$$

This shows that

$$f(x) = x(1 - x)^{-2} = \frac{x}{(1 - x)^2}$$

is the generating function for the given sequence 0, 1, 2, 3, …

Example 14.22

Find the generating function for the sequence $1^2, 2^2, 3^2, \ldots$

Solution: We know that

$$0 + 1.x + 2.x^2 + 3.x^3 + \ldots = x(1-x)^{-2} \qquad \ldots..(14.18)$$

Differentiating (14.18) w. r. t. x on both sides, we get that

$$1 + 2(2x) + 3(3x^2) + 4(4x^3) + \ldots = \frac{1+x}{(1-x)^3}$$

Therefore $f(x) = \dfrac{1+x}{(1-x)^3}$

is generating function for the given sequence $1^2, 2^2, 3^2, \ldots$

Example 14.23

Find the generating function for the sequence $0^2, 1^2, 2^2, \ldots$

Solution: We know that

$$0^2 + 1^2.x + 2^2.x^2 + \ldots = x[1^2 + 2^2.x + 3^2x^2 + \ldots]$$

$$= \frac{x(1+x)}{(1-x)^3} \qquad \ldots..(14.19)$$

Hence $f(x) = \dfrac{x(1+x)}{(1-x)^3}$

is generating function for the given sequence

$$0^2, 1^2, 2^2, \ldots$$

Example 14.24

Find the generating function for the sequence $0, 2, 6, 12, 20, \ldots$

Solution: Note that the rth term of the given sequence is of the form $r + r^2$ for $r = 0, 1, 2, \ldots$

We know that the generating function for the sequence $\{r\}$ is $0, 1, 2, \ldots$

$$0 + 1.x + 2.x^2 + \ldots$$

Hence $f(x) = \dfrac{x}{(1-x)^2}$

We know that the generating function for the sequence $\{r^2\} = (0^2, 1^2, 2^2, \ldots)$ is

$$g(x) = \frac{x(1+x)}{(1-x)^3}$$ (from above example)

Therefore the generating function for the given sequence is

$$f(x) + g(x) = \frac{x}{(1-x)^2} + \frac{x(1+x)}{(1-x)^3}$$

$$= \frac{2x}{(1-x)^3}$$

Example 14.25

Find a generating function for the sequence $\{b_m\}$ where

$$b_m = 1 \quad \text{for} \quad 0 \le m \le n$$

and $b_m = 0 \quad \text{for} \quad m \ge n+1$

Solution: For the given sequence $\{b_m\}$, the generating function is

$$1 + 1.x + 1.x^2 + \dots + 1.x^n + 0.x^{n+1} + 0.x^{n+2} + \dots$$

$$= 1 + x + x^2 + \dots + x^n$$

$$= \frac{1 - x^{n+1}}{1 - x}$$

Therefore $f(x) = \dfrac{1 - x^{n+1}}{1 - x}$

is generating function for the given sequence.

Example 14.26

Prove that $\dbinom{2n}{n}$ is the coefficient of x^n in the expansion of $(1+x)^{2n}$.

Solution: We know that

$$(1 + x)^{2n} = (1 + x)^n (1 + x)^n$$

$$= \sum_{r=0}^{n} \binom{n}{r} x^r . \sum_{s=0}^{n} \binom{n}{s} x^s$$ .....(14.20)

The coefficient of x^n in LHS of (14.20) is $\dbinom{2n}{n}$ .....(14.21)

The coefficient of x^n in the RHS of (14.20) is

$$= \binom{n}{0}\binom{n}{n} + \binom{n}{1}\binom{n}{n-1} + \dots + \binom{n}{n}\binom{n}{0}$$

$$= \sum_{r=0}^{n} \binom{n}{r}\binom{n}{n-r}$$

$$= \sum_{r=0}^{n} \binom{n}{r}\binom{n}{r}$$

$$= \sum_{r=0}^{n} \binom{n}{r}^2 \qquad\qquad .....(14.22)$$

From Eq. (14.21) and (14.22) we get that

$$\binom{2n}{n} = \sum_{r=0}^{n} \binom{n}{r}^2$$

Example 14.27

Find the coefficient of x^7 in $f(x) = (1 + x + x^2)(1 + x)^n$ where n is a positive integer.

Solution: Given that

$$f(x) = (1 + x + x^2)(1 + x)^n$$

$$= (1 + x + x^2).\sum_{r=0}^{n} \binom{n}{r} x^r$$

$$= \sum_{r=0}^{n} \binom{n}{r} x^r + \sum_{r=0}^{n} \binom{n}{r} x^{r+1} + \sum_{r=0}^{n} \binom{n}{r} x^{r+2} \qquad .....(14.23)$$

From eq. (14.23), we find that the coefficient of x^7 is $\dbinom{n}{7} + \dbinom{n}{6} + \dbinom{n}{5}$

Example 14.28

Find the coefficient of x^8 in $f(x) = (1 + x + x^2)(1 + x)^n$ where 'n' is a positive integer.

Solution: Given that

$$f(x) = (1 + x + x^2)(1 + x)^n$$

$$= (1 + x + x^2) \cdot \sum_{r=0}^{n} \binom{n}{r} x^r$$

$$= \sum_{r=0}^{n} \binom{n}{r} x^r + \sum_{r=0}^{n} \binom{n}{r} x^{r+1} + \sum_{r=0}^{n} \binom{n}{r} x^{r+2} \qquad \dots\dots(14.24)$$

From (14.24) we get that the coefficient of x^8 is,

$$\binom{n}{8} + \binom{n}{7} + \binom{n}{6}$$

Example 14.29

Find the coefficient of x^k for $0 \le k \le n + 2$, in $f(x) = (1 + x + x^2)(1 + x)^n$, where n is a positive integer.

Solution: Given that

$$f(x) = (1 + x + x^2)(1 + x)^n$$

$$= (1 + x + x^2) \cdot \sum_{r=0}^{n} \binom{n}{r} x^r$$

$$= \sum_{r=0}^{n} \binom{n}{r} x^r + \sum_{r=0}^{n} \binom{n}{r} x^{r+1} + \sum_{r=0}^{n} \binom{n}{r} x^{r+2} \qquad \dots\dots(14.25)$$

From (14.25), we find that the coefficient of x^k is,

$$\binom{n}{k} + \binom{n}{k-1} + \binom{n}{k-2}$$

Example 14.30

Find the coefficient of x^{27} in the function $(x^4 + 2x^5 + 3x^6 + \dots)^5$.

Solution: Given function is

$$= \left(x^4 + 2x^5 + 3x^6 + \dots\right)^5 = \left\{x^4\left(1 + 2x + 3x^2 + \dots\right)\right\}^5$$

$$= x^{20} \cdot \left(1 + 2x + 3x^2 + \dots\right)^5$$

$$= x^{20}\left((1-x)^{-2}\right)^{5}$$

$$= x^{20}.(1-x)^{-10}$$

$$= x^{20}.\sum_{r=0}^{\infty}\binom{a+r}{r}x^{r}$$

Therefore the required coefficient is given by

$$\binom{16}{7} = \frac{16!}{9!\,7!} = 11,440$$

Example 14.31

Find the coefficient of x^{27} in the function $(x^4 + x^5 + x^6 + ...)^5$.

Solution: Consider the given function

$$\left(x^{4} + x^{5} + x^{6} + ...\right)^{5} = \left[x^{4}\left(1 + x + x^{2} + ...\right)\right]^{5}$$

$$= x^{20}\left(1 + x + x^{2} + ...\right)^{5}$$

$$= x^{20}\left[(1-x)^{-1}\right]^{5}$$

$$= x^{20}.(1-x)^{-5}$$

$$= x^{20}.\sum_{r=0}^{\infty}\binom{4+r}{r}x^{r}$$

The coefficient of x^{27} in the given function is

$$\binom{11}{7} = \frac{11!}{7!\,4!} = 330$$

Example 14.32

Find the coefficient of x^{18} in the product $(x + x^2 + x^3 + x^4 + x^5)(x^2 + x^3 + x^4 + ...)^5$ of two functions.

Solution: Consider the given product

$$\left(x + x^{2} + x^{3} + x^{4} + x^{5}\right)\left(x^{2} + x^{3} + x^{4} + ...\right)^{5}$$

$$= \left\{x\left(1 + x + x^{2} + x^{3} + x^{4}\right)\right\}\left\{x^{2}\left(1 + x + x^{2} + ...\right)\right\}^{5}$$

$$=\left\{x\left(1+x+x^2+x^3+x^4\right)\right\}\left\{x^{10}\left(1+x+x^2+...\right)\right\}$$

$$=x^{11}\left(1+x+x^2+x^3+x^4\right)\left(1+x+x^2+...\right)^5$$

$$=x^{11}\left(1+x+x^2+x^3+x^4\right)\left\{(1-x)^{-1}\right\}^5$$

$$=x^{11}\left(1+x+x^2+x^3+x^4\right)(1-x)^{-5}$$

$$=x^{11}\left(1+x+x^2+x^3+x^4\right)\times\sum_{r=0}^{\infty}\binom{4+r}{r}x^r$$

The coefficient of x^{18} in the given product of functions is

$$\binom{11}{7}+\binom{10}{6}+\binom{9}{5}+\binom{8}{4}+\binom{7}{3}$$

Example 14.33

Find the coefficient of x^n in the function $(x^2+x^3+x^4+...)^4$.

Solution: Consider the given function

$$\left(x^2+x^3+x^4+...\right)^4=\left\{x^2\left(1+x+x^2+...\right)\right\}^4$$

$$=x^8\left(1+x+x^2+...\right)^4$$

$$=x^8\left((1-x)^{-1}\right)^4$$

$$=x^8(1-x)^{-4}$$

$$=x^8.\sum_{r=0}^{\infty}\binom{3+r}{r}x^r$$

Therefore, the coefficient of x^n in the given function is

$$\binom{3+n-8}{n-8}=\binom{n-5}{n-8}=\binom{n-5}{3}$$

Example 14.34

Use generating functions to prove Pascal Identity

$$C(n,\ r)=C(n-1,\ r)+C(n-1,\ r-1)$$

when 'n' and 'r' positive integers with $r<n$.   [JNTUH, June 2010, Set No. 1]

Solution: We know the generating function

$$(1+x)^n = \sum_{r=0}^{\infty} (n, r) \, x^r \qquad\qquad .....(14.26)$$

So $C(n, r)$ is the coefficient of x^r in the expansion of $(1 + x)^n$ .....(14.27)

Consider $(1 + x)^n = (1 + x) (1 + x)^{n-1}$

The coefficient of x^{r-1} in $(1 + x)^{n-1}$ is $(n - 1, r - 1)$ .....(14.28)

The coefficient of x^r in $(1 + x)^{n-1}$ is $(n - 1, r)$ .....(14.29)

The coefficient of x^r in $(1 + x) (1 + x)^{n-1}$

     = the coefficient of x^r in $(1 + x)^{n-1} + x(1 + x)^{n-1}$

     = [the coefficient of x^r in $(1 + x)^{n-1}$ + the coefficient of x^{r-1} in $(1 + x)^{n-1}$]

     = $(n - 1, r) + (n - 1, r - 1)$ [by eq. (14.29) and (14.28)]   .....(14.30)

Since $(1 + x)^n = (1 + x) (1 + x)^{n-1}$ and comparing the coefficients of x^r on both sides [use (14.27) and (14.30)] we get that $(n, r) = (n - 1, r) + (n - 1, r - 1)$.

Example 14.35

Find the generating function of $n^2 - 2$.   [JNTUH, Nov 2010, Set No.1]

Solution: Suppose $a_n = n^2 - 2$

Then $a_{n-1} = (n - 1)^2 - 2$

          $= n^2 - 2n + 1 - 2$

          $= n^2 - 2n - 1$

    $a_n - a_{n-1} = (n^2 - 2) - (n^2 - 2n - 1)$

           $= n^2 - 2 - n^2 + 2n + 1$

         $= 2n - 1$

So $a_n = a_{n-1} + 2n - 1$ is the required generating function with initial condition $a_0 = a$

Check: $a_1 = a_{1-1} + 2 - 1 = a_0 + 2 - 1 = 1 = 1^2$

        $a_2 = a_1 + (2 \times 2 - 1) = 1 + 3 = 4 = 2^2$

        $a_3 = a_2 + (2n - 1)$

           $= a_2 + (6 - 1) = 4 + 5 = 9 = 3^2$

and so on.

14.3   The Principle of Inclusion – Exclusion

Let us consider a finite set T containing 'm' number of elements. The number 'm' is called the order, (or size or the cardinality) of the set 'T' and it is denoted as o(T) (or n(T) (or) |T|.

For example if x = {a, b, c, d, e} then o(x) = | x | = 5

Note:

It is clear that $|\phi| = 0$ where ϕ is the empty set.

For every finite non-empty set T, we have that $|T| \geq 1$.

For any two finite sets X and Y, such that $X \subseteq Y$ we have $|X| \leq |Y|$.

If $X \subset Y$ then $|X| < |Y|$. If $X \subseteq Y$ then $X \subset Y$ if and only if $|X| < |Y|$.

Definition

If X is a subset of a finite universal set $\cup$, then the number of elements in the complement $\overline{X}$ is given by

$$\left|\overline{X}\right| = \left|\cup\right| - \left|X\right|$$

.....(14.31)

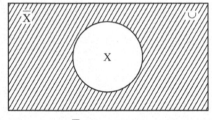

$\overline{X}$ (the shaded area)

14.3.1   Result

For any two sets X and Y we have that $\left|X \cup Y\right| = \left|X\right| + \left|Y\right| - \left|X \cap Y\right|$

Proof: Suppose we consider the union of two finite sets X and Y and wish to determine the number of elements in $X \cup Y$.

Since the elements of $X \cup Y$ consist of all elements which are either in X (or) in Y (or) in both X and Y, the number of elements in $X \cup Y$ is equal to the number of elements in X plus the number of elements in Y minus the number of elements (if any) that are common to both X and Y.

That is., $\left|X \cup Y\right| = \left|X\right| + \left|Y\right| - \left|X \cap Y\right|$

.....(14.32)

Note 1:

Consider the diagram given here.

A more explicit (visual) way of obtaining the above result is through the use of a Venn-diagram. Write $E_1 = X \setminus Y$, $E_2 = X \cap Y$ and $E_3 = Y \setminus X$.

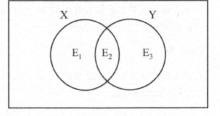

It is clear that $X \cup Y$ is equal to the disjoint union of E_1, E_2 and E_3.

Also note that $X = E_1 \cup E_2$, $Y = E_2 \cup E_3$,

$$|X| = |E_1| + |E_2|; \text{ and } |Y| = |E_2| + |E_3|.$$

From these facts, we get that

$$|X \cup Y| = |E_1| + |E_2| + |E_3|$$
$$= (|E_1| + |E_2|) + (|E_2| + |E_3|) - |E_2|$$
$$= |X| + |Y| - |X \cap Y|$$

Thus, for determining the number of elements in $X \cup Y$, we first include all elements in X and all elements in Y and then exclude all elements that are common to X and Y.

Note 2:

If $\cup$ is a finite universal set of which X and Y are subsets, then by virtue of the De'morgan law and the expansion (14.31) above, we have

$$\left(\overline{X} \cap \overline{Y}\right) = \overline{\left(X \cup Y\right)} = |\cup| - \left(X \cup Y\right)$$

Consider the Venn-diagram shown above.

In this diagram, the set X is made up of two disjoint parts E_1 and E_2, and the set Y is made up of two disjoint parts E_2 and E_3, where $E_2 = X \cap Y$. Also $X \cup Y$ is made up of parts E_1, E_2 and E_3.

Therefore,

$$|X| = \text{Number of elements in } E_1 + \text{Number of elements in } E_2$$
$$= |E_1| + |E_2|$$

Similarly, $|Y| = |E_2| + |E_3|$

$$|X \cap Y| = |E_2|$$

and $\quad |X \cup Y| = |E_1| + |E_2| + |E_3|$

with the use of formula (14.32) of above this becomes

$$\left|\overline{X} \cap \overline{Y}\right| = |\cup| - \left\{|X| + |Y| - |X \cap Y|\right\} = |\cup| - |X| - |Y| + |X \cap Y| \quad \ldots\ldots(14.33)$$

Expressions (14.32) and (14.33) are equivalent to one another.

Either of these is referred to as the Addition Principle (rule) or the Principle of "Inclusion – Exclusion" for two sets.

Note 3:

In the particular case: where X and Y are disjoint sets so that $X \cap Y = \phi$, the addition rule (14.32) becomes,

$$|X \cup Y| = |X| + |Y| - |\phi| = |X| + |Y| \qquad \qquad .....(14.34)$$

This is known as the Principle of Disjunctive Counting for two sets.

14.3.2   Principle of Inclusion–Exclusion for n–sets

The principle of inclusion–exclusion as given by expression (14.32) can be extended to 'n' sets, $n > 2$.

Let $\cup$ be a finite universal set and $X_1, X_2, ..., X_n$ be subsets of $\cup$. Then the principle of Inclusion–Exclusion for $X_1, X_2, ..., X_n$ states that

$$|X_1 \cup X_2 \cup ... \cup X_n| = \sum_{i=1}^{n} |X_i| - \sum |X_i \cap X_j|$$

$$+ \sum |X_i \cap X_j \cap X_k| + ... +$$

$$(-1)^{n-1} |X_1 \cap X_2 \cap ... \cap X_n|$$

Example 14.36

Suppose that 200 faculty members can speak English and 50 can speak Hindi while only 20 can speak both English and Hindi. How many faculty members can speak either English or Hindi?

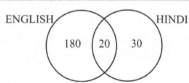

Solution: If E is the set of faculty who speak English and H is the set of faculty members who speak Hindi, then

$$|E| = 200$$

$$|H| = 50$$

$$|E \cap H| = 20$$

Now by known formula, we have

$$|E \cup H| = |E| + |H| - |E \cap H| = 200 + 50 - 20 = 230$$

14.3.3 Result

If X, Y, Z are any three subsets of the universal set $\cup$, then explain the formula for $|X \cup Y \cup Z|$ by using the Venn–diagram.

Proof: Let X, Y, Z are any 3 subsets of the universal set $\cup$.

We know that,

$$X = \left(X \cap \overline{Y} \cap \overline{Z}\right) \cup \left(X \cap Y \cap \overline{Z}\right) \cup \left(X \cap \overline{Y} \cap Z\right) \cup \left(X \cap Y \cap Z\right)$$

$$Y = \left(\overline{X} \cap Y \cap \overline{Z}\right) \cup \left(X \cap Y \cap \overline{Z}\right) \cup \left(\overline{X} \cap Y \cap Z\right) \cup \left(X \cap Y \cap Z\right)$$

$$Z = \left(\overline{X} \cap \overline{Y} \cap Z\right) \cup \left(X \cap \overline{Y} \cap Z\right) \cup \left(\overline{X} \cap Y \cap Z\right) \cup \left(X \cap Y \cap Z\right)$$

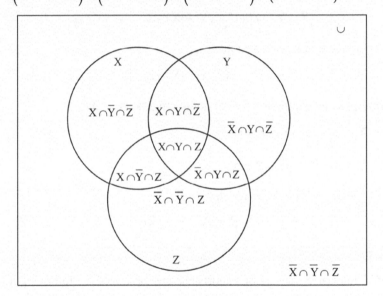

Since the above unions are disjoint unions, we can use the sum rule to get the following (14.35), (14.36) and (14.37).

$$|X| = X \cap \overline{Y} \cap \overline{Z} + X \cap Y \cap \overline{Z} + X \cap \overline{Y} \cap Z + X \cap Y \cap Z \qquad .....(14.35)$$

$$|Y| = \overline{X} \cap Y \cap \overline{Z} + X \cap Y \cap \overline{Z} + \overline{X} \cap Y \cap Z + X \cap Y \cap Z \qquad .....(14.36)$$

$$|Z| = \overline{X} \cap \overline{Y} \cap Z + X \cap \overline{Y} \cap Z + \overline{X} \cap Y \cap Z + X \cap Y \cap Z \qquad .....(14.37)$$

By adding eq. (14.35), (14.36), and (14.37), on both sides we get that

$$|X| + |Y| + |Z| = |X \cap \overline{Y} \cap \overline{Z}| + |X \cap Y \cap \overline{Z}| + |X \cap \overline{Y} \cap Z|$$

$$+ |X \cap Y \cap Z| + |\overline{X} \cap Y \cap \overline{Z}| + |\overline{X} \cap Y \cap Z|$$

$$+ |\overline{X} \cap \overline{Y} \cap Z| + |X \cap Y \cap \overline{Z}| + |X \cap Y \cap Z|$$

$$+ |X \cap \overline{Y} \cap Z| + |X \cap Y \cap Z| + |\overline{X} \cap Y \cap Z| \qquad \qquad .....(14.38)$$

The first 7 of these sets make up $X \cup Y \cup Z$, the next 2 make up $X \cap Y$ and the next 2 give $X \cap Z$.

Thus we have that

$$|X| + |Y| + |Z| = |X \cup Y \cup Z| + |X \cap Y| + |X \cap Z| + |\overline{X} \cap Y \cap Z|$$

By rearranging the terms, we have that

$$|X \cup Y \cup Z| = |X| + |Y| + |Z| - |X \cap Y| - |X \cap Z| - |\overline{X} \cap Y \cap Z| \qquad .....(14.39)$$

But we would like an expression free of complements, note that

$$|\overline{X} \cap Y \cap Z| + |X \cap Y \cap Z| = |Y \cap Z|$$

and so $\quad |\overline{X} \cap Y \cap Z| = |Y \cap Z| - |X \cap Y \cap Z| \qquad \qquad .....(14.40)$

From eq. (14.39) and (14.40), we get that

$$|X \cup Y \cup Z| = |X| + |Y| + |Z| - |X \cap Y| - |X \cap Z| - |Y \cap Z| + |X \cap Y \cap Z|$$

Example 14.37

A survey of 500 television viewers of a sports channel produced the following information: 285 watch baseball, 195 watch shuttle, 115 watch kabbadi, 45 watch baseball and kabbadi, 70 watch baseball and shuttle, 50 watch shuttle and kabbadi and 50 do not watch any of three kinds of games.

   (i)   How many viewers in the survey watch all three kinds of games?

   (ii)   How many viewers watch exactly one of the sports?

Solution: Suppose that U is the set of all viewers included in the survey.

Suppose X is the set of all viewers who watch baseball, Y denote the set of all viewers who watch shuttle and Z denote the set of all viewers who watch kabbadi.

Given that

$$|U| = 500$$

$$|X| = 285$$

$$|Y| = 195$$

$$|Z| = 115$$

$$|X \cap Z| = 45$$

$$|X \cap Y| = 70$$

$$|Y \cap Z| = 50$$

$$\left|\overline{X \cup Y \cup Z}\right| = 50$$

$$|X \cup Y \cup Z| = 500 - 50 = 450$$

Part (i): Let us consider the addition principle

$$|X \cup Y \cup Z| = |X| + |Y| + |Z| - |X \cap Y| - |X \cap Z| - |Y \cap Z| + |X \cap Y \cap Z|$$

We find that,

$$|X \cap Y \cap Z| = |X \cup Y \cup Z| - |X| - |Y| - |Z| + |X \cap Y| + |X \cap Z| + |Y \cap Z|$$

$$= 450 - 285 - 195 - 115 - 70 - 45 - 50$$

$$= 20$$

Thus the number of viewers who watch all three kinds of games is 20.

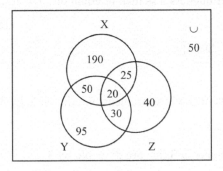

Part (ii): Let X_1 denote the set of viewers who watch only baseball, Y_1 denote the set of viewers who watch only shuttle and Z_1 denotes the set of viewers who watch only kabbadi.

Then,

$$X_1 = X - (Y \cup Z) = X - (X \cap Y) \cup (X \cap Z)$$

So $$|X_1| = |X| - |(X \cap Y) \cup (X \cap Z)|$$

$$= |X| - \{|X \cap Y| + |X \cap Z| - |X \cap Y \cap Z|\}$$

$$= |X| - |X \cap Y| - |X \cap Z| + |X \cap Y \cap Z|$$

$$= 285 - 70 - 45 + 20 = 190$$

Similarly, the number of viewers who watch only shuttle is

$$|Y_1| = |Y| - |Y \cap X| - |Y \cap Z| + |Y \cap Z \cap X|$$

$$= 195 - 70 - 50 + 20 = 95$$

The number of viewers who watch only kabbadi is

$$|Z_1| = |Z| - |Z \cap X| - |Z \cap Y| + |Z \cap Y \cap X|$$

$$= 115 - 45 - 50 + 20 = 40$$

From the above, we get that the number of viewers who watch exactly one of the games is

$$|X_1| + |X_2| + |X_3| = 190 + 95 + 40 = 325$$

Example 14.38

Find the number of positive integers less than or equal to 2076 and divisible by 3 or 4.

[JNTUH, Nov 2010, Set – 1]

Solution: Let S = {1, 2, ..., 2076}

Then $|S| = 2076$

Let A_1, A_2 be the subsets of S whose elements are divisible by 3, 4 respectively.

Then we have to find $|A_1 \cup A_2|$

   $|A_1|$ = No. of elements in S that are divisible by 3

   $= [2076/3] = 692$

   $|A_2|$ = No. of elements in S that are divisible by 4

   $= [2076/4] = 519$

$|A_1 \cap A_2|$ = No. of elements in S that are divisible by both 3 and 4

$$= [2076/(3 \times 4)] = [2076/12] = 173$$

By the principle of inclusion and exclusion,

$$|A_1 \cup A_2| = |A_1| + |A_2| - |A_1 \cap A_2|$$

$$= 692 + 519 - 173$$

$$= 1211 - 173 = 1038$$

Exercises

Binomial and Multinomial Coefficients

(i)   Prove that for a positive integer n:

$$^nC_0 - {}^nC_1 + {}^nC_2 - ... + (-1)^n \left({}^nC_n \right) = 0$$

(ii)   Find the coefficient of $x^9 y^3$ in the expansion $(2x - 3y)^{12}$.

Ans: 1946

(iii)   Determine the coefficient of
$X \, Y \, Z^2$ in $(2x - y - Z)^2$

Ans: (−24)

Generating Functions of Permutations and Combinations

(i)   Obtain the generating function of the numeric function $b_r = 3^{r+2}$, $r \geq 0$.

Ans: $\dfrac{9}{1 - 3x}$

(ii)   Show that $(1 - 4x)^{-\frac{1}{2}}$ generates the sequence ${}^{2r}C_r$

(iii)   Find the generating function for the sequence $<b_r>$ for $b_r = r + 1$.

Ans: $(1 - x)^{-2}$

The Principles of Inclusion and Exclusion

(i)   In how many ways 5 numbers of a's, 4 number of b's and 3 number of c's can be arranged so that all the identical letters are not in a single block?

Ans: 25762

(ii)   Verify the principle of inclusion – exclusion for

P = {a, b, c, d, e} and B = {c, e, f, h, k}

(iii) Find the number of integers between 1 and 1000 that are not divisible by 2, 3, 5 or 7?

Ans: 228

CHAPTER - 15

Recurrence Relations

The expressions for permutations, combinations and partitions (that were developed in the earlier chapters) are the most fundamental tools for counting the elements of finite sets.

An important approach uses "recurrence relations" (sometimes called difference equations) to define the term of a sequence.

Many combinatorial problems can be solved by reducing them to analogous problems involving a smaller number of objects, and important feature of the recurrence relations is to express one term of a collection of numbers as a function of preceding terms of the collection.

With the help of a recurrence relation, we can reduce a problem involving "n" objects to one involving n − 1 objects, then to one involving n − 2 objects and so on.

By successive reduction of the number of terms involved, we hope to eventually end up with a problem that can be solved by earlier techniques.

Before going to recurrence relations in detail, we re-collect the concept of generating functions and study, understand the problems that are dealt with the partial fractions, and calculating coefficients of Generating Functions.

15.1   Generating Function of Sequences

15.1.1   Generating Functions

Consider a sequence of real numbers $b_0, b_1, \ldots$ Let us denote this sequence by $\{b_n\}$, $n = 0, 1, 2, \ldots$ or just $\{b_n\}$.

The function

$$f(x) = b_0 + b_1 x + b_2 x^2 + \ldots + b_n x^n + \ldots = \sum_{k=0}^{\infty} b_k x^k \qquad \ldots (15.1)$$

is called a generating function for the sequence $\{b_n\}$. Note that b_k is the coefficient of x_k in the expansion of $f(x)$ in a series of powers of x.

If $f(x)$ is the generating function of the sequence $\{b_n\}$, then we say that $f(x)$ generates the sequence $\{b_n\}$.

The series on the RHS of the eq. (15.1) is known as the power series expansion of $f(x)$.

Example 15.1

Find a generating function for the sequence, 0, 2, 6, 12, 20, 30, 42, ...

Solution: Given sequence is

$$0, 2, 6, 12, 20, 30, 42, \ldots$$

Here $b_0 = 0 = 0 + 0$

$b_1 = 2 = 1 + 1$

$b_2 = 6 = 2 + 2^2$

$b_3 = 12 = 3 + 3^2$

$b_4 = 20 = 4 + 4^2$

...

so on

so that

$$b_k = k + k^2 \quad \text{for } k = 0, 1, 2, \ldots$$

We know that a generating function for the sequence $\{k\} = 0, 1, 2, \ldots$ is

$$f(x) = \frac{x}{(1-x)^2} \text{ and}$$

a generating function for the sequence $\{k^2\} = 0^2, 1^2, \ldots$ is

$$g(x) = \frac{x(1+x)}{(1-x)^3}$$

Therefore a generating function for the given sequence is

$$f(x)+g(x)=\frac{x}{(1-x)^2}+\frac{x(1+x)}{(1-x)^3}$$

$$=\frac{2x}{(1-x)^3}$$

Example 15.2

The generating functions

$$P(X)=\sum_{r=0}^{\infty}2^r\times x^r$$

$$Q(X)=2x^5+2x^6+2x^7+2x^8+2x^9+3x^{10}+4x^{11}$$

$$R(X)=\sum_{r=0}^{\infty}(r+1)x^r$$

$$S(X)=\sum_{r=0}^{\infty}(r^2)x^r$$

generates the sequences P, Q, R and S where

$$P=\{2^r\}_{r=0}^{\infty}$$

Q be the finite sequence 2, 2, 2, 2, 2, 3, 4.

$$R=\{r+1\}_{r=0}^{\infty}=\{n\}_{n=1}^{\infty}$$

$$S=\{r^2\}_{r=0}^{\infty}$$

Example 15.3

(i) The generating function of the sequence 1, 2, 3, ... of natural numbers is

$$f(x)=1+2x+3x^2+\ldots$$

(ii) The generating function of the arithmetic sequence 1, 4, 7, 10, ... is

$$f(x)=1+4x+7x^2+10x^3+\ldots$$

Definition

Let $P(X)=\sum_{r=0}^{\infty}a_r x^r$, $Q(X)=\sum_{s=0}^{\infty}b_s x^s$ be two formal power series. Then define the following concepts.

Equality: $P(X) = Q(X)$ if and only if $a_n = b_n$ for each $n \geq 0$.

Multiplication by a scalar number C: $d.P(X) = \sum_{r=0}^{\infty} (da_r) x^r$

Sum: $P(X) + Q(X) = \sum_{n=0}^{\infty} (a_n + b_n) x^n$

Product: $P(X) \, Q(X) = \sum_{n=0}^{\infty} P_n X^n$

where $P_n = \sum_{j+k=n} a_j b_k$

Example 15.4

Let $S_1(x) = 5 + 2x + 3x^2$ and $S_2(x) = 2 + 3x + 4x^2$. Then

(i) $S_1(x) + S_2(x) = (5 + 2x + 3x^2) + (2 + 3x + 4x^2) = 7 + 5x + 7x^2$.

(ii) $3S_1(x) = 3(5 + 2x + 3x^2) = 15 + 6x + 9x^2$.

(iii) $S_1(x)S_2(x) = (5 + 2x + 3x^2)(2 + 3x + 4x^2)$

$$= 5(2 + 3x + 4x^2) + 2x(2 + 3x + 4x^2) + 3x^2(2 + 3x + 4x^2)$$

$$= 10 + 15x + 20x^2 + 4x + 6x^2 + 8x^3 + 6x^2 + 9x^3 + 12x^4$$

$$= 10 + 19x + 32x^2 + 17x^3 + 12x^4$$

Note that in (iii): Coefficient of $x^0 = 10$

Coefficient of $x^1 = 19$

Coefficient of $x^2 = 32$

Coefficient of $x^3 = 17$

Coefficient of $x^4 = 12$

Note:

Let $f(x) = a_0 + a_1 x + a_2 x^2 + \ldots$ and $g(x) = b_0 + b_1 x + b_2 x^2 + \ldots$ be two generating sequences, then $f(x) + g(x) = (a_0 + b_0) + (a_1 + b_1) x + (a_2 + b_2) x^2 + \ldots$ and $f(x)g(x) = (a_0 b_0) + (a_1 b_0 + a_0 b_1)x + (a_0 b_2 + a_1 b_1 + a_2 b_0)x^2 + \ldots$, the coefficient of x^n in the product $f(x)g(x)$ is the finite sum: $a_0 b_n + a_1 b_{n-1} + a_2 b_{n-2} + \ldots + a_n b_0$.

Example 15.5

If $f(x) = 1 + x + x^2 + \ldots + x^n + \ldots$ and $g(x) = 1 - x + x^2 - x^3 + \ldots + (-1)^n x^n + \ldots$,

then $f(x) + g(x) = (1 + 1) + (1 - 1)x + (1 + 1)x^2 + \ldots + (1 + (-1)^n)x^n + \ldots$

$$= 2 + 2x^2 + 2x^4 + \ldots$$

$$f(x) \cdot g(x) = 1 + [1(-1) + 1(1)]x + [1(1) + 1(-1) + 1(1)]x^2 + \ldots$$

$$= 1 + x^2 + x^4 + x^6 + \ldots$$

Note:

If $S(x)$ is a generating function for a sequence $\{b_k\}$ and $T(x)$ is a generating function for a sequence $\{c_k\}$, then $d_1 S(x) + d_2 T(x)$ is a generating function for the sequence $\{d_1 b_k + d_2 c_k\}$, where d_1 and d_2 are any two real numbers.

15.2   Partial Fractions (Definition)

If $P(X)$ and $Q(X)$ are power series, we say that $P(X)$ divides $Q(X)$ if there is a formal power series $R(X)$ such that $Q(X) = P(X) R(X)$ and we write

$$R(X) = \frac{Q(X)}{P(X)}$$

15.2.1   Partial Fractions

If $P(X)$ and $Q(X)$ are polynomials, we discuss how to compute $\dfrac{Q(X)}{P(X)}$ by using the partial fractions.

   $P(X)$ is a product of Linear factors, $P(X) = a_n (x - \alpha_1)^{r_1} (x - \alpha_2)^{r_2} ... (x - d_n)^{r_n}$ and if $Q(X)$ is any polynomial of degree less than the degree of $P(X)$ then $\dfrac{Q(X)}{P(X)}$ can be written as the sum of elementary fractions as it is given below:

$$\frac{Q(X)}{P(X)} = \frac{k_{11}}{(x - \alpha_1)^{r_1}} + \frac{k_{12}}{(x - \alpha_2)^{r_1 - 1}} + ... + \frac{k_1 r_1}{(x - \alpha_1)}$$

$$+ \frac{k_{21}}{(x - \alpha_2)^{r_2}} + ... + \frac{k_2 r_2}{(x - \alpha_2)} + ... +$$

$$\frac{k_{n1}}{(x - \alpha_n)^{r_n}} + \frac{k_{n2}}{(x - \alpha_n)^{r_n - 1}} + ... + \frac{k_{nn}}{(x - \alpha_n)} \qquad .....(15.2)$$

   To find the numbers k_{11}, k_{12}, ..., k_{nn} we multiply both sides of the eq. (15.2) by $(x - \alpha_1)^{r_1} (x - \alpha_2)^{r_2} ... (x - \alpha_n)^{r_n}$ to clear of denominators and then we equate coefficients of the same powers of X. Then the required coefficients can be solved from the resulting system of equations.

Example 15.6

Calculate $B(X) = \sum_{r=0}^{\infty} b_r x^r = \dfrac{1}{x^2 - 5x + 6}$

Solution: Since

$\quad x^2 - 5x + 6 = (x - 3)(x - 2)$, we write that $\dfrac{1}{x^2 - 5x + 6} = \dfrac{A}{x - 3} + \dfrac{B}{x - 2}$

Thus $1 = A(x - 2) + B(x - 3)$ .....(15.3)

Substitute $X = 2$ in (15.3)

to get $B = -1$

Substitute $X = 3$ in (15.3)

to get $A = 1$

Thus $\dfrac{1}{x^2 - 5x + 6} = \dfrac{1}{x - 3} - \dfrac{1}{x - 2}$

$$= \dfrac{1}{-3\left(1 - \dfrac{x}{3}\right)} - \dfrac{1}{-2\left(1 - \dfrac{x}{2}\right)}$$

$$= \dfrac{-1}{3\left(1 - \dfrac{x}{3}\right)} + \dfrac{1}{2\left(1 - \dfrac{x}{2}\right)}$$

$$= -\dfrac{1}{3}\left(1 - \dfrac{x}{3}\right)^{-1} + \dfrac{1}{2}\left(1 - \dfrac{x}{2}\right)^{-1}$$

$$= -\dfrac{1}{3}\sum_{r=0}^{\infty}\left(\dfrac{1}{3}\right)^r x^r + \dfrac{1}{2}\sum_{r=0}^{\infty}\left(\dfrac{1}{2}\right)^r x^r$$

$$\left(\text{Since } \left(1 - \dfrac{x}{a}\right)^{-1} = \sum_{r=0}^{\infty}\left(\dfrac{1}{a}\right)^r \cdot x^r\right)$$

$\Rightarrow \quad B(X) = \sum \left[-\dfrac{1}{3^{r+1}} + \dfrac{1}{2^{r+1}}\right] x^r$

Therefore for each r we have that

$$b_r = -\dfrac{1}{3^{r+1}} + \dfrac{1}{2^{r+1}} \text{ is the coefficient of } X^r \text{ in } B(X)$$

Thus

$$\frac{x^5}{x^2 - 5x + 6} = x^5 \sum_{r=0}^{\infty} \left[-\frac{1}{3^{r+1}} + \frac{1}{2^{r+1}} \right] x^r$$

$$= \sum \left[-\frac{1}{3^{r+1}} + \frac{1}{2^{r+1}} \right] x^{r+5}$$

and if we make the substitution $k = r = 5$ we see that

$$\frac{x^5}{x^2 - 5x + 6} = \sum_{r=0}^{\infty} \left[-\frac{1}{3^{r+1}} + \frac{1}{2^{r+1}} \right] x^r = \sum_{r=0}^{\infty} d_r x^r$$

and what this final equality says that $B(X) = \sum_{r=0}^{\infty} b_r x^r = \frac{1}{x^2 - 5x + 6}$

Example 15.7

Resolve $\dfrac{9}{(x-1)(x+2)^2}$ into partial fractions.

Solution: Let

$$\frac{9}{(x-1)(x+2)^2} = \frac{A}{x-1} + \frac{B}{x+2} + \frac{C}{(x+2)^2}$$

$$\Rightarrow \qquad 9 = A(x+2)^2 + B(x-1)(x+2) + C(x-1) \qquad \qquad \dots\dots(15.4)$$

Substitute $x = 1$ in (15.4), then

$$9 = 9A \quad \Rightarrow \quad A = 1$$

Substitute $x = -2$ in (15.4), then

$$9 = -3C \quad \Rightarrow \quad C = -3$$

If $x = 0$, then (15.4) becomes

$$4A - 2B - C = 9 \quad \Rightarrow \quad 4(1) - 2B - (-3) = 9$$

$$\Rightarrow \qquad 2B = -2 \quad \Rightarrow \quad B = -1$$

Therefore

$$\frac{9}{(x-1)(x+2)^2} = \frac{1}{x-1} - \frac{1}{x+2} - \frac{3}{(x+2)^2}$$

Example 15.8

Resolve $\dfrac{x^2+1}{\left(x^2+4\right)(x-2)}$ into partial fractions.

Solution: Let

$$\frac{x^2+1}{\left(x^2+4\right)(x-2)}=\frac{Ax+B}{x^2+4}+\frac{C}{x-2}$$

$$=\frac{(Ax+B)(x-2)+C\left(x^2+4\right)}{\left(x^2+4\right)(x-2)}$$

$$\Rightarrow\quad x^2+1=(Ax+B)(x-2)+C\left(x^2+4\right)$$

$$=(A+C)x^2+(-2A+B)x+(-2B+4C)$$

$$\Rightarrow\quad A+C=1 \qquad\qquad\qquad\qquad\qquad .....(15.5)$$
$$\qquad -2A+B=0 \qquad\qquad\qquad\qquad\quad .....(15.6)$$
$$\qquad -2B+4C=1 \qquad\qquad\qquad\qquad\quad .....(15.7)$$

By (15.6) $B = 2A$

Substitute this in (15.7), then we get

$$-4A+4C=1 \qquad\qquad\qquad\qquad\qquad .....(15.8)$$

By solving (15.5) and (15.8), we get $C=\dfrac{5}{8},\ A=\dfrac{3}{8}$

So $\quad B=2A=\dfrac{2.3}{8}=\dfrac{6}{8}=3/4$

Therefore $\dfrac{x^2+1}{\left(x^2+4\right)(x-2)}=\dfrac{3x+6}{8\left(x^2+4\right)}+\dfrac{5}{8(x-2)}$

Example 15.9

Resolve $\dfrac{x^2+5x+7}{(x-1)^3}$ into partial fractions.

Solution: write $\quad x-1=y$

Then $\qquad\qquad x=y+1$

Therefore

$$\frac{x^2 + 5x + 7}{(x-1)^3} = \frac{(y+1)^2 + 5(y+1) + 7}{y^3}$$

$$= \frac{y^2 + 7y + 13}{y^3}$$

$$= \frac{y^2}{y^3} + \frac{7y}{y^3} + \frac{13}{y^3}$$

$$= \frac{1}{y} + \frac{7}{y^2} + \frac{13}{y^3}$$

$$= \frac{1}{x-1} + \frac{7}{(x-1)^2} + \frac{13}{(x-1)^3}$$

15.3   Calculating Coefficient of Generating Functions

In this section we have been interested to develop algebraic techniques for calculating the coefficients of generating functions.

Also we study the concept of division of power series.

Definition

Let $S(x) = \Sigma a_r x^r$ be a formal power series. A formal power series $T(x) = \Sigma b_k x^k$ is said to be multiplicative inverse of $S(x)$ if $S(x)T(x) = 1$. In this case, we say that the formal power series $S(x)$ have a multiplicative inverse $T(x)$.

Note:

In particular if $S(x)$ has a multiplicative inverse, then $a_0 b_0 = 1$, and so a_0 must be non-zero.

The converse is also true.

If $a_0 \neq 0$, then we can determine the coefficients of $T(x)$ by writing down the coefficients of successive powers of x in $S(x)T(x)$ by definition of two power series, and equating these to the coefficient of like powers of x in the power series $S(x)$.

Then we have

$$a_0 b_0 = 1 \qquad\qquad\qquad\qquad .....(i)$$
$$a_0 b_1 + a_1 b_0 = 0 \qquad\qquad\qquad .....(ii)$$
$$a_0 b_2 + a_1 b_1 + a_2 b_0 = 0 \qquad\qquad .....(iii)$$
$$.$$
$$.$$
$$.$$
$$a_0 b_n + a_1 b_{n-1} + ... + a_n b_0 = 0 \qquad .....(n)$$

By solving equation (i), we get $b_0 = 1/a_0$

By solving equation (ii), we get $b_1 = (-a_1b_0)/a_0 = (-a_1a_0^2)$ (by (i))

By solving equation (iii), we get $b_2 = (-a_1b_1 - a_2b_0)/a_0 = (a_1^2 - a_2a_0)/a_0^3, \ldots,$

Continuing in this manner, we can solve for each coefficient of T(x).

15.3.1   Remark

From the above discussion, we conclude that the given formal power series $S(x) = \sum a_r x^r$ has a multiplication inverse if and only if the constant term a_0 is different from 0 (that is, $a_0 \neq 0$).

Definition

If S(x) and T(x) are power series, we say that S(x) divides T(x) if there is a power series V(x) such that T(x) = S(x)V(x). In this case, we write V(x) = T(x)/S(x).

Note:

If $A(x) = \sum a_r x^r$ and $a_0 = 0$, but some coefficient of A(x) is not zero, then let a_k be the first non-zero coefficient of A(x) and $A(x) = x^k A_1(x)$ where a_k,(the constant term of $A_1(x)$) is non-zero.

Then in order for A(x) to divide C(x) it must be true that x^k is also a factor of C(x).

That is, $C(x) = x^k C_1(x)$, where $C_1(x)$ is a formal power series.

If this is the case, then cancel the common powers of x from both A(x) and C(x) and then we can find $C(x)/A(x) = C_1(x)/A_1(x)$ by using the multiplicative inverse of $A_1(x)$.

15.3.2   Some Formulas (Special Cases of Binomial Theorem)

If 'n' is a positive integer, then we have the following:

(i) $\quad (1+x)^n = 1 + \binom{n}{1}x + \binom{n}{2}x^2 + \ldots + \binom{n}{n}x^n$

(ii) $\quad (1+x^k)^n = 1 + \binom{n}{1}x^k + \binom{n}{2}x^{2k} + \ldots + \binom{n}{n}x^{nk}$

(iii) $\quad (1-x)^n = 1 - \binom{n}{1}x + \binom{n}{2}x^2 + \ldots + (-1)^n\binom{n}{n}x^n$

(iv) $\quad (1-x^k)^n = 1 - \binom{n}{1}x^k + \binom{n}{2}x^{2k} + \ldots + (-1)^n\binom{n}{n}x^{nk}$

The above formulas (i) – (iv) are some special cases of the Binomial Theorem.

15.3.3   Geometric Series

Now we determine the multiplicative inverse for $P(x) = 1 - x$.

Let $\quad Q(x) = \dfrac{1}{P(x)} = \displaystyle\sum_{n=0}^{\infty} C_n x^n$

Solving successively for C_0, C_1, ..., we get that

$$C_0 = \frac{1}{p_0} = 1$$

$$C_1 = \frac{-p_1 q_0}{p_0} = \frac{-(-1)(1)}{1} = 1$$

$$C_2 = \frac{-p_1 q_1 - p_2 q_0}{p_0} = \frac{-(-1)(1) - (0)(1)}{1} = 1$$

$$C_3 = \frac{-p_1 q_2 - p_2 q_1 - p_3 q_0}{p_0} = 1$$

and so on. So each $C_i = 1$

Therefore

$$\frac{1}{1-x} = \sum_{r=0}^{\infty} x^r \qquad\qquad\qquad .....(15.9)$$

If we replace in (15.9) the term x by kx, where k is a real number, then we get that

$$\frac{1}{1-kx} = \sum_{r=0}^{\infty} k^r x^r \qquad\qquad\qquad .....(15.10)$$

(15.10) is a geometric series (with common ratio k).

In particular, take $a = -1$, then we get

$$\frac{1}{1+x} = \sum_{r=0}^{\infty} (-1)^r x^r = 1 - x + x^2 - ... \qquad\qquad .....(15.11)$$

This is an alternating geometric series.

In the same way, we get that

$$\frac{1}{1+kx} = \sum_{r=0}^{\infty} (-1)^r k^r x^r \qquad\qquad\qquad .....(15.12)$$

Let us take a positive integer n.

If $Q_1(x)$, $Q_2(x)$, ..., and $Q_n(x)$ are the multiplicative inverses of $P_1(x)$, $P_2(x)$, ..., and $P_n(x)$ respectively, then $Q_1(x).Q_2(x), ... Q_n(x)$ is the multiplicative inverse of $P_1(x).P_2(x) ... P_n(x)$. (since $P_i(x) Q_i(x) = 1$ for each i)

In particular, if $Q(x)$ is the multiplicative inverse of $P(x)$, then $(Q(x))^n$ is the multiplicative inverse of $(P(x))^n$. Now we apply this observation to $P(x) = 1 - x$.

For a positive integer n, we have

$$\frac{1}{(1-x)^n} = \left[\sum_{k=0}^{\infty} x^k\right]^n \qquad \text{(by (15.9))}$$

$$= \sum_{r=0}^{\infty} C(n-1+r, r) x^r \qquad \qquad .....(15.13)$$

This equality $\dfrac{1}{(1-x)^n} = \sum_{r=0}^{\infty} C(n-1+r, r) x^r$

can be obtained by using mathematical induction and the identity

$$C(n-1, 0) + C(n+1, 2) + ... +$$

$$C(n+r-1, r) = C(n+r, r)$$

By replacing x by –x in the above (15.13), we get the following identity (15.14):
For a positive integer n we get that

$$\frac{1}{(1+x)^n} = \sum C(n-r+r, r) . (-1)^r x^r \qquad .....(15.14)$$

Following this pattern, replace X by KX in (15.13) and (15.14) to get

$$\frac{1}{(1-kx)^n} = \sum_{r=0}^{\infty} C(n-1+r, r) k^r x^r \qquad .....(15.15)$$

$$\frac{1}{(1+kx)} = \sum C(n-1+r, r) (-k)^r . x^r \qquad .....(15.16)$$

By replacing x by x^k in (15.9), we get that (for a positive integer k)

$$\frac{1}{(1-x^k)} = \sum_{k=0}^{\infty} x^{kr} = 1 + x^k + x^{2k} + ... \qquad .....(15.17)$$

and

$$\frac{1}{1+x^k} = \sum (-1)^n x^{br} \qquad .....(15.18)$$

If k is a non zero real number, then we get that

$$\frac{1}{k-x} = \frac{1}{k}\left(\frac{1}{1-\frac{x}{k}}\right) = \frac{1}{k}\sum \frac{x^r}{k^r} \qquad \qquad .....(15.19)$$

and $\qquad \dfrac{1}{x-k} = -\dfrac{1}{k-x} = -\dfrac{1}{k}\sum_{r=0}^{\infty} \dfrac{x^r}{k^r}$ .....(15.20)

If 'n' is a positive integer

$$1 + x + x^2 + ... + x^n = \frac{1-x^{n+1}}{1-x} \qquad \qquad .....(15.21)$$

because $\left(1 + x + x^2 + ... + x^n\right)\left(1-x\right) = \left(1-x^{n+1}\right)$

Hence we have the following list of formulas.

15.3.4 A List of Formulas

Example 15.10

Compute the coefficient of

$$\sum_{k=0}^{\infty} c_k x^k = \frac{x^2 - 5x + 3}{x^4 - 5x^2 + 4}$$

Solution: we know that

$$x^4 - 5x^2 + 4 = (x^2 - 1)\,(x^2 - 4)$$

$$= (x-1)\,(x+1)\,(x-2)\,(x+2)$$

Now we write

$$\frac{x^2 - 5x + 3}{x^4 - 5x^2 + 4} = \frac{A}{x-1} + \frac{B}{x+1} + \frac{C}{x-2} + \frac{D}{x+2} \qquad .....(15.22)$$

By multiplying (15.22) by $x^4 - 5x^2 + 4$ on both sides, we get that

$$x^2 - 5x + 3 = A\left(x+1\right)\left(x-2\right)\left(x+2\right) + B\left(x-1\right)\left(x-2\right)\left(x+2\right)$$

$$+ C\left(x-1\right)\left(x+1\right)\left(x+2\right) + D\left(x-1\right)\left(x+1\right)\left(x-2\right) \qquad .....(15.23)$$

If x = 1, then all the terms of the RHS of (15.23) that involve the factor x − 1 will vanish, and so we have $-1 = -6\,A$ and $A = \dfrac{1}{6}$

Similarly taking $x = -1$, $x = 2$ and $x = -2$, we get that $B = \dfrac{3}{2}$, $C = -\dfrac{1}{4}$ and $D = -\dfrac{17}{12}$

Therefore

$$\frac{x^2 - 5x + 3}{x^4 - 5x^2 + 4} = \frac{1}{6(x-1)} + \frac{3}{2(x+1)} - \frac{1}{4(x-2)} - \frac{17}{12(x+2)}$$

$$= \frac{1}{2}\left[\frac{-1}{3(1-x)} + \frac{3}{1+x} + \frac{1}{4\left(1-\dfrac{x}{2}\right)} - \frac{17}{12\left(1+\dfrac{x}{2}\right)}\right]$$

$$= \frac{1}{2}\left[-\frac{1}{3}\sum_{r=0}^{\infty} x^r + 3\sum_{r=0}^{\infty}(-1)^r x^r + \frac{1}{4}\sum_{r=0}^{\infty}\left(\frac{1}{2^r}\right)x^r - \frac{17}{12}\sum_{r=0}^{\infty}\left(-\frac{1}{2}\right)^r x^r\right]$$

$$= \frac{1}{2}\sum_{r=0}^{\infty}\left[\left(-\frac{1}{3}\right) + 3(-1)^r + \frac{1}{4}\cdot\frac{1}{2^r} - \frac{17}{12}\left(-\frac{1}{2}\right)^r\right]x^r$$

Therefore,

$$d_r = \frac{1}{2}\left[-\frac{1}{3} + 3(-1)^r + \frac{1}{2^{r+2}} - \frac{17}{3}(-1)^r\frac{1}{2^{r+2}}\right]$$

After necessary simplification, we get that

$$d_r = \frac{1}{2}\left[-\frac{1}{3} + 3 + \frac{1}{2^{r+2}}\left(1 - \frac{17}{3}\right)\right] = \frac{1}{3}\left(4 - \frac{14}{2^{r+3}}\right) \quad \text{if r is even}$$

$$= \frac{1}{2}\left[-\frac{1}{3} - 3 + \frac{1}{2^{r+2}}\left(1 + \frac{17}{3}\right)\right] = \frac{1}{3}\left(-5 + \frac{5}{2^{r+1}}\right)$$

If r is odd.

Example 15.11

Find the coefficient of x^{20} in the expansion of

$$\left(x^3 + x^4 + x^5 + \ldots\right)^5$$

Solution: Simplify the expression by extracting x^3 from each factor.

Thus

$$\left(x^3 + x^4 + ...\right)^5 = \left[x^3\left(1 + x + ...\right)\right]^5$$

$$= x^{15}\left(1 + x + x^2 + ...\right)^5$$

$$= x^{15}\left[\sum_{r=0}^{\infty} x^r\right]^5$$

$$= \frac{x^{15}}{\left(1 - x\right)^5} = x^{15}\sum_{r=0}^{\infty} C(5 - 1 + r,\ r)x^r$$

Observe that the coefficient of x^{20} in the original expression becomes the coefficient of x^5 in $\sum_{r=0}^{\infty} C(4 + r,\ r)x^r$

Therefore the coefficient we required is (when $r = 5$ in the last power series), $C(4 + 5, 5) = C(9, 5)$.

15.4 Recurrence Relations and Formation of Recurrence Relation

Definition

A recurrence relation is a formula that relates for any integer $n = 1$, the nth term of the sequence $x = \{x_r\}_{r=1}^{n}$ to one or more of the terms $x_1, x_2, ...$

15.4.1 Examples of Recurrence Relations

(i) Let 'S_n' denotes the sum of the first 'n' positive integers, then

$$S_n = n + S_{n-1}$$

(ii) Suppose, 'd' is a real number, then the nth term of an arithmetic progression with common difference 'd' satisfies the relation

$$b_n = b_{n-1} + d$$

(iii) If P_n denotes the nth term of a geometric progression with common ratio 'r', then

$$P_n = r\,P_{n-1}$$

15.4.2 Some More Examples

Suppose that a_n is the nth term of a sequence

(i) $a_n - 3a_{n-1} + 2a_{n-2} = 0$

(ii) $a_n - 3a_{n-1} + 2a_{n-2} = n^2 + 1$

(iii) $a_n^2 + \left(a_{n-1}\right)^2 = -1$

Definition

(i) Let 'n' and 'k be two non-negative integers. The recurrence relation of the given form

$$b_0(n)a_n + b_1(n)a_{n-1} + \dots + b_k(n)a_{n-k} = f(n)$$

for $n \geq k$, where f(n), $b_0(n)$, $b_1(n)$, …, $b_k(n)$ are functions of 'n', is said to be a linear recurrence relation.

(ii) In case, if $b_0(n)$ and $b_k(n)$ are not identically zero, then it is said to be a linear recurrence relation of degree k.

(iii) If $b_0(n)$, $b_1(n)$, …, $b_k(n)$ are constants, then the recurrence relation is known as a recurrence relation with constant coefficients.

(iv) If f(n) is identically equal to zero, then the recurrence relation is said to be a homogeneous recurrence relation. Otherwise, it is said to be inhomogeneous recurrence relation.

15.4.3   Solutions of Recurrence Relations

(i) The process of determining 'a_n' from a recurrence relation is called "solving" of the recurrence relation.

(ii) A value 'a_n' that satisfies a recurrence relation is called its "general solution".

(iii) If the values of some particular terms of the sequence are specified, then by making use of these values in the general solution, we can obtain the "particular solution" that uniquely determines the sequence.

Example 15.12

Consider the sequence $x = \left\{x_n\right\}_{n=0}^{\infty}$ where $x_n = 2^n$.

This satisfies the recurrence relation $x_n = 2x_{n-1}$ where n ranges over all the set of integers.

Note:

To solve the linear homogeneous recurrence relations, we shall follow the following three methods to solve the linear homogeneous recurrence relations: (we discuss these methods in the next sections)

(i) Substitution method.

(ii) By using generating functions.

(iii) The method of characteristic roots.

Example 15.13

Find a recurrence relation for a_n, the number of different ways to distribute either a Rs 3 bill, a Rs 5 bill, a Rs 7 bill or a Rs 12 bill on successive days until a total of 'n' rupees has been distributed.

Solution: If on the first day, we distribute Rs 3 bill, then we are to distribute $(n - 3)$ rupees on the succeeding day and there are a_{n-3} ways to do that.

If on the other hand, the first day distribution was Rs 5 bill, there remains the problem of distributing $n - 5$ rupees; this can be done a_{n-5} ways and so on.

Thus finally we get that

$$a_n = a_{n-3} + a_{n-5} + a_{n-7} + a_{n-12}$$

Example 15.14

Find the recurrence relation for number of ways to climb "n" stairs if the person climbing the stairs can take one, two, or three stairs at a time.

[JNTUH, Nov 2008, Set No. 3]

Solution: Write

a_n = the number of ways the person can climb 'n' stairs for $n \geq 1$

Note that $a_1 = 1$, $a_2 = 2$, $a_3 = 3$

Let us solve for a_n in terms of a fewer number of stairs.

Suppose the person takes only 1 stair on the first stride, then there are remaining $n - 1$ stairs to climb for which there are a_{n-1} ways to climb them.

On the other hand, if the person took 2 stairs in the first stride, then there are $n - 2$ steps left for which there are a_{n-2} ways to climb.

Similarly, if the person took 3 stairs in the first stride, then there are $n - 3$ stairs left for which there are a_{n-3} ways to climb.

Then it should be clear that

$$a_n = a_{n-1} + a_{n-2} + a_{n-3}$$

where each summand is determined by whether the first stride talcs 1, 2 or 3 stairs.

Example 15.15

What are the initial conditions? How many ways can this person climb a flight of 8 stairs?

[JNTUH, Nov 2008, Set No. 3]

Solution:

Part 1: Finding initial conditions:

The person (climbing stairs) can take one, two or three stairs at a time.

So $a_1 = 1$ (because one stair can take in one step)

$\quad a_2 = 2$

(Possibility-1: taking one stair each time

Possibility-2: taking two stairs in one time)

$$a_3 = 4$$

(Possibility-1: taking one stair each time

Possibility-2: taking two stairs first time and one stair second time

Possibility-3: taking one stair first time and two stairs second time

Possibility-4: taking three stairs at a time)

The initial conditions obtained are:

$$a_1 = 1$$
$$a_2 = 2$$
$$a_3 = 4$$

***Part* 2:** To find recurrence relation

From 15.4.9, we get a recurrence relation

$$a_n = a_{n-1} + a_{n-2} + a_{n-3}$$

***Part* 3:** To find the number of ways:

$$a_n = a_{n-1} + a_{n-2} + a_{n-3}$$

Take $n = 4$, then

$$a_4 = a_3 + a_2 + a_1 = 4 + 2 + 1 = 7$$

Take $n = 5$, then

$$a_5 = a_4 + a_3 + a_2 = 7 + 4 + 2 = 13$$

Take $n = 6$, then

$$a_6 = a_5 + a_4 + a_3 = 13 + 7 + 4 = 24$$

Take $n = 7$, then

$$a_7 = a_6 + a_5 + a_4 = 24 + 13 + 7 = 44$$

Take $n = 8$, then

$$a_8 = a_7 + a_6 + a_5 = 44 + 24 + 13 = 81$$

Exercises

Generating Function of Sequences

1. Find the generating functions of the following:

    (i)   2, 4, 8, 16, 32, ...   (ii)   2, –2, 2, –2, ...

    (iii)   1, 0, 1, 0, ...   (iv)   0, 1, –2, 4, –8, ...

Ans: (i) $\dfrac{2}{1-2x}$, $|2x| < 1$; (ii) $\dfrac{2}{1+x}$, $|-x| < 1$; (iii) $\dfrac{1}{1-x^2}$, $|x^2| < 1$; (iv) $\dfrac{x}{1+2x}$, $|2x| < 1$

2. Find the generating function of an $= 2n + 3$, $n = 0, 1, 2, \ldots$

Ans: $\dfrac{3-x}{(1-x)^2}$

3. Find the sequences generated by the following:

   (i) $\dfrac{1}{1-x} + 2x^3$

   (ii) $3x^3 + e^{2x}$

Ans: (i) $(1, 1, 1, 3, 1, 1, 1, \ldots)$ (ii) $(1, \dfrac{2}{1!}, \dfrac{2^2}{2!}, 3 + \dfrac{2^3}{3!}, \ldots)$

Partial Fractions

1. Use partial fractions to compute

$$\dfrac{1}{1-7x+12x^2}$$

Ans: $\displaystyle\sum_{r=0}^{\infty} \left(4^{r+1} - 3^{r+1}\right) x^r$

2. Solve the recurrence relation $a_n - 4a_{n-1} = 0$, $n \geq 1$, $a_0 = 1$ by using generating functions (Partial functions)

Ans: $a_n = 4^n$

3. Solve the recurrence relation $a_n = a_{n-1} + \dfrac{1}{n(n+1)}$, $n \geq 1$, $a_0 = 1$ by using generating functions (Partial functions)

Ans: $a_n = 2 - \dfrac{1}{n+1}$

Calculating Coefficient of Generating Functions

1. Determine the coefficient of
    x^{12} in $x^3 (1-2x)^{10}$

Ans: -5120

2. Find the coefficient of x^n in the function
    $(1 + x^2 + x^4 + \ldots)^7$

Ans: when 'n' is odd coefficient of x^n is '0'.

when 'n' is even, the coefficient of x^n is $\left[\,^{6+\frac{n}{2}}C_6 \right]$

3. Determine the coefficient of
x^{30} in $(x^3 + x^4 + x^5 + x^6 + \ldots)^6$

Ans: $\left({}^{17}C_{12} \right) - 6\left({}^{13}C_8 \right) + \left({}^{6}C_2 \right)\left({}^{9}C_4 \right) - \left({}^{6}C_3 \right)$

Recurrence Relations and Formulation of Recurrence Relations

1. Solve the recurrence relation $a_n = 7a_{n-1}$, where $n \geq 1$ given that $a_2 = 98$

Ans: $a_n = 2 \times 7^n$ for $n \geq 1$

2. Find the recurrence relation for the sequence
$0, 2, 6, 12, 20, 30, 42, \ldots$

Ans: $a_n - a_{n-1} = 2n$ for $n \geq 0$.

3. The number of virus affected files in a system is 1000 (to start with) and this increases 250% every two hours. Use a recurrence relation to determine the number of virus affected files in the system after one day?

Ans: recurrence relation: $a_{n+1} = 3.5\, a_n$ and $1000 \times (3.5)^{12}$

CHAPTER - 16

Some Methods of Solving Recurrence Relations

From the earlier chapter, we came to know that we study the following three methods to solve the linear homogeneous recurrence relations.

(i) Substitution method

(ii) By using generating functions

(iii) The method of characteristic roots

16.1 Solving Linear Homogeneous Recurrence Relations by Substitution Method

LEARNING OBJECTIVES

♦ *to understand the concept of Solving Linear Homogeneous Recurrence Relations by using Substitution Method*

♦ *to find the Characteristics Roots of generating function*

♦ *to find the Solutions of inhomogeneous Recurrence Relations*

16.1.1 Substitution Method

In this substitution method, the recurrence relation for a_n is used repeatedly to solve for the general expression (for the terms a_n to express in terms of n), we wish that this expression involves no other terms of the sequence except those given by the boundary conditions.

Example 16.1

Solve the recurrence relation $b_n = b_{n-1} + g(n)$ for $n \geq 1$ by substitution method.

Solution: Given recurrence relation is

$$b_n = b_{n-1} + g(n) \text{ for } n \geq 1 \qquad .....(16.1)$$

$$\Rightarrow \quad b_1 = b_0 + g(1)$$

$$b_2 = b_1 + g(2) = b_0 + g(1) + g(2)$$

$$b_3 = b_2 + g(3)$$

$$= b_0 + g(1) + g(2) + g(3)$$

$$b_n = b_0 + g(1) + g(2) + \ldots + g(n)$$

$$= b_0 + \sum_{k=1}^{n} g(k)$$

This shows that b_n is just the sum of the $g(k)$'s plus b_0.

Note:

Observe the above example. In general, if k is a constant, then we can solve the relation

$$b_n = k \, b_{n-1} + g(n) \quad \text{for } n \geq 1$$

In a similar way,

$$b_1 = k \, b_0 + g(1)$$

$$b_2 = k \, b_1 + g(2)$$

$$= k \, (kb_0 + g(1)) + g(2)$$

$$= k^2 a_0 + k \, g(1) + g(2)$$

Similarly, $a_3 = k^3 \, a_0 + k^2 \, g(1) + k \, g(2) + g(3)$

.

.

.

$$a_n = k \, a_{n-1} + f(n)$$

$$= k^n \, a_0 + k^{n-1} \, g(1) + k^{n-2} \, g(2) + \ldots + k \, g(n-1) + g(n)$$

$$= k^n a_0 + \sum_{k=1}^{n} k^{n-k} g(k)$$

16.1.2   First Order Recurrence Relation

Let us consider the recurrence relation, of the form

$$a_n = c \, a_{n-1} + f(n) \quad \text{for } n \geq 1 \quad\quad\quad \ldots..(16.2)$$

Now we wish to solve the relation (16.2)

By replacing n by (n + 1) we get that

$$a_{n+1} = c \, a_n + f(n + 1) \quad\quad\quad\quad\quad \text{for } n \geq 0$$

For   n = 0, 1, 2, 3, …

$$a_1 = c \, a_0 + f(1)$$

$$a_2 = c\, a_1 + f(2)$$
$$= c\,[c\, a_0 + f(1)] + f(2)$$
$$= c^2 a_0 + c\, f(1) + f(2)$$

and so on.

In general,

$$a_n = c^n a_0 + c^{n-1} f(1) + c^{n-2} f(2) + \ldots + c\, f(n-1) + f(n)$$

$$= c^n a_0 + \sum_{k=1}^{n} c^{n-k} f(k), \qquad\qquad n \geq 1$$

This is the general solution for (16.2).

Example 16.2

Solve the difference equation

$$u_n - 2\, u_{n-1} = 5(2)^n \text{ using generating functions.}$$

[JNTUH, Nov 2010, Set No. 3]

Solution: The given recurrence relation is,

$$u_n - 2\, u_{n-1} = 5\,(2)^n \qquad\qquad\qquad \ldots..(16.3)$$

$$n \geq 1$$

Replace n by n+1 in (16.3), to get

$$u_{n+1} = 2\, u_n + 5\,(2)^{n+1} \qquad\qquad\qquad \ldots..(16.4)$$

$$n \geq 0$$

$$= 2\, u_n + f\,(n+1) \qquad\qquad \text{where } f(n) = 5\,(2^n)$$

The general solution of (16.4) is in the form (refer 16.1.5)

$$u_n = 2^n u_0 + \sum_{r=1}^{n} 2^{n-r} f(r)$$

Example 16.3

Define recurrence relation? Show that $\{a_n\}$ is a solution of recurrence relation

$$a_n = -3a_{n-1} + 4a_{n-2} \quad \text{if } a_n = 1$$

[JNTUH, June 2010, Set No. 3]

[JNTUH, Nov 2008, Set No. 1]

Solution: Given recurrence relation is

$$a_n = -3a_{n-1} + 4a_{n-2}$$

$\Rightarrow \quad a_n + 3a_{n-1} - 4a_{n-2} = 0$ .....(16.5)

Characteristic equation of (16.5) is

$$k^2 + 3k - 4 = 0$$

$\Rightarrow \quad k^2 + 4k - k - 4 = 0$

$\Rightarrow \quad k(k+4) - 1(k+4) = 0$

$\Rightarrow \quad (k-1)(k+4) = 0$

$\Rightarrow \quad k = 1, -4$

Therefore the general solution is

$$a_n = (1)^n + (-4)^n = 1 + (-4)^n$$

Example 16.4

What is solution of the recurrence relation?

$\qquad a_n = a_{n-1} + 2a_{n-2}$ with $a_0 = 2$ and $a_1 = 7$ [JNTUH, Nov 2008, Set No.1]

[JNTUH, June 2010, Set No. 3]

Solution: Given recurrence relation is

$$a_n = a_{n-1} + 2\, a_{n-2}$$ .....(16.6)

with $a_0 = 2$ and $a_1 = 7$

Put $a_n = r^n$ in (16.6), then we get

$$r^n - r^{n-1} - 2r^{n-2} = 0, \ r > 0$$

$\Rightarrow \quad r^{n-2}(r^2 - r - 2) = 0$

$\Rightarrow \quad r^2 - r - 2 = 0$ .....(16.7)

This is the characteristic equation.

$\Rightarrow \quad (r+1)(r-2) = 0$

$\Rightarrow \quad r = -1, 2$

So the characteristic roots are $r = -1, 2$ which are real and distinct.

So the solution is of the form

$$a_n = c_1(-1)^n + c_2.2^n \qquad\qquad .....(16.8)$$

where c_1, c_2 are arbitrary constants.

By the given initial conditions

i.e., $a_0 = 2$ and $a_1 = 7$

From (16.8) we get

$$a_0 = c_1 + c_2 = 2 \qquad\qquad .....(16.9)$$

$$a_1 = -c_1 + 2c_2 = 7 \qquad\qquad .....(16.10)$$

$(16.9) + (16.10) \quad \Rightarrow \qquad 3c_2 = 9$

$\qquad\qquad\qquad\quad \Rightarrow \quad c_2 = 3$

by (16.9) $c_1 = 2 - c_2 = 2 - 3 = -1$

Therefore $a_n = (-1)(-1)^n + 3.2^n$

$\qquad\qquad\qquad\qquad = (-1)^{n+1} + 3.2^n$

Example 16.5

Find the generating function of $(n-1)^2$ [JNTUH, Nov 2010, Set No. 3]

Solution: write $\{a_n\}_{n=1}^{\infty} = \{(n-1)^2\}_{n=0}^{\infty}$

For the given sequence $\{a_n\}$, we note that

$$\sum_{n=1}^{\infty} a_n x^n = \sum_{n=1}^{\infty} (n-1)^2 x^n$$

$$= x^2 + 4x^3 + 9x^4 + 16x^5 + ...$$

$$= x^2(1 + 4x + 9x^2 + 16x^3 + ...)$$

$$= x^2(1^2 + 2^2 x + 3^2 x^2 + 4^2 x^3 + ...) \qquad\qquad .....(16.11)$$

We know that

$$1.x + 2x^2 + 3x^2 + ... = \frac{x}{(1-x)^2} \qquad\qquad .....(16.12)$$

Differentiating (16.12) w.r.t. x on both sides, we get

$$1 + 2(2x) + 3(3x)^2 + \ldots = \frac{d}{dx}\left[\frac{x}{(1-x)^2}\right]$$

$$= \frac{1+x}{(1-x)^3}$$

$$\Rightarrow \quad 1 + 4x + 9x^2 + \ldots = \frac{1+x}{(1-x)^3} \qquad \qquad \ldots\ldots(16.13)$$

So by (16.11) and (16.13) we get that

$$x^2(1 + 4x + 9x^2 + \ldots) = x^2\left[\frac{(1+x)}{(1-x)^3}\right]$$

is a generating function.

16.2   Generating Functions and the Method of Characteristic Roots

Before going to the method "Solutions by Generating Functions": let us understand the following basic property.

16.2.1   The Shifting Properties of Generating Functions

(i)   We know that $P(X) = \sum\limits_{n=0}^{\infty} b_n X^n$ generates the sequence $\{b_0, b_1, \ldots\}$ then $X\,P(X)$ generates the sequence $\{0, b_0, b_1, \ldots\}$; $X^2 P(X)$ generates $\{0, 0, b_0, b_1, \ldots\}$, and in general $X^k P(X)$ generates $\{0, 0, \ldots, 0, b_0, b_1\}$ in which there are k – zeros before b_0.

So if P(X) is the generating function for the sequence $\{b_0, b_1, \ldots\}$ and if we multiply P(X) by X will amounts to shifting the sequence one place to the right and inserting a zero in the beginning. Multiplying P(X) by X^k results the shifting of the sequence k positions to the right and inserting K zeros in the beginning.

(ii)   This process may be described by a change in the dummy variable in the formal power series expressions as follows:

$$X^k P(X) = X^k \sum_{n=0}^{\infty} b_n X^n$$

$$= \sum_{n=0}^{\infty} b_n X^{n+k} \qquad \qquad \ldots\ldots(16.14)$$

In the last expression (16.14), if we replace n + k by 'r', and then taking n = r − k then the expression becomes to $\sum\limits_{r=k}^{\infty} b_{r-k} X^r$.

In this form, the expression $\sum\limits_{r=k}^{\infty} b_{r-k} x^r$ generates the sequence $\{C_r\}_{r=0}^{\infty}$ with

$0 = C_0 = C_1 = \ldots = C_{k-1}, C_k = b_0, C_{k+1} = b_1$. In general $C_r = b_{r-k}$ if $r \geq k$.

Therefore the nth term in the new sequence is obtained from the old sequence by replacing b_n by b_{n-k} (if $n \geq k$) and '0' (if $n < k$).

(iii) For example, consider the known equation: $\dfrac{1}{1-X} = \sum\limits_{n=0}^{\infty} X^n$ which generates the

sequence (1, 1, ...). This is the sequence $\{b_n\}_{n=0}^{\infty}$ where $b_n = 1$ for each $n \geq 0$.

Therefore

$$\frac{X}{(1-X)} = \sum_{n=0}^{\infty} X^{n+1}$$

$$= \sum_{r=1}^{\infty} X^r \qquad \qquad \ldots\ldots(16.15)$$

generates (0, 1, 1, ...) and

$$\frac{X^2}{(1-X)} = \sum_{n=0}^{\infty} X^{n+2} = \sum_{r=2}^{\infty} X^r \qquad \qquad \ldots\ldots(16.16)$$

generates $\{0, 0, 1, 1, \ldots\}$.

Similarly, the relation

$$\frac{1}{(1-X)^2} = \sum_{n=0}^{\infty} \left({}^{n+1}C_n \right) X^n$$

$$= \sum_{n=0}^{\infty} (n+1) X^n \qquad \qquad \ldots\ldots(16.17)$$

generates the sequence $\{1, 2, 3, 4, \ldots\}$ and the relation

$$\frac{X}{(1-X)^2} = \sum_{n=0}^{\infty} (n+1) X^{n+1}$$

$$= \sum_{r=1}^{\infty} r X^r \qquad \qquad \ldots\ldots(16.18)$$

generates the sequence $\{r\}_{r=0}^{\infty} = \{0, 1, 2, ...\}$. Here we understand that the expression $\sum_{r=1}^{\infty} r\, X^r$ describes that the coefficient of X^0 is '0' (because the sum is taken from $r = 1$ to ∞) but the form of the coefficients provides the same conclusion even if r is allowed to equal '0'.

Both the expressions mean that the coefficient of X^0 is zero. Similarly, the relation

$$\frac{X^2}{(1-X)^2} = \sum_{n=0}^{\infty} (n+1)\, X^{n+2}$$

$$= \sum_{r=2}^{\infty} (r-1)\, X^r \qquad\qquad .....(1619)$$

generates the sequence $\{0, 0, 1, 2, 3, ...\}$, (that is, the sequence $\{C_r\}_{r=0}^{\infty}$, where $C_r = r - 1$ if $r \geq 2$, but $0 = C_0 = C_1$).

Since the expression $C_r = r - 1$ equals to zero when $r = 1$, we have that

$$\frac{X^2}{(1-X)^2} = \sum_{r=2}^{\infty} (r-1)\, X^r$$

$$= \sum_{r=1}^{\infty} (r-1)\, x^r$$

(iv)   In the same way, we can get that the relation

$$\frac{1}{(1-X)^3} = \sum_{n=0}^{\infty} \left({}^{n+2}C_n \right) X^n$$

$$= \sum_{n=0}^{\infty} \frac{(n+2)(n+1)}{2}\, X^n \qquad\qquad .....(16.20)$$

generates the sequence

$$\left\{ \frac{(n+2)(n+1)}{2} \right\}_{n=0}^{\infty} = \left\{ \frac{1.2}{2}, \frac{2.3}{2}, \frac{3.4}{2}, ... \right\}$$

Therefore $\dfrac{2}{(1-X)^3} = \sum_{n=0}^{\infty} (n+2)(n+1)\, X^n$

generates the sequence $\{(n+2)(n+1)\}_{n=0}^{\infty} = \{1.2, 2.3, 3.4, ...\}$.

(v)   It follows that the relation

$$\frac{2X}{(1-X)^3} = \sum_{n=0}^{\infty} (n+2)(n+1) \, X^{n+1}$$

$$= \sum_{r=1}^{\infty} (r+1)(r) \, x^r \qquad\qquad .....(16.21)$$

generates the sequence $\{0, 1.2, 2.3, 3.4, ...\}$

Since $C_r = (r+1) \, r$ equal 0 when $r = 0$, we understand that we can write

$$\frac{2X}{(1-X)^3} = \sum_{r=0}^{\infty} (r+1)(r) \, X^r$$

$$= \sum_{r=0}^{\infty} (r+1) \, r \, X^r$$

So $\dfrac{2x}{(1-x)^3}$ generates $\{(r+1) \, r\}_{r=0}^{\infty}$

(vi)   Similarly the relation

$$\frac{2X^2}{(1-X)^2} = \sum_{n=0}^{\infty} (n+2)(n+1) \, X^{n+2}$$

$$= \sum_{r=2}^{\infty} (r)(r-1) \, X^r$$

$$= \sum_{r=0}^{\infty} (r)(r-1) \, X^r$$

generates the sequence $\{0, 0, 1.2, 2.3, ...\}$ and the last sum can be taken from 0 to ∞ (because the coefficient $r(r-1)$ is '0' when $r = 0, 1$).

Gluing all these results, we can find the generating function for some other sequences.

(vii)   For example

$$\frac{2X}{(1-X)^3} = \sum_{n=0}^{\infty} \left({}^{n+3}C_n \right) X^n$$

$$= \sum_{n=0}^{\infty} \frac{(n+3)(n+2)(n+1)}{6} X^n$$

generates the sequence

$$\left\{(r+1)(r)-r\right\}_{r=0}^{\infty} = \left\{r^2\right\}_{r=0}^{\infty}$$

$$= \{0, 1, 4, 9, \ldots\}$$

Also we can verify that

$$\frac{1}{(1-X)^4} = \sum_{n=0}^{\infty} \left(^{n+3}C_n\right)X^n$$

$$= \sum_{n=0}^{\infty} \frac{(n+3)(n+2)(n+1)}{6}X^n$$

generates $\left\{\dfrac{(n+3)(n+2)(n+1)}{6}\right\}_{n=0}^{\infty}$

The relation $\dfrac{6}{(1-X)^4}$ generates $\left\{(n+3)(n+2)(n+1)\right\}_{n=0}^{\infty}$

The relation

$$\frac{6X}{(1-X)^4} = \sum_{n=0}^{\infty} (n+3)(n+2)(n+1)X^{n+1}$$

$$= \sum_{r=1}^{\infty} (r+2)(r+1)r\ X^r$$

$$= \sum_{r=0}^{\infty} (r+2)(r+1)(r)\ X^r$$

generates $\left\{(r+2)(r+1)\ r\right\}_{r=0}^{\infty}$

(viii) The relation

$$\frac{6X^2}{(1-X)^4} = \sum_{n=0}^{\infty} (n+3)(n+2)(n+1)\ X^{n+2}$$

$$= \sum_{r=2}^{\infty} (r+1)(r)(r-1)\ X^r$$

$$= \sum_{r=0}^{\infty} (r+1)(r)(r-1)\ X^r$$

generates $\left\{(r+1)(r)(r-1)\right\}_{r=0}^{\infty}$

(ix)   Since $(r+3)(r+2)(r+1) = r^3 + 6r^2 + 11r + 6$

We have that

$$r^3 = (r+3)(r+2)(r+1) - 6r^2 - 11r - 6$$

and $\{r^3\}_{r=0}^{\infty}$ is generated by the relation

$$\frac{6}{(1-x)^4} - \frac{6(x)(1+X)}{(1-X)^3} - \frac{11.X}{(1-X)^2} - \frac{6}{(1-x)} = \frac{X(1+4X+X^2)}{(1-X)^4}$$

In a similar way, we can get generating functions for the sequences such as $\{r^4\}_{r=0}^{\infty}$, $\{r^5\}_{r=0}^{\infty}$.

(x) In solving recurrence relations by generating functions, we use these shifting properties so frequently. We now, list some equivalent expressions for ready reference

If $P(X) = \sum_{n=0}^{\infty} b_n X^n$ then

(a) $\displaystyle\sum_{n=k}^{\infty} b_n X^n = P(X) - b_0 - b_1 X - \cdots - b_{k-1} X^{k-1}$

(b) $\displaystyle\sum_{n=k}^{\infty} b_{n-1} X^n = X\{P(X) - b_0 - b_1 X - \cdots - b_{k-2} X^{k-2}\}$

(c) $\displaystyle\sum_{n=k}^{\infty} b_{n-2} X^n = X^2\{P(X) - b_0 - b_1 X - \cdots - b_{k-3} X^{k-3}\}$

$\vdots$

$$\sum_{n=k}^{\infty} b_{n-k} X^n = X^k P(X)$$

16.2.2   Sketch of the Method of Generating Functions

Let us discuss the sketch of the method of generating functions.

(i) Given a Linear recurrence relation with constant coefficients of degree 'k', without loss of generality, we assume that the given linear recurrence relation has the form:

$$b_n + m_1 b_{n-1} + m_2 b_{n-2} + \ldots + m_k b_{n-k} = 0,$$

where $m_1, m_2 \ldots, m_k$ are constants, $m_k \neq 0$ and $n \geq k$.

(ii)   Suppose that $P(X) = \sum_{n=0}^{\infty} b_n X^n$. Multiply each term of the recurrence relation by X^n. Sum from k to ∞. Replace all infinite sums by the equivalent expressions. Then the recurrence relation is transformed into an algebraic equation of the form:

$$P(X) = \frac{Q(X)}{R(X)}$$

where $Q(X)$ and $R(X)$ are given by

$$Q(X) = b_0 + (b_1 + m_1 b_0)X + (b_2 + m_1 b_1 + m_2 b_0)X^2 + ... +$$

$$(b_{k-1} + m_1 b_{k-2} + ... + m_{k-1} b_0)x^{k-1}$$

$$R(X) = 1 + m_1 X + m_2 X^2 + ... + m_k X^k$$

(iii)   Knowing $Q(X)$ and $R(X)$; we transform $P(X)$ back to get the coefficients b_n (call this as "performing the inverse transformation"). This can be done in one of several known ways.

(iv)   Suppose that we know the factorization of $R(X)$. Say, $R(X) = (1 - r_1 x)(1 - r_2 x) ...$ $(1 - r_k x)$.

Then use partial fractions method and also some necessary identities for familiar generating functions to get $P(X)$ as the sum of some familiar series. Then we get b_n as the sum of the coefficients of known series.

(v)   In case, if we cannot factor the $R(X)$ and if we are given initial conditions, then we can solve for as many coefficients of $P(X)$ as we desire, by long division [or by finding the multiplicative inverse of $R(X)$ (because constant term of $R(X)$ is non zero)].

Example 16.6

Solve $b_n - 6b_{n-1} + 12b_{n-2} - 8b_{n-3} = 0$ by generating functions.

Solution: We follow the method.

Given that

$$b_n - 6b_{n-1} + 12b_{n-2} - 8b_{n-3} = 0 \qquad\qquad .....(16.22)$$

Suppose

$$P(X) = \sum_{n=0}^{\infty} b_n X^n = \frac{Q(X)}{R(X)}$$

$$= \frac{b_0 + (b_1 - 6b_0)X + (b_2 - 6b_1 + 12a_0)X^2}{1 - 6X + 12X^2 - 8X^3} \qquad\qquad .....(16.23)$$

Since
$$1 - 6X + 12X^2 - 8X^3 = (1 - 2X)^3, \text{ we use partial fractions.}$$

In this case, we know that there exist constants d_1, d_2, d_3 such that

$$P(X) = \frac{d_1}{1 - 2X} + \frac{d_2}{(1 - 2X)^2} + \frac{d_3}{(1 - 2X)^3}$$

$$= d_1(1 - 2X)^{-1} + d_2(1 - 2X)^{-2} + d_3(1 - 2X)^{-3}$$

$$= d_1 \sum_{n=0}^{\infty} 2^n X^n + d_2 \sum_{n=0}^{\infty} \binom{n+1}{n} 2^n X^n + d_3 \sum_{n=0}^{\infty} \binom{n+2}{n} 2^n X^n$$

$$= \overline{Z} \left[d_1 2^n + d_2(n+1)2^n + \frac{d_3(n+1)(n+2)}{2} 2^n \right] X^n$$

Therefore $b_n = d_1 2^n + d_2(n+1)2^n + d_3 \dfrac{(n+1)(n+2)}{2} 2^n$

Example 16.7

Solve the recurrence relation

$$a_n + a_{n-1} - 6\,a_{n-2} = 0 \qquad\qquad \text{for } n \geq 2$$

with the initial conditions: $a_0 = -1$ and $a_1 = 8$

Solution: The recurrence relation given is

$$a_n + a_{n-1} - 6\,a_{n-2} = 0 \qquad\qquad \text{for } n \geq 0 \qquad\qquad \text{.....(16.24)}$$

Suppose that

$$P(X) = \sum_{n=0}^{\infty} a_n X^n = \frac{Q(X)}{R(X)}$$

$$= \frac{a_0 + (a_1 + a_0)X}{1 + X - 6X^2} \qquad\qquad \text{.....(16.25)}$$

Since $1 + X - 6X^2 = 1 + 3X - 2X - 6X^2$

$$= 1(1 + 3X) - 2X(1 + 3X)$$

$$= (1 - 2X)\,(1 + 3X)$$

We conclude that $1 + X - 6X^2$ can be factored. In this case, we know that there exist constants k_1, k_2 satisfying

$$P(X) = \frac{k_1}{(1 - 2X)} + \frac{k_2}{(1 + 3X)}$$

Now

$$P(X) = k_1(1-2X)^{-1} + k_2(1+3X)^{-1}$$

$$= k_1 \sum_{n=0}^{\infty} 2^n X^n + k_2 \sum_{n=0}^{\infty} (-1)^n 3^n X^n$$

$$= \sum_{n=0}^{\infty} \left[k_1 2^n + k_2 (-1)^n 3^n \right] X^n$$

Therefore, we get that

$$a_n = k_1 2^n + k_2 (-1)^n 3^n \qquad \qquad .....(16.26)$$

The initial conditions given are

$$a_0 = -1, \text{ and } a_1 = 8$$

By (16.26) $a_0 = k_1 + k_2 = -1$ $\qquad \qquad .....(16.27)$

$a_1 = 2k_1 - 3k_2 = 8$ $\qquad \qquad .....(16.28)$

By multiplying (16.27) with 2 we get

$$2k_1 + 2k_2 = -2 \qquad \qquad .....(16.29)$$

We solve (16.28) and (16.29)

By subtraction, we get that

$$5k_2 = -10 \text{ and so}$$

$$k_2 = -2$$

Also $k_1 = -1 - k_2 = -1 + 2 = 1$

Therefore $a_n = 2^n + (-2)(-1)^n 3^n = 2^n - 2(-1)^n 3^n$

16.2.3   The Method of Characteristics Roots

This new method is nothing but the synthesis of all that we know from the method of generating functions.

(i)   To solve $b_n + d_1 b_{n-1} + ... + d_k b_{n-k} = 0$ where $d_k \neq 0$, we find $P(X) = \dfrac{Q(X)}{R(X)}$ where

$R(X)$ is a polynomial of degree k. The factor of $R(X)$ completely determine the form of the coefficient of $P(X)$.

(ii)   Let us take a polynomial where we replace X in $R(X)$ by $\dfrac{1}{t}$ and multiply by t^k to get the polynomial.

$$C(t) = t^k R\left(\frac{1}{t}\right) \qquad \qquad .....(16.30)$$

The polynomial C(t) is known as the "characteristic polynomial" of the given recurrence relation.

(iii) It is important to understand that if the given recurrence relation is $b_n + d_1 b_{n-1} + d_2 b_{n-2} + ... + d_k b_{n-k} = 0$ for $n \geq k$, where $d_k \neq 0$, then the characteristic polynomial for this recurrence relation is

$$C(t) = t^k + d_1 t^{k-1} + ... + d_k$$

This, in turn, equals to $t^k R\left(\dfrac{1}{t}\right)$ where $R(X) = 1 + d_1 X + d_1 X^2 + ... + d_k X^k$.

If C(t) factors as $(t - \alpha_1)^{r_1} (t - \alpha_2)^{r_2} ... (t - \alpha_s)^{r_s}$, in such a case, R(X) factors as

$$(1 - \alpha_1 X)^{r_1} (1 - \alpha_2 X)^{r_2} ... (1 - \alpha_s X)^{r_s}$$

(iv) *Distinct Roots:* Suppose that the characteristic polynomial has distinct roots α_1, α_2, ..., α_k. In this case, the general form of the solutions for the homogenous equation is $b_n = d_1 \alpha_1^n + ... + d_k \alpha_k^n$ where d_1, d_2, ..., d_k are constants that may be choosen to satisfy any initial conditions.

Example 16.8

Solve $a_n - 7a_{n-1} + 12a_{n-2} = 0$ for $n \geq 2$.

Solution: The recurrence relation given is

$$a_n - 7a_{n-1} + 12a_{n-2} = 0 \qquad\qquad .....(16.31)$$

Here $k = 2$

Suppose that $P(X) = \displaystyle\sum_{n=0}^{\infty} b_n X^n \qquad\qquad .....(16.32)$

Multiply each term in the recurrence relation by X^n and sum from $k = 2$ to ∞.
Then

$$P(X) = \sum_{n=2}^{\infty} b_n X^n - 7\sum_{n=2}^{\infty} b_{n-1} x^n + 12\sum_{n=2}^{\infty} b_{n-2} X^n = 0$$

We know that,

$$\sum_{n=2}^{\infty} b_n X^n = P(X) - b_0 - b_1 X$$

$$\sum_{n=2}^{\infty} b_{n-1} X^n = X[P(X) - b_0]$$

$$\sum_{n=2}^{\infty} b_{n-2} X^n = X^2[P(X)]$$

Therefore

$$[P(X) - b_0 - b_1 X] - 7X[P(X) - b_0] + 12X^2 P(X) = 0$$

$$\Rightarrow \quad P(X)[1 - 7X + 12X^2] = b_0 + b_1 X - 7Xb_0$$

$$\Rightarrow \quad P(X) = \frac{b_0 + b_1 X - 7b_0 X}{1 - 7X + 12X^2}$$

Note that the denominator is

$$R(X) = 1 - 7X + 12X^2$$

Replace X by $\dfrac{1}{t}$ in R(X) and multiply by t^2, to get

$$C(t) = t^2 \, R\!\left(\frac{1}{t}\right) = t^2\left[1 - 7.\frac{1}{t} + 12.\frac{1}{t^2}\right]$$

$$= t^2\left[\frac{t^2 - 7t + 12}{t^2}\right]$$

$$= t^2 - 7t + 12 = t^2 - 4t - 3t + 12$$

$$= (t - 4)t - 3(t - 4)$$

$$= (t - 4)(t - 3)$$

Therefore t = 4, t = 3 are the real and distinct roots.

Hence the general solution is given by

$$b_n = d_1 3^n + d_2 4^n$$

16.3   Solving Inhomogeneous Recurrence Relations

In this section, we present a method of solving second and higher order linear inhomogeneous recurrence relations with constant coefficients.

16.3.1   Method

The general form we considered is given below:

$$m_n a_n + m_{n-1} a_{n-1} + m_{n-2} a_{n-2} + \ldots + m_{n-k} a_{n-k} = f(n) \quad \text{for } n \ge k \ge 2 \qquad \ldots\text{(16.33)}$$

Here m_n, m_{n-1}, m_{n-2}, …, m_{n-k} are real constants with $m_n \ne 0$ and f(n) is a given real-valued function of n.

A general solution for the recurrence relation (16.33) is given below

$$a_n = a_n^{(h)} + a_n^{(p)} \qquad \ldots\text{(16.34)}$$

Here $a_n^{(h)}$ is a general solution of the homogenous part of the relation (16.33) namely the relation (16.33) with $f(n) = 0$.

The term $a_n^{(p)}$ is a particular solution of the relation (16.33).

We can get the part $a_n^{(h)}$ of solution (16.34) by the methods (second order linear homogeneous recurrence relation and third and higher-order linear homogeneous recurrence relations). But the determination of $a_n^{(p)}$ for arbitrary $f(n)$ is a little difficult job. In some special cases, only, we have to find $a_n^{(p)}$ in a straight forward way. In the following we present some of such special cases.

(i) Suppose that $f(x)$ is a polynomial of degree q and 1 is not a root of the characteristic equation of the homogeneous part of the relation (16.33).

In such case, $a_n^{(p)}$ is taken in the form

$$a_n^{(p)} = B_0 + B_1 n + B_2\ n^2 + ... + B_q\ n^q \qquad\qquad .....(16.35)$$

Here $B_0, B_1, ..., B_q$ are constants to be evaluated by using the fact that $a_n = a_n^{(p)}$ satisfies the relation given by (16.33).

(ii) Suppose that $f(n)$ is a polynomial of degree q and 1 is a root of multiplicity m of the characteristic equation of the homogeneous part of the relation given in (16.33).

In such a case $a_n^{(p)}$ is taken in the form

$$a_n^{(p)} = n^m \left\{ B_0 + B_1 n + B_2 n^2 + ... + B_q n^q \right\} \qquad\qquad .....(16.36)$$

Here $B_0, B_1, B_2, ..., B_q$ are constants to be evaluated by using the fact that $a_n = a_n^{(p)}$ satisfies the relation given in (16.33).

(iii) Suppose that $f(n) = \alpha b^n$, where α is a constant and b is not a root of the characteristic equation of the homogeneous part of the relation given in (16.33).

Then $a_n^{(p)}$ is taken the form

$$a_n^{(p)} = B_0\ b^n \qquad\qquad .....(16.37)$$

Here B_0 is a constant to be evaluated by using the fact $a_n = a_n^{(p)}$ satisfies the relation given in (16.33).

(iv) Suppose that $f(n) = \alpha\ b^n$, where α is a constant and b is a root of multiplicity m of the characteristic equation of the homogeneous part of the relation given in (16.33).

Now $a_n^{(p)}$ is taken in the form

$$a_n^{(p)} = B_0\ n^m b^n \qquad\qquad .....(16.38)$$

Here B_0 is a constant to be evaluated by using the fact that $a_n = a_n^{(p)}$ satisfies the relation given in (16.33).

Example 16.9

Solve the recurrence relation

$$b_n + 4b_{n-1} + 4b_{n-2} = 8 \quad \text{for } n \geq 2, \text{ and } b_0 = 1, b_1 = 2$$

Solution: The recurrence relation given is

$$b_n + 4b_{n-1} + 4b_{n-2} = 8 \qquad \ldots\ldots(16.39)$$

for $n \geq 2$ and the initial conditions $b_0 = 1, b_1 = 2$.

For the homogeneous part $b_n + 4b_{n-1} + 4b_{n-2} = 0$, the characteristic equation is

$$k^2 + 4k + 4 = 0$$

$$\Rightarrow \quad (k+2)^2 = 0$$

$$\Rightarrow \quad k = -2, -2$$

Therefore the roots of the characteristic equation are $-2, -2$.

The roots are real and are equal.

Therefore the general solution of the homogeneous part is given by

$$b_n = (C + D_n)(-2)^n \qquad \ldots\ldots(16.40)$$

where C and D are two arbitrary constants.

To find the particular solution, put

$$b_n = A_0$$

where A_0 is a constant to be evaluated, and which satisfies the relation (16.33).

$$A_0 + 4A_0 + 4A_0 = 8 \quad \Rightarrow \quad 9A_0 = 8$$

$$\Rightarrow \quad A_0 = \frac{8}{9} \qquad \ldots\ldots(16.41)$$

Therefore the general solution for a_n is given by

$$b_n = (C + D_n)(-2)^n + \frac{8}{9} \qquad \ldots\ldots(16.42)$$

Given initial conditions are $b_0 = 1$ and $b_1 = 2$.

Using these initial conditions in (16.42), we get that

$$b_0 = (C + D.0)(-2)^0 + \frac{8}{9} = 1$$

$$\Rightarrow \quad C + \frac{8}{9} = 1$$

$$\Rightarrow \quad C = 1 - \frac{8}{9} = \frac{9-8}{9} = \frac{1}{9}$$

and $\quad b_1 = (C + D.1)(-2)^1 + \frac{8}{9} = 2$

$$\Rightarrow \quad \left(\frac{1}{9} + D\right)(-2) = 2 - \frac{8}{9} = \frac{10}{9}$$

$$\Rightarrow \quad \frac{1}{9} + D = \frac{10}{9} \times \frac{-1}{2} = \frac{-5}{9}$$

$$\Rightarrow \quad D = \frac{-5}{9} - \frac{1}{9} = \frac{-6}{9} = \frac{-2}{3}$$

Therefore the general solution is given by

$$b_n = \left[\frac{1}{9} - \frac{2}{3}n\right](-2)^n + \frac{8}{9}$$

which is the required solution.

Example 16.10

Solve the recurrence relation

$$b_{n+2} + 3b_{n+1} + 2b_n = 3^n \quad \text{for } n \geq 0$$

The initial conditions:

$$b_0 = 0, \ b_1 = 1$$

Solution: The recurrence relation given is

$$b_{n+2} + 3b_{n+1} + 2b_n = 3^n \qquad\qquad\qquad .....(16.43)$$

for $n \geq 0$, $b_0 = 0$, $b_1 = 1$

For the homogeneous part of the given relation, the characteristic equation is given by

$$k^2 + 3k + 2 = 0$$

$$\Rightarrow \quad (k+1)(k+2) = 0$$

$$\Rightarrow \quad k = -1, -2$$

The roots are real and distinct.

Therefore $b_n = C(-2)^n + D(-1)^n \qquad\qquad\qquad .....(16.44)$

where C and D are two constants.

To find the particular solution, write

$$b_n = A_0 \times 3^n \text{ in } (16.43)$$

which satisfies the (16.43). Then

$$A_0 \times 3^{n+2} + 3A_0 \times 3^{n+1} + 2A_0 \times 3^n = 3^n$$

$$\Rightarrow \quad (A_0 \times 3^2) + (3A_0 \times 3) + 2A_0 = 1$$

$$\Rightarrow \quad 9A_0 + 9A_0 + 2A_0 = 1$$

$$\Rightarrow \quad 20\,A_0 = 1$$

$$\Rightarrow \quad A_0 = \frac{1}{20}$$

Substituting this value, into b_n, we get that

$$b_n = \frac{1}{20}\,3^n$$

Hence, the general solution is of the form

$$b_n = C(-2)^n + D(-1)^0 + \frac{1}{20} \times 3^0 = 0 \qquad\qquad \dots\dots(16.45)$$

By the given initial conditions, we have that

$$b_0 = C(-2)^0 + D(-1)^0 + \frac{1}{20} \times 3^0 = 0$$

$$\Rightarrow \quad C + D + \frac{1}{20} = 0$$

$$\Rightarrow \quad C + D = -\frac{1}{20} \qquad\qquad \dots\dots(16.46)$$

$$b_1 = C(-2)^1 + D(-1)^1 + \frac{1}{20} \times 3^1 = 1$$

$$\Rightarrow \quad -2C - D = 1 - \frac{3}{20} = \frac{17}{20} \qquad\qquad \dots\dots(16.47)$$

By adding (16.46) and (16.47), we get

$$-C = -\frac{1}{20} + \frac{17}{20} = \frac{16}{20}$$

$$\Rightarrow \quad C = -\frac{16}{20} = -\frac{4}{5}$$

$$D = -\frac{1}{20} - C$$

$$= -\frac{1}{20} - \left(-\frac{4}{5}\right) = -\frac{1}{20} + \frac{4}{5}$$

$$= \frac{15}{20} = \frac{3}{4}$$

Therefore the general solution is given by

$$b_n = \left(-\frac{4}{5}\right)(-2)^n + \frac{3}{4}(-1)^n + \frac{1}{20} \times (3)^n$$

which is the required solution.

Example 16.11

Find all solutions of the recurrence relation

$$a_n = 5a_{n-1} - 6a_{n-2} + 7^n \qquad\qquad \text{[JNTUH, June 2010, Set No. 4]}$$

Solution: The recurrence relation given is

$$a_n = 5a_{n-1} - 6a_{n-2} + 7^n$$

$$\Rightarrow \quad a_n - 5a_{n-1} + 6a_{n-2} = 7^n \qquad\qquad\qquad .....(16.48)$$

write $\quad A(X) = \sum_{n=0}^{\infty} a_n X^n \qquad\qquad\qquad\qquad .....(16.49)$

Then,

$$A(X) = \sum_{n=2}^{\infty} \frac{7^n X^n}{1 - 5X + 6X^2} + \frac{a_0(a_1 - 5a_0)X}{1 - 5X + 6X^2} \qquad .....(16.50)$$

Now $\quad \sum_{n=2}^{\infty} 7^n X^n = 7^2 X^2 \sum_{n=2}^{\infty} 7^{n-2} X^{n-2}$

which by a change of dummy variable becomes,

$$7^2 X^2 \sum_{n=0}^{\infty} 7^n X^n = 7^2 \frac{X^2}{1 - 7^x}$$

Therefore we have that

$$A(X) = \frac{7^2 X^2}{(1 - 7X)(1 - 5X + 6X^2)} + \frac{a_0 + (a_1 - 5a_0)X}{1 - 5X + 6X^2}$$

We know that

$$1 - 5X + 6X^2 = (1 - 2X)(1 - 3X)$$

So we see that the homogeneous solutions have the form

$$C_1 2^n + C_2 3^n$$

Similarly,

$$\frac{7^2 X^2}{(1-7X)(1-2X)(1-3X)} = \frac{C}{1-7X} + \frac{D}{1-2X} + \frac{E}{1-3X}$$

so that $\dfrac{7^2 X^2}{(1-7X)(1-2X)(1-3X)}$ generates a sequence

$\{b_n\}_{n=0}^{\infty}$

where $b_n = C\ 7^n + D.2^n + E\ 3^n$ for some constants C, D and E.

Note that

$D.2^n + E.3^n$ is also a solution of homogeneous recurrence relations

$$a_n - 5a_{n-1} + 6a_{n-2} = 0$$

So the only new information gained is the part $C\ 7^n$.

When we compare this with the original function $f(n) = 7^n$, we see that this function has almost reproduced itself.

Therefore at least $\displaystyle\sum_{n=k}^{\infty} \frac{f(n)x^n}{Q(X)}$ has generated a sequence

$$b_n = c\ f(n) + h(n)$$

where h(n) is a solution of (1).

Example 16.12

Find the general solution for the recurrence relation

$a_n - a_{n-1} = 4(n + n^3)$ where $n \geq 1$ and $a_0 = 5$.   [JNTUH, Nov 2010, Set No. 2]

Solution: Given recurrence relation is

$$a_n - a_{n-1} = 4(n + n^3)\ \ \forall\, n \geq 1 \qquad \text{and} \qquad a_0 = 5 \qquad\qquad ..... (16.51)$$

$\Rightarrow \quad a_n = a_{n-1} + 4(n + n^3)\ $ for all $n \geq 1$

From (2) we get the following:

$$a_1 = a_0 + 4(1 + 1^3)$$

$$a_2 = a_1 + 4(2 + 2^3)$$

$$= a_0 + 4(1 + 1^3) + 4(2 + 2^3)$$

$$= a_0 + 4\left\{(1+2) + (1^3 + 2^3)\right\}$$

$$a_3 = a_2 + 4(3 + 3^3)$$

$$= a_0 + 4(1 + 2 + 3 + 1^3 + 2^3 + 3^3)$$

$$a_n = a_{n-1} + 4(n + n^3)$$

$$= a_0 + 4\{(1 + 2 + \dots + n) + (1^3 + 2^3 + \dots + n^3)\}$$

$$a_n = a_0 + 4\left\{\frac{n(n+1)}{2} + \frac{n(n+1)^2}{2}\right\}$$

$$= a_0 + 4\left\{\frac{n^2 + n}{2} + \frac{(n^2 + n)^2}{4}\right\}$$

$$= a_0 + 4\left\{\frac{n^2 + n}{2} + \frac{n^4 + 2n^3 + n^2}{4}\right\}$$

$$= a_0 + 2(n^2 + n) + (n^4 + 2n^3 + n^2)$$

$$= 5 + (2n^2 + 2n) + (n^4 + 2n^3 + n^2) \qquad \text{(since } a_0 = 5)$$

$$= n^4 + 2n^3 + 3n^2 + 2n + 5$$

Example 16.13

Find the form of particular solution to

$$a_n - 5a_{n-1} + 6a_{n-2} = n^2 4^n \quad \text{for } n \geq 2.$$

[JNTUH, June 2010, Set No. 2]

Solution: Given linear recurrence relation is

$$a_n - 5a_{n-1} + 6a_{n-2} = n^2 4^n \quad \text{for } n \geq 2 \qquad \qquad \dots\dots(16.52)$$

write

$$A(X) = \sum_{n=0}^{\infty} a_n X^n \qquad \qquad \dots\dots(16.53)$$

Then

$$A(X) = \frac{\sum_{n=2}^{\infty} n^2 4^n X^n + a_0 + (a_1 - 5a_0)X}{1 - 5X + 6X^2}$$

Now

$$\sum_{n=2}^{\infty} n^2 4^n X^n = 4^2 x^2 \sum n^2 4^{n-2} X^{n-2} \qquad \qquad .....(16.54)$$

We write $r = n - 2$. Then we have

$$\sum_{n=2}^{\infty} n^2 4^n X^n = n^2 X^2 \sum_{r=0}^{\infty} (r+2)^2 4^r X^r$$

Write $(r+2)^2 = r^2 + 4r + 4$ as

$$2\,C(r+2,\,r) + C\,(r+1,\,r) + C\,(r,\,r)$$

So that

$$4^2 X^2 \sum_{r=0}^{\infty} (r+2)^2 4^r X^r = 4^2 X^2 \left[2\sum_{r=0}^{\infty} C(r+2,\,r)4^r X^r + \sum_{r=0}^{\infty} C(r+1,\,r)4^r X^r \right.$$

$$\left. + \sum_{r=0}^{\infty} C(r+1,\,r)4^r X^r + \sum_{r=0}^{\infty} 4^r X^r \right]$$

$$= 4^2 X^2 \left[\frac{2}{(1-4X)^3} + \frac{1}{(1-4X)^2} + \frac{1}{1-4X} \right]$$

$$= \frac{4^2 X^2 \left[2 + (1-4X) + (1-4X)^2 \right]}{(1-4X)^3}$$

Therefore

$$A(X) = \frac{4^2 X^2 \left[2 + (1+4X) + (1-4X)^2 \right] + (1-4X)^3 P(X)}{(1-4X)^3 (1-5X+6X^2)}$$

where $\quad P(X) = a_0 + (a_1 - 5a_0)X$

Thus we have that

$$A(x) = \frac{F(X)}{(1-4X)^3 (1-2X)(1-3X)}$$

where $F(x)$ is a polynomial of 4 or less.

By partial fractions, we get that

$$A(X) = \frac{A}{(1-4X)^3} + \frac{B}{(1-4X)^2} + \frac{C}{1-4X} + \frac{D}{1-2X} + \frac{E}{(1-3X)} \qquad .....(16.55)$$

Clearly $\dfrac{D}{1-2X} + \dfrac{E}{1-3X}$ satisfies the homogeneous recurrence relation.

The series is given by

$$\frac{A}{(1-4X)^3} + \frac{B}{(1-4X)^2} + \frac{C}{1-4X}$$

$$= \sum [A.C(n+2,\ n) + B.C(n+1,\ n) + C] 4^n X^n$$

So that a particular solution has the form $[A\ C(n+2,\ n) + B\ C(n+1,\ n) + C] 4^n$.

But after expanding the binomial coefficients, we get that the above solution becomes

$$(P_0 + P_1.n + P_2.n^2) 4^n$$

Therefore $f(n) = n^2 4^n$ determines a particular solution of the form: a polynomial of degree 2 times 4^n.

Example 16.14

Solve the $a_n - 6a_{n-1} + 8a_{n-2} = n4^n$

where $a_0 = 8,\ a_1 = 22$

[JNTUH, June 2010, Set No. 2]

Solution: Given recurrence relation is

$$a_n - 6a_{n-1} + 8a_{n-2} = n.4^n \qquad \qquad .....(16.56)$$

where the initial condition are: $a_0 = 8,\ a_1 = 22$

Here the particular solution takes the form $n(A_0 + A_1 n)\, 4^n$ since 4 is a root of characteristic polynomial of multiplicity 1.

By substituting this expression into the recurrence relation, we get that

$$n(A_0 + A_1^n)\, 4^n - 6(n-1)(A_0 + A_1(n-1)) 4^{n-1}$$

$$+ 8(n-2)\ (A_0 + A_1(n-2)) 4^{n-2} = n4^n \qquad \qquad .....(16.57)$$

By cancelling the common term 4^{n-2}, we have

$$16_n (A_0 + A_1 n) - 24(n-1)\ (A_0 + A_1(n-1))$$

$$+ 8(n-2)\ (A_0 + A_1(n-2)) = 16n$$

Now this is an expression that holds for all values of n. In particular for $n = 0$, we get the simplified equation.

$$A_0 + A_1 = 0 \text{ and then for } n = 1$$

Now we obtain $A_0 + 3A_1 = 2$

These equations have the unique solution.

$$A_0 = -1 \text{ and } A_1 = 1$$

Therefore

$$\left.\begin{array}{c} a_n^p = n(-1+n)\,4^n \\ = n(n-1)\,4^n \end{array}\right\} \qquad \qquad \dots\dots(16.58)$$

which is a particular solution.

But $a_n = n(n-1)4^n + C_1 4^n + C_2 2^n$ .....(16.59)

is the general solution of the relation.

By substituting the initial conditions in (16.59), we get that

$$C_1 = 3 \text{ and } C_2 = 5$$

Hence $a_n = n(n-1)\,4^n + (3)\,4^n + (5)\,2^n$ is the unique solution to the given recurrence relation with the initial conditions mentioned.

Example 16.15

Solve $a_n = a_{n-1} + n$ where $a_0 = 2$ by substitution.   [JNTUH, Nov 2010, Set No. 1]

Solution: Given recurrence relation is,

$$a_n = a_{n-1} + n \qquad \text{where } a_0 = 2 \qquad \dots\dots(16.60)$$

From the given relation, we find the following:

$$a_1 = a_0 + 1$$
$$a_2 = a_1 + 2 = (a_0 + 1) + 2 = a_0 + (1+2)$$
$$a_3 = a_2 + 3 = (a_0 + 1 + 2) + 3$$
$$\quad = a_0(1+2+3)$$
$$\vdots$$
$$a_n = a_0 + (1+2+3+\dots+n)$$
$$\quad = a_0 + \frac{n(n+1)}{2}$$

Since we know that

$$\left[1+2+3+\dots+n = \frac{n(n+1)}{2} \right]$$

Substituting $a_0 = 2$, we get that

$$a_n = \frac{n(n+1)}{2} + 2$$

Example 16.16

Find a general expression for a solution to the recurrence relation

$$a_n - 5a_{n-1} + 6a_{n-2} = n(n-1) \quad \text{for } n > 2 \qquad \text{[JNTUH, Nov 2008, Set No. 2]}$$

Solution: Given recurrence relation is

$$a_n - 5a_{n-1} + 6a_{n-2} = n(n-1) \quad \text{for } n > 2 \qquad \qquad .....(16.61)$$

Let $\quad A(X) = \sum_{n=0}^{\infty} a_n X^n \qquad \qquad .....(16.62)$

Multiply (16.61) on both sides by X^n and take sum from 2 to ∞, then we get

$$\sum_{n=2}^{\infty} a_n X^n - 5\sum_{n=2}^{\infty} a_{n-1} X^n + 6\sum_{n=2}^{\infty} a_{n-2} X^n$$

$$= \sum_{n=2}^{\infty} n(n-1) X^n \qquad \qquad .....(16.63)$$

Replacing the infinite sums by equivalent expressions, we have

$$(A(X) - a_0 - a_1 X) - 5X(A(X) - a_0) + 6x^2 A(X) = \sum_{n=2}^{\infty} n(n-1) X^n \qquad .....(16.64)$$

By using the shifting properties of generating functions, we get that

$$\sum_{n=2}^{\infty} n(n-1) X^n = \frac{2X^2}{(1-X)^3}$$

Thus

$$A(X)(1 - 5X + 6X^2) = a_0 + (a_1 - 5a_0)X + \frac{2X^2}{(1-X)^3}$$

and

$$A(X) = \frac{a_0 + (a_1 - 5a_0)X}{(1 - 5X + 6X^2)} + \frac{2X^2}{(1-X)^3(1 - 5X + 6X^2)} \qquad .....(16.65)$$

Exercises

Solving Linear Homogeneous Recurrence Relation by Substitution

1. Solve $a_n = a_{n-1} + n^3$ where $a_0 = 5$ by using substitution method.

    Ans: $a_n = \dfrac{n^2(n+1)^2}{4} + 5$

2. Solve $a_n = a_{n-1} + 3^n$ where $a_0 = 1$ by using substitution method.

    Ans: $a_n = \dfrac{3^{n+1} - 1}{2}$

3. Show by substitution that $C(2n-2, n-1)\,((n-1)!)$ is the solution of the relation
    $a_n = (4n-6)a_{n-1}$ for $n \geq 2$ where $a_1 = 1$

Generating Functions and the Method of Characteristic Roots

1. Solve the recurrence relation
    $a_{n+2} = 4(a_{n+1} - a_n)$, $n \geq 0$, $a_0 = 1$, $a_1 = 3$
    by using generating functions (characteristic roots)

    Ans: $a_n = 2^n + n(2^{n-1})$

2. Solve the recurrence relation
    $F_{n+2} = F_{n+1} + F_n$ for $n \geq 0$, $F_0 = 0$, $F_1 = 1$
    by using generating functions (characteristic roots)

    Ans: $F_n = \dfrac{1}{\sqrt{5}}\left[\left(\dfrac{1+\sqrt{5}}{2}\right)^n - \left(\dfrac{1-\sqrt{5}}{2}\right)^n\right]$

3. Solve $a_{n+2} + a_n = 0$, $n \geq 0$, $a_0 = 0$, $a_1 = 3$
    by using generating functions (characteristic roots)

    Ans: $a_n = 3\sin\left[\dfrac{n\pi}{2}\right]$

Solving Inhomogeneous Recurrence Relations

1. Solve the recurrence relation
    $a_{n+2} - 4a_{n+1} + 3a_n = -200$, $n \geq 0$ and $a_0 = 3000$, $a_1 = 3300$

    Ans: $a_n = 2900 + 100 \times 3^n + 100_n$

2. Solve $a_n - 6a_{n-1} + 8a_{n-2} = 9$, $n \geq 2$, $a_0 = 10$, $a_1 = 25$

$$\textit{Ans: } a_n = 4 \times 4^n + 3 \times 2^n + 3$$

3. Solve the recurrence relation

$a_{n+2} - 2a_{n+1} + a_n = 2^n$, $n \geq 0$ and $a_0 = 1$, $a_1 = 2$ by the method of generating function.

$$\textit{Ans: } a_n = 2^n$$

Bibliography

1. Barker S. F. The Elements of Logic, 5th Ed., McGraw-Hill, New York, 1989.

2. Beineke L. W., Wilson R.J., and Camerong P.J. 'Topics in Algebraic Graph Theory', Cambridge University Press (2004).

3. Bondy J. A. and U.S.R. Murthy, Graph Theory with Applications, American Elsevier, New York.

4. Grewel B.S. "Higher Engineering Mathematics" Khanna Publishers, 42nd edition.

5. Fraleigh J.B. "A First Course in Abstract Algebra", Narosa Pub. House , New Delhi, 1992.

6. Graham R. L., Knuth, D. E., and Patashnik O. Concrete Mathematics: A Foundation for Computer Science, 2nd Ed., Addison Wesley, 1994.

7. Halmos P. R. "Naive Set Theory", Springer-Verlag, New York, 1974.

8. Herstein I. N. "Topics in Algebra", New York, Blaisdell, 1964.

9. Hungerford T.W. "Algebra" Rinehart and Winston. Inc., New York, 1974.

10. Mott J.L., Kandel A. and Baker T.P. Discrete Mathematics for Computer Scientist and Mathematicians-Prentice Hall of India, 2nd Edn. 1999.

11. Lipschutz S. Schaum's Outline of Theory and Problems of Set Theory and Related Topics, 2nd Ed., McGraw-Hill, New York, 1968.

12. Liu. Elements of Discrete Mathematics.

13. Liu. C. L. Introduction to Combinatorial Mathematics, McGraw-Hill, New York, 1968.

14. Narsing Deo. "Graph Theory with Applications to Engineering and Computer Science", Prentice-Hall of India Pvt. Ltd., 1997.

15. N. P. Bali and Manish Goyal. "A Textbook of Engineering Mathematics" by Lakshmi Publications, Delhi.

16. Bali N. P., Satyanarayana Bhavanari and Indrani Promod Kelkar. "A Textbook of Engineering Mathematics" - Sem I (JNTU, Kakinada) 2012 Mathematics by Lakshmi Publications, Delhi.

17. Richard Johnsonbaugh. "Discrete Mathematics" Pearson Education Asia, 2001.

18. Rosen K. H. Hand Book of Discrete and Combinatorial Mathematics, CRC Press, 1999.

19. Satyanarayana Bhavanari. "Contributions to Near-ring Theory", VDM Verlag Dr Muller, Germany, 2010 (ISBN 978-3-639-22417-7).

20. Satyanarayana Bhavanari and Syam Prasad Kuncham. "An isomorphism theorem on Directed Hypercubes of Dimension n", Indian J. Pure & Appl. Math (2003).

21. Satyanarayana Bhavanari and Syam Prasad Kuncham. "Nearrings, Fuzzy Ideals and Graph Theory" Taylor and Francis Series (CRC Press, UK) 2013 (ISBN: 13:978-1-4398-7310-6).

22. Satyanarayana Bhavanari, Syam Prasad Kuncham and Nagaraju Dasari. "Prime Graph of a Ring", Journal of Combinatorics, Information and System Sciences, Vol. 35, 1-12, (2010).

23. Satyanarayana Bhavanari, Syam Prasad Kuncham, Pradeep Kumar T.V, Madhavi Latha, T. and Padmavathi, A. "Discrete Mathematics", (for *BCA /MCA*), Maruthi Publishers, Guntur.

24. Satyanarayana Bhavanari, Pradeep Kumar T.V. and Srinivasulu D. "Liniar Algebra and Vector Calculus" Studera Press, New Delhi, in progress.

25. Satyanarayana Bhavanari and Syam Prasad Kuncham. "Discrete Mathematics with Graph Theory" PHI (Printice Hall of India), New Delhi, 2009 (ISBN: 978-81-203-3842-5).

26. Satyanarayana Bhavanari and Syam Prasad Kuncham. "Discrete Mathematics with Graph Theory" PHI (Printice Hall of India), New Delhi, 2014 Second Edition (ISBN: 978-81-203-4948-3).

27. Trembly, J.P., and Manohar. R. "Discrete Mathematical Structures with Applications to Computer Science", Mc-Graw Hill, 1975.

28. Wilson R.J. Introduction to Graph Theory, 3rd ed., Longman, 1984.

29. Zadeh L. A. "Fuzzy Sets", Information and Control, 8, 338-353, 1965.